Undergraduate Topics in Computer Science

Undergraduate Topics in Computer Science (UTiCS) delivers high-quality instructional content for undergraduates studying in all areas of computing and information science. From core foundational and theoretical material to final-year topics and applications, UTiCS books take a fresh, concise, and modern approach and are ideal for self-study or for a one- or two-semester course. The texts are all authored by established experts in their fields, reviewed by an international advisory board, and contain numerous examples and problems. Many include fully worked solutions.

More information about this series at http://www.springer.com/series/7592

Bernhard Reus

Limits of Computation

From a Programming Perspective

 Springer

Bernhard Reus
Department of Informatics
School of Engineering and Informatics
University of Sussex
Brighton
UK

Series editor
Ian Mackie

ISSN 1863-7310   ISSN 2197-1781   (electronic)
Undergraduate Topics in Computer Science
ISBN 978-3-319-27887-2   ISBN 978-3-319-27889-6   (eBook)
DOI 10.1007/978-3-319-27889-6

Library of Congress Control Number: 2015960818

This Springer imprint is published by SpringerNature
The registered company is Springer International Publishing AG Switzerland

Foreword

Computer Science centers on questions about computational problems, and computer *programs* to solve problems:

- *What* is a computational problem, what is an algorithm, what does it mean to have solved a problem, are some problems unsolvable by any computing device, are there problems intrinsically difficult or impossible to solve automatically, are there problems intrinsically easier to solve than others, ...?
- *How* can a machine solve a problem, how to build such a machine, how to specify an algorithm for computer execution, how to design good algorithms, which "language" can a human use to direct the computer, how to design and "debug" programs for computer execution, how to build good programs, ...?

Good news: rapid progress has been made in both areas; we stand on the shoulders of giants in both theory and practice. The questions above have many and various answers. The first questions led in mathematical directions to foundational studies: the theories of computability, recursive functions, automata theory and more. The second led in engineering directions: computer architectures, the architectures of programming languages, the art or discipline of programming, software engineering and more.

Since the 1930s our understanding of both areas has developed hand in hand, led by theoreticians such as Kleene, Church, and Gödel; by hardware and software inventors such as Babbage, Von Neumann, and McCarthy; and by Alan Turing's genius at the borderline between the two areas.

This book focuses on the first question area by an approach near the borderline: the theory of computability and complexity (C&C for short) is presented by using a simple programming language. In this language one is able to perform the (many) program constructions needed for the theory. This is done abstractly enough to reveal the great breadth and depth of C&C. Further, it is done concretely and precisely enough to satisfy practice-oriented readers about the constructions' feasibility. Effect: a reader can see the relative efficiency of the constructions and understand the efficiency of what is constructed.

My 1997 C&C book was a step in this direction, but suffers from several problems: a scope too great for a single-semester university course; sections that (without warning) require more mathematical maturity than others; too few exercises; and too few historical and current references to research contexts.

Bernhard Reus has succeeded very well in overcoming these problems, and writing a deep, interesting, up-to-date, and even entertaining book on computability and complexity. I recommend it highly, both for systematic study and for spot reading.

Neil D. Jones
DIKU, University of Copenhagen

Preface

About 12 years ago a student[1] asked me after one of my lectures in *Computability and Complexity* why he had to write tedious Turing machine programs, given that everyone programmed in languages like Java. He had a point. What is the best way to teach a *Computability and Complexity* module in the twenty-first century to a cohort of students who are used to programming in high level languages with modern tools and libraries; to students who live in a world of smart phones and 24h connectivity, and, even more importantly maybe, who have not been exposed to very much formal reasoning and mathematics?

Turn the clock back only two or three decades. Then, a first year in a computer science Bachelor degree mainly consisted of mathematics (analysis and linear algebra, later discrete maths, numerical analysis, and basic probability theory). When computability and complexity was taught in the second or third year, students were already acquainted with a formal and very mathematical language of discourse, maybe because computer science lecturers in the 80s were usually mathematicians by trade. Things have changed significantly. Curriculum designers for Bachelor degrees are under pressure to push more and more new exciting material into a three-year degree program that should prepare students for their lives as working IT professionals. Any new module moved into the curriculum necessarily forces another one out. Often, allegedly "unpopular" modules, including formal theory and mathematics, are the victims. As a consequence, computer science students have to a degree lost the skills to digest material presented in an extremely formal and symbolic fashion while, at the same time, they are prolific programmers and quite knowledgeable in the use of tools.

So, *Computability and Complexity*, do they really have to be taught using Turing machines or μ-recursive functions? Do they have to be presented in the style a logician or mathematician would prefer? Seeking for alternatives, I eventually stumbled across Neil Jones' fantastic book *Computability and Complexity—From a Programming Perspective*. The subtitle already gives away the book's philosophy.

---

[1] Alexis Petrounias.

The leitmotif of Neil's textbook is to present the most important results in computability and complexity theory "using programming techniques and motivated by programming language theory" as well as "using a novel model of computation, differing from traditional ones in crucial aspects".[2] The latter, WHILE, is a simple imperative language with one datatype of lists (s-expressions) à la LISP. Admittedly, this language is not Java, but it has the hallmarks of a modern high level language and is infinitely more comfortable to program in than Turing machines or Gödel numbers. Java, or any similar powerful language, would be impractical for our purposes "since proofs about them would be too complex to be easily understood".[3]

So when rebranding the module under the name *Limits of Computation* in 2008, I adopted Neil's book as course textbook. Delivering an introductory, one semester final-year module, I picked the most important and appealing chapters. This was easy as the design of the book was exactly made to mix and match.[4] Soon, however, it turned out that students found Neil's book tough going. In fact, this became more apparent as the years went past. There were several factors. First of all, Neil's students would have had ML, a functional language with built in list type, as a first programming language, whereas our students were raised on Java. Yet, the datatype of WHILE is a functional one, and this caused more problems to the students than anticipated. Second, and more importantly, I had not put enough attention to the prerequisites. Neil expected readers of his book to be senior undergraduate students "with good mathematical maturity."[5] It turned out that not all the third-year students had this maturity (given the heterogeneity of backgrounds and reduction of maths teaching in the undergraduate years one and two).

As a response to mitigate the issues above, I started writing explicit notes to accompany my slides, intended as additional comments and explanations for the selected book chapters. I ended up adding more and more new material and rearranging it. The results of this effort are the 23 individual lectures of this book. A (British) semester is 12 weeks long, which usually requires 24 lectures to be delivered. The shortfall of one lecture is intentional, it acts as a buffer (in case things take more time) and also allows for extra events like invited talks or in-class tests.

This book was heavily influenced by Neil Jones's textbook, which is clearly visible in some chapters. To pay homage to his book, its telling subtitle "From a Programming Perspective" has been adopted.

Brighton   Bernhard Reus
November 2015

---

[2]Preface of Neil's book, page x.

[3]The book "Understanding Computation From Simple Machines to Impossible Programs" by Tom Stuart, published by O'Reilly in 2013, appears to follow the same idea and philosophy, using Ruby as programming language. It does not deal with complexity however.

[4]Neil's Preface, page xii.

[5]Neil's Preface, page xiii.

For Tutors

The students using this book are expected to be senior undergraduates who can master at least one imperative programming language. They are supposed to know arithmetic, Boolean algebra, graphs and some basic graph algorithms. Similarly, knowledge of basic set theory, function, and relations is needed, but Chap. 2 contains a short summary of basic definitions used in the book. Some exposition to formal reasoning would be helpful as well, but is not strictly required. To understand the probabilistic complexity classes in Chap. 21 some basic knowledge of probability theory will be needed.

This book is divided into 23 chapters with the intention that each chapter corresponds to one lecture. If not all chapters can be delivered, the following chapters can be omitted without interrupting the natural narrative:

- Chapter 10, *Self-referencing Programs*, but the self-producing program is a brain teaser which turns out to be very popular with students;
- Chapter 21, *How to Solve NP-complete Problems*, which may, however, be the chapter that has the most impact on students' future projects;
- The last two chapters *Molecular Computing* (Chap. 22) and *Quantum Computing* (Chap. 23), but students find this material particularly exciting.

A few chapters are significantly longer than others. These are Chap. 11, *The Church-Turing Thesis*, Chap. 16, *Famous Problems in P*, Chap. 20 *Complete Problems*, Chap. 21 *How Solve NP-complete Problems?*, and Chap. 22 *Molecular Computing*. The longer chapters allow the tutor to pick some of the sections and present them in the lecture, leaving the remaining ones for self-study or exercises.

Acknowledgements

This book would not have been possible without the help and influence of so many people. They all deserve a big thank you.

First of all, I wish to thank Neil Jones for his book on *Computability and Complexity*, which has been inspirational. The results of many brilliant researchers have been reported in this textbook. Many of them have been mentioned and referenced but, this being an introductory textbook, some of them may have been left out. My sincerest apologies to those who were omitted.

I am grateful for the feedback I have received over the years from my teaching assistants Matthew Wall, Jan Schwinghammer, Cristiano Solarino, Billiejoe Charlton, Ben Horsfall, and Shinya Sato.

I enjoyed and benefited from talking to many students on the *Limits of Computation* course I taught at Sussex. Thanks go in particular to Alexis Petrounias, Marek Es, Thomas Weedon Hume, Alex Jeffery, Susan Coleman, Sarah Aspery, Benjamin Hayward, Jordan Hobday, and Lucas Rijllart. The latter five also gave feedback on early versions of various chapters.

I also benefited from discussions with Thomas Streicher and my colleagues at Sussex: thanks go to Martin Berger, Matthew Hennessy, Julian Rathke, Des Watson, and particularly, George Parisis. Des and George also provided most welcome proof-reading services.

Luca Cardelli was kind enough to discuss current topics in molecular computing and gave valuable pointers regarding Chap. 22. Neil Jones and Matthew Hennessy provided comments on an entire draft of this book which helped improve the presentation. All remaining errors are of course solely mine.

I would also like to thank series editor Ian Mackie for his encouragement, Helen Desmond from Springer for her continuous support and Divya Meiyazhagan from the production team for all the last minute edits. The wonderful *LaTeX* typesetting and the TikZ (PGF) vector graph drawing systems have been used. Thanks to all those people who contributed to their development.

My sister's family and my friends deserve acknowledgment for moral support, and for being there for me when it counted.

Finally, I would like to dedicate this book to the memory of my parents and my brother. His interest in the sciences and in computing aroused my curiosity already at a young age.

Contents

Chapter 1
Limits? What Limits?

What kinds of limits are we studying and why?

As final year undergraduate (or postgraduate) students will have learned how a modern digital computing device works. They will know about these gadgets' architecture, their operating systems, their input/output devices, their networking capabilities. They will appreciate the importance of compilers that can be used to translate programs in high-level languages into a form that the underlying hardware of the computer in use can interpret as instructions. This code the students will have learned to write in modern high-level programming languages like Java, Haskell or Scala. They also practise, sometimes quite painfully, how to debug their programs to find and eliminate bugs so that in the end the system designed and implemented does the job required.

All these skills are very important (also for future employability) and building systems is, of course, the creative "fun part" of being a computer scientist. Yet, as the term "scientist" suggests, there is more to being a computer scientist than being able to write nifty programs and debug them (which is the "engineering" aspect of computing).

Rocket scientists who design space ships need to know the laws of physics and the *limits* described by those laws. For instance, they need to know that nothing can go faster than the speed of light. Similarly, a computer scientist needs to know the *limits* of computation. These limits, however, should not just be the ones obtained by resource constraints of a given computing device. Since there are so many different devices, we are striving for limits that are more abstract, independent of whether one has a computer of brand A or B.

Therefore, this chapter right at the beginning will stake the book's territory. We have to answer some important questions: What kind of limits of computation will we discuss and what kind do we leave out? In Sect. 1.1 we briefly discuss

© Springer International Publishing Switzerland 2016
B. Reus, *Limits of Computation*, Undergraduate Topics in Computer Science,
DOI 10.1007/978-3-319-27889-6_1

some physical limitations, discussing several types of physical limits in relation to computing that are well known. Then we give an overview of the content of the book (Sect. 1.2).

1.1   Physical Limits of Computation

When asked about limitations in the context of computing one most likely will think about the limits of the devices we use: the limited memory and limited clock speed. Or one might think about the limits memory or clock speed can theoretically reach in the production of computing devices. As this book is intended for *Computer Science* students with a focus on software and not hardware, we do *not* really address physical limits.[1] In this section, however, we give a short overview discussing several types of physical limits in relation to computing that are well known (and usually one needs to understand quite some physics to get deeper into the subject).

1.1.1   Fundamental Engineering Constraints to Semiconductor Manufacturing and Scaling

- Limits on manufacturing
  There are limits to layering material on silicon on an extremely small scale using "precision optics and photochemical processes". Lithography is limited but new technologies emerge and there is "additional progress in multiple patterning and directed self-assembly" which "promises to support photolithography beyond the 10 nm technology node" [5].
- Limits on interconnects
  "Metallic wires can be either fast or dense, but not both at the same time—smaller cross-section increases electrical resistance, while greater height or width increase parasitic capacitance with neighboring wires" [5]. The solution is to build so called 'interconnect-stacks' with varying thickness of wires according to purpose. New technology uses alternatively optical interconnects but this also is restricted by Maxwell's equation on propagation speed of electromagnetic waves.
- Limits on conventional transistors
  "Transistors are limited by their tiniest feature—the width of the gate dielectric,[2]—which recently reached the size of several atoms" [5].

---

[1]Although we briefly discuss some issues related to miniaturisation in Chap. 23.

[2]This is the insulator (usually silicon dioxide) between gate and substrate of a metal-oxide semi-conductor field-effect transistor (MOSFET). When the gate of the transistor is positive the dieletric is responsible for inducing a conducting channel between the source and drain of the transistor due to the so-called field effect.

- Limits on design effort
  With the rising complexity of integrated circuits (ICs) in order to achieve higher
  speed at smaller scale and lower power usage, computer-aided design (CAD) is
  unavoidable and each new design required new CAD software. Clever algorithms
  for optimisations are required and hardware as well as software for the ICs need to
  be co-developed and verified. The limitations of the CAD software has therefore
  a direct limiting effect on the produced ICs.

1.1.2 Fundamental Limits to Energy Efficiency

- Limits on energy consumption, power supply and cooling
  In modern times of cloud computing and computing as a service, data centres
  play an important role. In the US they "consumed 2.2% of total U.S. electricity
  in 2011" [5]. One can improve transmission and power conversion in datacenters,
  but on-chip power management soon reaches some limits. "Modern IC power
  management includes clock and power gating, per-core voltage scaling, charge
  recovery and, in recent processors, a CPU core dedicated to power scheduling. IC
  power consumption depends quadratically on supply voltage, which has decreased
  steadily for many years, but recently stabilized at 0.5–2 V.... Cooling technologies
  can improve too, but fundamental quantum limits bound the efficiency of heat
  removal" [5].

1.1.3 Fundamental Physical Constraints on Computing in General

Generally speaking, energy limits speed and entropy limits memory. In [4] a hypothet-
ical "computer with a mass of 1 kg and a volume of 1 l, operating at the fundamental
limits of speed and memory capacity fixed by physics" has been analysed. It is called
the 'ultimate' laptop and serves to expose the limits of physical computation. Whereas
current machines perform roughly (in terms of order of magnitude) 10^{10} operations
per second on 10^{10} bits,[3] "the ultimate laptop performs $2mc^2 \frac{1}{\pi \hbar} = 5.4258 \times 10^{50}$
logical operations per second on approx 10^{31} bits" [4] where m is the mass of the
computer, c light speed, and $\hbar$ is the Planck constant. However, "the ultimate laptop's
memory looks like a thermonuclear explosion or a little piece of the Big Bang!" [4].
Moreover, one achieves only an "overall processing rate of $\simeq 10^{40}$ operations per
second in ordinary matter" [4]. Those computed limits are based on the speed of
light, the quantum scale, and the gravitational constant (see also [1]).

---

[3]This was at the time of [4] in 2000 and has increased somewhat by now.

1.2 The Limits Addressed

The above mentioned physical limits are not only interesting but also most significant for the semiconductor and computing industry. However, we are interested in limitations inherent to computing independent of the hardware used to compute. In the first part, computability, we allow unlimited memory and time and investigate whether nonetheless there remain any limits to what one can compute. In the second part, complexity, we restrict the time allowed for computation, abstracting again from concrete details. We will be looking at the *asymptotic behaviour of programs*: how much does the computation time increase when we increase the size of the input. In this book we focus on time complexity only due to limitations of a 12 week semester module. It is expected that the diligent reader, once familiar with the concepts of computational time complexity, is able to read independently the literature on space complexity which addresses the size of memory needed for computation. The short last part of the book is dedicated to new emerging paradigms of computing: molecular computing and quantum computing and the question whether they can extend those limits.

It might come as a surprise that the computational limits we will discuss, despite the idealisation, appear to be deeply connected with limits in our physical reality.

1.2.1 Computability Overview

Finding the limits of computation requires a clear-cut definition of what computing means. The origins of what is called "computability theory" go back to Alan M. Turing,[4] a British mathematician, who achieved much more than cracking the German code used by the *Enigma* machines in WWII, but also wrote a world famous paper that, for the very first time, gave a definition of what it means "to compute". He defined a computation device on paper that is now called *Turing machine* in his memory.

At the time the term *computer* still referred to a human being carrying out a computation. Turing showed that, based on this definition, there is a function[5] that cannot be computed in the proposed way. And this was in 1936 before the invention of the general purpose digital computer. We will look at this seminal paper [7] a bit later.

Also, when searching for the limits of computation, we will simplify the kind of problems we consider to be computed. This is fine since if we have found limitations for a restricted class of problems we of course have also found limitations of a larger class of problems.

After we have defined an adequate class of problems (Chap. 2) we will idealise our computation devices and investigate what cannot be computed using unbounded resources.

---

[4]To commemorate Turing's 100th birthday, 2012 was the Alan Turing Year, see http://www.mathcomp.leeds.ac.uk/turing2012/.

[5]At least one such function.

So the first big quest is to find a problem that cannot be computed even with unlimited time (and memory).

To be able to do this ourselves (in Chap. 8), we need to work with a particular model of computation. Of course we would like to work with a language that is similar to our modern high-level languages and not, for instance, with fiddly (Turing) machine programs. But on the other hand, we also want to be able to program an interpreter for the chosen language. Thus, the language should not be too complicated. We follow Neil Jones's inspired choice from his textbook [3] and (in Chap. 3) pick WHILE, a simple "while language" with variables, assignment, while-loops and with binary trees as data type. We fix programs to have exactly one input and one output value. Again, this is a simplification but good enough to successfully go on our quest of non-computable problems. In Chap. 4 we ensure that WHILE programs are actually effectively computable and in Chap. 5 we present some additional language features, which facilitate programming further.

Then we will learn how to use programs as objects (Chap. 6) such that we can write a WHILE interpreter in WHILE (Chap. 7).

This will help us to further distinguish certain kinds of non-computable problems. We will then also meet several other (famous) undecidable problems in Chap. 9. We will learn that, roughly speaking, any interesting statement referring solely to the semantics of programs is undecidable. Historically, Hilbert's *Entscheidungsproblem*[6] is of interest. Hilbert and many others believed that there is a decision procedure for first-order logic sentences with axioms, for instance, arithmetic (number theory). Such a procedure could, given any arithmetic sentence, compute whether this given sentence is true or not. In 1936 this was disproved by Gödel, Turing and Church [2].

In Chap. 10 it is discussed how **programs that are recursively defined, i.e. *self-referencing programs*, can be interpreted despite the fact that WHILE does not possess any recursive features**. The corresponding programming principle is called *reflection*.

Next, we will have to ensure that our definition of computation is robust (Chap. 11). That means that our result does not depend on the choice of computing device or notion of computation. Otherwise, somebody might say that the problem we have found might have been computable with a different (e.g. more "powerful") computing device.

We will show that what is computable with unlimited resources does not depend on the computing device. The device must be of course reasonably powerful. There is a generalised thesis stating this called the *Church-Turing-Thesis*, and we will give evidence for it. The evidence will be the fact that all commonly used models of computation are of equivalent power. We can show this by writing compilers that translate programs or machines from one kind into another. This translation has to preserve the semantics of the original program or machine of course.

---

[6]German for "decision problem".

1.2.2   Complexity Overview

In the complexity part we investigate what can be computed with limited time resources, defining time complexity classes (those should already be known from an "Algorithms and Data Structures" module). Complexity classes for programs (or algorithms) are classes of programs that can be computed with a certain time bound. That means, we consider how much time is needed to compute a program in relation to the size of the input data (Chap. 12). It is clear that bigger input computations will usually take more time.

We can then also define complexity classes for problems (Chap. 13). An interesting question is at what point the computations become *intractable*, i.e. too long to be useful in practice. By the way, we usually consider worst case time complexity when we consider complexity classes. The complexity class is a program (or problem) which depends on the worst input for the program (or decision procedure for the problem). "The worst input" denotes the input for which the run time is the longest.

In Chap. 14 we address again the issue of *robustness* but now taking into consideration time consumption. In other words we show that **with respect to polynomial time computation it does not matter which notion of computability we use**. We will *extend* the Church-Turing thesis accordingly (and discuss variations such as Cook's Thesis).

To show that a problem is in a certain complexity class, it suffices to produce an algorithm that solves it and to show that the algorithm's time bound (its runtime complexity) is within the class. It is much less straightforward though to show that a problem is *not* in a complexity class since one needs to prove that no algorithm solving the problem has the required complexity. A technique to show exactly that and in fact a hierarchy that shows that **with more time one can solve more problems** is presented in Chap. 15.

We will look at some of those problems that are very much "real life" **combinatorial optimisation problems, which are important in practical applications in industry and transport.** The problems presented in Chap. 16 can all be solved in polynomial time but the ones in Chap. 17 are "hard" in the sense that the only algorithms known to solve them exactly have exponential complexity. This is not good enough since for medium-sized input the runtime complexity becomes astronomically large (this is down to what is called the "combinatorial explosion"). For instance, consider the problem of finding the shortest tour passing through a list of given cities. The number of possible tours to search through grows exponentially with the number of cities. There is no problem if the number of cities is small. For three cities there exists only one[7] tour and for four cities there are only three. For 16 cities, however, there exist already several hundred billion possible tours and for 64 cities there exist orders of magnitude more tours than there are atoms in the known

---

[7]We disregard whether travelling clockwise or counterclockwise.

universe.[8] It is impossible to straightforwardly check which tour is the shortest. However, we don't know yet whether there are some very clever algorithms that have acceptable runtime that just nobody has discovered yet.[9]

In order to define yet **another complexity class based on verifying solutions rather than finding them**. It contains all those optimisation problems encountered earlier, for which we don't know yet whether they have polynomial solutions. **Whether this class is equal to the class of problems decidable in polynomial time is the most famous open problem in theoretical computer science** (Chap. 18). Of this class we can prove some more interesting properties in Chap. 19. It is possible to define the hardest problems in this class in a precise way by means of reduction. The resulting class does contain all the "hard" problems that we encounter in applications and, very remarkably, it is the case that either all problems in this class are tractable (i.e. have polynomial complexity) or none. Such problems can then be rightfully called *intractable* or *infeasible*.

In Chap. 20 we then **present even more problems that are as "hard" as the optimisation problems encountered in Chap. 17, among them are some famous games and puzzles** as well as problems regarding database queries or computer networks.

Regarding space complexity (which we will not cover in this book) the question is still open whether with polynomial space we can solve actually more problems than with polynomial time. Note that the complexity class, which was mentioned earlier and contains all the hard problems, lies in between these two classes. Again nobody knows if any of these classes (and if so which ones) are actually equal.

After all this we seem to be left with a big dilemma: **if there are no known feasible solutions to so many practical (combinatorial) problems what do we do?** This is the topic of Chap. 21. One could be led to think that it helps to look for approximative solutions rather than optimal ones. Instead of looking for the optimal solution, one could just be as happy with computing one that is at most a certain percentage worse that the optimal solution. Alas, it turns out that computing these approximative solutions is as hard as computing the real thing most of the time.[10] **In some cases, however, changing the problem slightly might give rise to approximative solutions that can be computed in polynomial time**.

Another fix for this dilemma could be the usage of **many parallel computers or processors**. This will certainly speed up matters. But does it make otherwise intractable problems tractable? Also this is an open problem and subject of ongoing research. A big problem here is that one would need super-polynomially many processors to get a real polynomial speed-up, and communication between those is likely to take super-polynomial time.

---

[8]"It is estimated that the there are between 10^{78} to 10^{82} atoms in the known, observable universe. In layman's terms, that works out to between ten quadrillion vigintillion and one-hundred thousand quadrillion vigintillion atoms." [8]

[9]There are, of course, clever algorithms that work reasonably well for reasonably large input and we will get back to this in Chap. 21.

[10]This is still an active research area.

Another question we will briefly address is this. **Do probabilistic methods help?** There are two versions of probabilistic algorithms: one is always correct but only probably fast. The other is always fast but only probably correct. Of course one needs to obtain a high probability to be able to use the latter kind of algorithms. In the 1970s very good polynomial time algorithms, e.g. for primality testing, were developed and the research area of probabilistic algorithms was established. However, recently it has been discovered that primality testing is polynomial even deterministically (the famous AKS algorithm after the initials of the inventors which will be discussed in Sect. 16.4). Anyway, **it is still unknown whether probabilistic algorithms resolve our dilemma and give us good complexity algorithms for our hard problems**.

Another hope to achieve feasibility of hard problems is to consider **new models of computation.** One of those is the so-called *molecular computing* (also called DNA computing) discussed in Chap. 22. Molecular computing has been pioneered in the 1990s by Len Adleman[11]. However, like quantum computing, this is still technically difficult (one needs the right experiment set up in a lab). The massive parallelism in those molecular computations stems from the massive parallelism in interaction in the "molecular soup". Other fascinating aspects of DNA computing are that it can be carried out in vivo (inside living cells), that DNA information can be highly packed, and that it consumes very little energy. The latter two facts have been used already to successfully store information in DNA. It is currently unknown whether molecular computing could be used to resolve intractability. This is unlikely, as one needs too large a volume of DNA to obtain such dramatic speed-ups. However, molecular computing is a hot research area, due to the possible medical applications it could give rise to.

Quantum computers are able to do massively parallel computations via so-called superposition of states in the world of quantum physics. The problem here is that it is extremely difficult to build quantum computers, even quantum storage, as on the quantum scale the environment can easily interfere with the quantum effects, from the casing to the wires. Moreover, it is difficult to harness the parallelism of quantum computers as any observation of the result state produces just one result and the magic of the superposition is subsequently lost. No more results can be observed. The area of quantum computing is an active and exciting research area. On the theoretical front, some breakthroughs have been already achieved. Shor's[12] algorithm [6] is a quantum algorithm for prime factoring. This algorithm can be executed on a quantum computer in polynomial time. But it should be pointed out that we still don't know for sure whether factorisation is computable in polynomial time, i.e. feasibly, at least on "standard" hardware anyway. This will be covered in more detail in Chap. 23.

---

[11] Leonard Max Adleman (born December 31, 1945) is an American theoretical computer scientist, best known for being a co-inventor of the RSA cryptosystem and molecular computing. He received the Turing award in 1977.

[12] Peter Williston Shor (born August 14, 1959) is an American professor of applied mathematics at MIT. He won the Gödel Prize for his work on quantum computing.

The problem whether the class of problems that can be **solved in polynomial time** is equal to the class of problem whose solutions of which can be **verified in polynomial time** appears to be strongly connected with the world of physics.

In any case, whether it's quantum computing or molecular computing, none of them would allow us to solve more problems than we already can with conventional computers. **This fact, also known as the Church-Turing-thesis, is another major topic** (see Chap. 11).

What Next?

We have determined what limits we will investigate and we have completed an overview of the topics of this book. Therefore we are prepared and ready to start with the first part about computability. The obvious first question here is: what does computable actually mean? We give an answer in the next chapter.

References

1. Bremermann, D.: Optimization through evolution and recombination. In: Yovits, M.C., Jacobi, G.T., Goldstein, G.D. (eds.) Self-Organizing Systems, pp. 93–106. Spartan Books, Washington D.C. (1962)
2. Church, A.: A note on the Entscheidungsproblem. J. Symb. Log. **1**(1), 40–41 (1936)
3. Jones, N.D.: Computability and complexity: From a Programming Perspective. MIT Press, Cambridge (1997). Available online at http://www.diku.dk/~neil/Comp2book.html
4. Lloyd, S.: Ultimate physical limits to computation. Nature **406**(6799), 1047–1054 (2000)
5. Markov, I.L.: Limits on fundamental limits to computation. Nature **512**(7513), 147–154 (2014)
6. Shor, P.W.: Polynomial-time algorithms for prime factorization and discrete logarithms on a quantum computer. SIAM J. Comput. **26**(5), 1484–1509 (1997)
7. Turing, A.M.: On computable numbers, with an application to the Entscheidungsproblem. J. Math. **58**, 345–363 (1936)
8. Villanueva, J.C.: How Many Atoms Are There in the Universe? Universe Today, 30 July 2009. Available via DIALOG http://www.universetoday.com/36302/atoms-in-the-universe/. Accessed 30 June 2015

Part I
Computability

Chapter 2
Problems and Effective Procedures

What does computable mean? What problems do we consider?

To a computer science student the question *"what does computable mean?"* might appear frivolous. We use computers every day, so don't they—obviously—"compute things" for us, although one might say, very often they mainly retrieve and display information, for instance when we are browsing web pages or watching videos. And "computable" means "being able to be computed", so what is the point?

The ACM[1] Computing Curricula [11, Sect. 2.1] states "computing to mean any goal-oriented activity requiring, benefiting from, or creating computers. Thus, computing includes ... processing, structuring, and managing various kinds of information; finding and gathering information relevant to any particular purpose, and so on. The list is virtually endless, and the possibilities are vast." This is uncontroversial but quite generic and does not really define what "computable" means.

It should be clear that we need to pin down what computable means precisely and formally if we want to explore the limits of computation in a scientific manner. The same issue arises with the definition of problems. Everybody has their own understanding of what a problem is: from not being able to pay the rent to finding the shortest path in a graph. Also in that respect we will have to restrict the definition in order to be able to apply formal reasoning so that we can prove results. It is also important to understand the difference between a problem and a program. Computable problems will be the ones for which there are programs that "solve" them. All these concepts will be carefully defined below.

We begin with a very short historical perspective (Sect. 2.1) introducing the notion of "effective procedure". Sets and structures on sets, i.e. relations and functions,

---

[1]The Association for Computing Machinery (ACM), "the world's largest educational and scientific computing society, delivers resources that advance computing as a science and a profession" [1]. It was founded in 1947 and has its headquarters in New York.

© Springer International Publishing Switzerland 2016
B. Reus, *Limits of Computation*, Undergraduate Topics in Computer Science,
DOI 10.1007/978-3-319-27889-6_2

together with their basic operations, are defined in Sect. 2.2 and some basic reasoning principles recalled. Finally, we define precisely what we mean by a "problem" in Sect. 2.3.

2.1 On Computability

In order to define the term "computable" we need to have a look at *what* is to be computed and *how* "computed" is actually defined. *What* is to be computed is generically called a problem. As computing in the 21st century is ubiquitous and microprocessors are not only in computers, but also in games consoles, mobile phones, music players and all kinds of consumer products, even washing machines, we need to restrict the problem domain since the "problem" in the context of washing will be very different from the "problem" in the context of game playing, or other areas.[2]

But first we address the "computable" question as the kind of problems we will look at depends on this definition.

2.1.1 *Historical Remarks*

Kleene wrote that the origin of algorithms[3] goes back at least to Euclid[4] ca. 330 B.C. according to [10] which provides an excellent historic overview. There have been many machines designed for calculation, from Gottfried Leibniz[5] to Charles Babbage[6] who wanted to automate calculations done in analysis.

The *Entscheidungsproblem*, the *decision problem for first order logic*, was raised in the 1920s by David Hilbert[7] and was described in [6]. The problem is to give a decision procedure "that allows one to decide the validity (respectively satisfiability) of a given logical expression by a finite number of operations" [6, pp. 72–73]. For Hilbert this was a fundamental problem of mathematical logic and played an important part

---

[2]And every reader may have their own problems, i.e. their own idea of what a "problem" is.

[3]The name "algorithm" dates back to the name of the ninth century Persian mathematician Al-Khwarizmi.

[4]Euclid (ca. 300 BC) was a Greek mathematician often called the "Father of Geometry" who also worked in number theory. He is the inventor of the common divisor algorithm.

[5]Gottfried Wilhelm von Leibniz (July 1, 1646–November 14, 1716) was a German mathematician and philosopher, credited with the independent invention of differential and integral calculus. He also invented calculating machines.

[6]Charles Babbage, (December 26, 1791–October 18, 1871) was an English mathematician, philosopher, inventor and mechanical engineer (and a Fellow of the Royal Society). He is also known for originating the concept of a programmable calculating machine. The London Science Museum has constructed two of his machines where they are on display.

[7]David Hilbert (January 23, 1862–February 14, 1943) was a world-renowned German mathematician.

in his program of finding a finite axiomatisation of mathematics that is consistent, complete and decidable (in an automatic way).

Gödel[8] then proved in 1931 that no axiomatic system of arithmetic can exist that is consistent and complete. This result proves a significant inherent limitation of mathematical logic and deductive systems. He gave the definition of *general recursive functions* on natural numbers based on previous work by Herbrand, Skolem, Hilbert, and Péter.

At the age of only 22 and still a student, Alan Turing . . .

> . . . worked on the problem for the remainder of 1935 and submitted his solution to the incredulous Newman on April 15, 1936. Turing's monumental paper *1936* was distinguished because: (1) Turing analyzed an idealized human computing agent (a computer) which brought together the intuitive conceptions of a function produced by a mechanical procedure which had been evolving for more than two millenia from Euclid to Leibniz to Babbage and Hilbert; (2) Turing specified a remarkably simple formal device (Turing machine) and proved the equivalence of (1) and (2); (3) Turing proved the unsolvability of Hilberts *Entscheidungsproblem* which established mathematicians had been studying intently for some time; (4) Turing proposed a universal Turing machine, . . . an idea which was later to have great impact on the development of high speed digital computers and considerable theoretical importance. [10, Sect. 3]

Turing used the term *a*-machine for his theoretical computing device but we now call them Turing machines in his honour.

Independently, Alonzo Church[9] proposed Church's Thesis "which asserts that the effectively calculable functions should be identified with the recursive functions" [10].[10] Church had initially intended this to be the definition of "effectively computable". Nowadays one uses the term *"Church-Turing thesis"* which amalgamates both theses, identifying all the intuitive notions of computation and all the various formal definitions. What is to be subsumed under the notion of "intuitively computable", is obviously up to interpretation. There have been suggestions that there are computational models much more powerful than Turing machines, called *hypercomputation*, but this is currently hotly debated. We will discuss this in more detail in Chap. 11 dedicated to the Church-Turing thesis.

Despite general recursive functions and Turing machines being the first formal definitions of computability, this book will not use the former and only briefly look at the latter. The reason is that the former needs some mathematical background and the latter is tedious to program. We follow the idea of Neil Jones [7] and use a high-level programming language which will be introduced in the next chapter. Thus, we need to justify that our language qualifies as "intuitive notion of computability". We must therefore understand what is required for such an intuitive notion of computation.

---

[8] Kurt Friedrich Gödel (April 28, 1906–January 14, 1978) was an Austrian (and American) logician, mathematician, and philosopher and is considered one of the most significant logicians in history. He proved many important results, relevant here is the *Incompleteness Theorem*.

[9] Alonzo Church (June 14, 1903–August 11, 1995) was an important American mathematician and logician.

[10] Church's first version that the computable functions are those definable by λ-terms [3] was initially rejected.

2.1.2 Effective Procedures

Effective Procedures, or effective algorithms are the programs that we understand to perform computations. The naming goes back to *Alan Turing*: "*A function is said to be effectively calculable if its values can be found by some purely mechanical process. . . . We may take this statement literally, understanding by a purely mechanical process one which could be carried out by a machine* [12, p. 166]. Note that Turing uses the word "calculable" here. In the 1930s computations usually referred to mathematical calculations. The machines he suggested have been called Turing machines and we will look at them more carefully in Chap. 11.

So what is an effective procedure? Copeland gives the following definition in [4]:

'Effective' and its synonym 'mechanical' . . . do not carry their everyday meaning. A method, or procedure, M, for achieving some desired result is called 'effective' or 'mechanical' just in case

1. M is set out in terms of a finite number of exact instructions (each instruction being expressed by means of a finite number of symbols);
2. M will, if carried out without error, always produce the desired result in a finite number of steps;
3. M can (in practice or in principle) be carried out by a human being unaided by any machinery save paper and pencil;
4. M demands no insight or ingenuity on the part of the human being carrying it out.

The instructions of an effective procedure must therefore be executable in a mechanical way. This means that instructions (or commands) in programs must not be "vague". For instance "find a number that has property P" which cannot be carried out effectively. How do we find the number? We need instructions that produce a number effectively such that it has the desired property. Therefore we cannot use oracles or choice axioms in our effective procedures. Moreover, we must be able to carry out the procedures in a finite amount of time. Infinite computations are by definition not effective. However, all notions of computation allow the definition of infinite computations as well. It will become clear in Chap. 8 why it is difficult to separate finite from infinite computations.

The exact meaning of "intuitive" computable is to a certain degree subject to interpretation. Some researchers insist that the mechanical computability by Turing machines does not include the so-called interactive computation, where humans (or other potentially non-computable oracles) interact with the program (see [5]). This appears to be equivalent to Turing's *o*-machines, Turing machines with an "oracle tape", an extra tape on which the Turing machine can write a word w and then ask the environment, the oracle, to answer whether w is in a certain set A which can be arbitrarily complicated (in particular it does not have to be decidable by an *a*-machine). The resulting definition of computability by *o*-machines is called *relative computability*. Relative computability will *not* be covered in this introductory book. Also we will not discuss computability of infinite objects (e.g. real number computation).

Turing's machine model extends the concept of a finite state automaton with extra memory. This memory is organised as a tape on which symbols can be written and read sequentially by a head that moves along the tape and that is controlled by the finite state automaton. Programming Turing machines is therefore a tedious and error-prone undertaking. For this reason, we don't want to use them to prove anything in this book, but rather use a programming language close to what we use on a daily basis. In Chap. 3, a more convenient notion of "effective procedure" will thus be presented. Turing machines will, for the sake of completeness and historical importance, be presented in detail in Sect. 11.3.

The following definition will be useful to compare languages later (for instance in Chap. 11 and Sect. 10.2).

2.2 Sets, Relations and Functions

Before we continue and define problems and solutions more formally, we recall some basic definitions that allow us to make formal statements throughout this book. Readers well familiar with those concepts can skip this section. We will discuss sets and structures on sets, namely relations and functions. We introduce operations on sets, fix notation, and recall some basic reasoning principles, which will be used throughout the book. A proper introduction to sets and logic for computing can be found e.g. in [8].

2.2.1 Sets

Sets are collections of objects. The collections can be finite or infinite. We will usually only consider *homogeneous* sets which means that the objects in a set are all of the same type.[11] For each element of this type one must be able to say whether the element is in the given set or not. An example of a set of natural numbers is the set S_{10} containing the numbers from 1 to 10. In this case, number 3 is in the set S_{10} but number 42 is not. It is important to observe that one does not care how many times the objects appears in the set as one would do in a list or an array. A set thus abstracts away from the number of occurrences. An object simply is either in or out. If we have such knowledge for all objects of the given underlying type we have uniquely defined a set.

Let us now fix some notation:

Definition 2.1 (*Sets*) A finite set containing n different objects $e_1, e_2, \ldots, e_n$ is written

$$\{e_1, e_2, \ldots, e_n\}$$

---

[11] This type may be a set again.

We call those objects contained in a set, the *elements* of this set. The *empty set* is the unique set that contains no elements at all and is usually denoted {} or ∅.

Let A be a set of elements of type T. The *elementhood operation* is a statement

$$x \in A$$

stating that element x is in set A ("belongs to A", "is contained in A"). If the set is infinite, we cannot write down all the elements. In this case we usually write the "law" that states which elements are in the set as follows (which can also be used to describe finite sets). If S is a type and $P(x)$ denotes a condition on variable x then

$$\{x \in S \mid P(x)\}$$

describes the set of all elements of type S that have property P. The type of all natural numbers is denoted $\mathbb{N}$ (which contains 0), the integer numbers is denoted $\mathbb{Z}$ and the real numbers is denoted $\mathbb{R}$. The type of Boolean values {true, false} is denoted $\mathbb{B}$.

Example 2.1 Here are some examples of finite sets with objects (elements) in $\mathbb{N}$:

1. {1, 10, 100}: the set of natural numbers containing the three elements 1, 10 and 100.
2. ∅: the empty set containing no natural number.
3. $\{x \in \mathbb{N} \mid x \text{ is even}\}$: the infinite set of all even natural numbers, which is the set {0, 2, 4, 6, 8, 10, 12, ...}. The notation with ... followed by a closing } is sometimes used to indicate an infinite set when the condition P used to define it is clear from the context. Note that in this example the condition $P(x)$ is "x is even".
4. $\{x \in \mathbb{N} \mid x = 10^n, 0 \leq n \leq 2\}$: the finite set containing the first three powers of 10, namely $1 = 10^0$, $10 = 10^1$ and $100 = 10^2$. So in fact this set is equal to the first. More about equality of sets follows these examples.

Definition 2.2 (*Set equality and subsets*) Let S_1 and S_2 be two sets ranging over the same type T of objects. We say that two sets S_1 and S_2 are *equal*, short $S_1 = S_2$, if, and only if, they contain exactly the same elements. This confirms that it is enough to know which elements are in the set and which are not to uniquely define a set.

We say that a set S_1 is a *subset* of a set S_2 (or S_1 is contained in S_2) if, and only if, every element of S_1 is also an element of S_2. More formally we can also write

$$S_1 \subseteq S_2 \iff \forall x \in T. \; x \in S_1 \Rightarrow x \in S_2$$
$$S_1 = S_2 \iff \forall x \in T. \; x \in S_1 \Leftrightarrow x \in S_2$$

where $\Rightarrow$ denotes implication and $\Leftrightarrow$ denotes equivalence and $\forall x \in T. \; P$ denotes universal quantification over all elements of type T.

In the above definition we used the phrase *"if, and only if"* (in the formal version $\Leftrightarrow$) and not just *"if"* (formally $\Leftarrow$) for a good reason. For a definition, it is important to cover all cases exactly. Consider the following statement: "Sets A and B (over natural numbers) are equal if S and T are both the empty set." This is obviously a correct statement about equality of sets A and B. But it is far from a definition of equality. The statement does not specify anything about the equality of non-empty sets. Clearly, its contraposition "if A and B are equal sets then A and B are both empty" is wrong. Thus the statement "sets A and B (over natural numbers) are equal if, and only if, S and T are both the empty set." is equally wrong.

As explained above the use of phrase "if, and only if" is important and we will encounter it often throughout the book. Therefore, we sometimes abbreviate it and simply write "iff" instead of "if, and only if".

In order to show equality of two sets, an important reasoning principle is often used:

Proposition 2.1 *Let S_1 and S_2 be sets of objects in T, then $S_1 = S_2$ if, and only if, $S_1 \subseteq S_2$ and $S_2 \subseteq S_1$. In other words, S_1 equals S_2 if, and only if, S_1 is a subset of S_2 and vice versa.*

Proof We need to show the two directions of the "if, and only if". The "only if" ($\Rightarrow$) and the "if" ($\Leftarrow$) direction.

"$\Rightarrow$": If $S_1 = S_2$ then by definition $S_1 \subseteq S_2$, as being equal is a special (degenerated) case of being a subset of. Analogously, $S_2 \subseteq S_1$.

"$\Leftarrow$": Assume $S_1 \subseteq S_2$ and $S_2 \subseteq S_1$. To show that both sets are equal we must show that they contain exactly the same elements, i.e. for all $x \in T$ it must hold that $x \in S_1$ iff $x \in S_2$. Unfolding the meaning of "iff" we get two conditions for all $x \in T$, namely $x \in S_1 \Rightarrow x \in S_2$ and $x \in S_2 \Rightarrow x \in S_1$. We can move the quantifier $\forall x$ around both conditions separately without changing the meaning of the formula, so it suffices to show:

$$\forall x \in T.\ x \in S_1 \Rightarrow x \in S_2 \qquad \text{and}$$
$$\forall x \in T.\ x \in S_2 \Rightarrow x \in S_1$$

and thus by Definition 2.2 that $S_1 \subseteq S_2$ and $S_2 \subseteq S_1$ which were our assumptions.

Definition 2.3 (*Set operations*) We will use the following standard operations on sets: union ($S_1 \cup S_2$) intersection ($S_1 \cap S_2$) and set difference ($S_1 \backslash S_2$). They are defined as follows:

$$x \in S_1 \cup S_2 \iff x \in S_1 \vee x \in S_2$$
$$x \in S_1 \cap S_2 \iff x \in S_1 \wedge x \in S_2$$
$$x \in S_1 \backslash S_2 \iff x \in S_1 \wedge \neg(x \in S_2)$$

where $\vee$ denotes logical disjunction ("or"), $\wedge$ denotes logical conjunction ("and"), and $\neg$ denotes logical negation ("not"). If x is not contained in A, we usually abbreviate $\neg(x \in A)$ by simply writing

$$x \notin A.$$

If S is a set of elements of type T, we call $T \backslash S$ the *complement* of S, which is sometimes also abbreviated $\overline{S}$.

Example 2.2 Here are some concrete examples of set operations and their results:

$$\{3, 5, 7\} \cup \{2, 4, 6, 8\} = \{2, 3, 4, 5, 6, 7, 8\}$$
$$\{3, 5, 7\} \cap \{2, 4, 6, 8\} = \{\}$$
$$\{3, 5, 7\} \backslash \{3, 5, 8, 16\} = \{7\}$$
$$\mathbb{N} \backslash \{x \in \mathbb{N} \mid x \text{ is even}\} = \{x \in \mathbb{N} \mid x \text{ is odd}\}$$

Definition 2.4 (*Cartesian product*) Let S_1 and S_2 be sets. Then $S_1 \times S_2$, the *Cartesian product*[12] of S_1 and S_2, is the set of pairs (i.e. tuples) (s, t) where $a \in S_1$ and $b \in S_2$. In other words:

$$S_1 \times S_2 = \{(s, t) \mid s \in S_1 \wedge t \in S_2\}$$

Based on the cartesian product one can also form sets of tuples of length k over a given set or type:

Definition 2.5 (*k-tuples*) Let S be a set. Then S^k, the *k-tuples* over S, is defined as follows:

$$S^k = \underbrace{S \times S \times S \ldots \times S}_{k \text{ times}}$$

such that elements of S^k are tuples of length k, i.e. $(s_1, s_2, \ldots, s_{k-1}, s_k)$ where $s_i \in S$ for all $1 \leq i \leq k$.

If we consider sets from a programming perspective as a kind of datatype then we would like to nest the set data type constructor. In other words, we would like to have set of sets, and so on.

Definition 2.6 (*Powerset*) Let S be a set. Then $Set(S)$, the *powerset* of S, denotes the set of all subsets of S, including the empty set, and the set S.

$$Set(S) = \{\text{set } s \mid s \subseteq S\}$$

Example 2.3

$$Set(\{1, 10, 100\}) = \{\emptyset, \{1\}, \{10\}, \{100\}, \{1, 10\}, \{1, 100\}, \{10, 100\}, \{1, 10, 100\}\}$$
$$Set(\mathbb{N}) \qquad = \{\text{set } s \mid \forall x \in s . \ x \in \mathbb{N}\}$$

---

[12]The Cartesian product is named in honour of Reneé Descartes (31 March 1596–11 February 1650), a French mathematician and philosopher, who spent most of his life in the Netherlands, and is famous for his saying "I think therefore I am" as well as for the development of (Cartesian) analytical geometry. He was invited to the court of Queen Christina of Sweden in 1649. "In Sweden—where, Descartes said, in winter men's thoughts freeze like the water—the 22-year-old Christina perversely made the 53-year-old Descartes rise before 5:00 am to give her philosophy lessons, even though she knew of his habit of lying in bed until 11 o'clock in the morning" [9]. Consequently, Descartes caught pneumonia and died.

Another standard set used in this book is the set of words over a finite alphabet Σ as used for instance by *finite state automata*. These words are just finite strings of letters of the alphabet, including the empty string.

Definition 2.7 (*Set of words*) Let Σ be a finite alphabet of symbols (or letters). Then we define a new set Σ^* by providing the rules to generate elements of this set and state that these are the only rules to generate elements of the set. The rules are as follows:

$$\varepsilon \quad \in \Sigma^*$$
$$aw \in \Sigma^* \ \ \text{if} \ \ a \in \Sigma \wedge w \in \Sigma^*$$

This is an *inductive* definition. The first rule states that the empty word ε is a word which provides the termination case of the induction. The second rule describes how to generate new elements from already generated ones.

Example 2.4 For alphabet $\Sigma = \{0, 1\}$ here are some examples of words in Σ^*:

- ε (the empty word)
- 0
- 01
- 11111001

Finally, for the definition of partial functions in Sect. 2.2.4 we need a way to extend a set by a unique new symbol.

Definition 2.8 (*One-point-extension*) Let S be a set over type T. Then we define a new set $S_\perp$ in $Set(T)$ by adding to S a new element $\perp$ (called[13] "undefined") that is assumed to be different from all elements in S.

$$S_\perp = \{x \in T \cup \{\perp\} \mid x \in S \ \vee \ x = \perp\}$$

2.2.2 Relations

Relations are special sets. We have already seen some relations in the previous section, the equality and subset relation are both binary relations on sets. A binary relation R is simply a set of pairs and we say that two elements s and t are in this relation R if the pair (s, t) is in the set R.

Definition 2.9 (*Relations*) Relations are sets. We define:

- a *unary relation* R over elements of type T is a subset $R \subseteq T$. An object $t \in T$ is said to be *in relation* R iff $t \in R$.
- a *binary relation* R over elements of type $S \times T$ is a subset $R \subseteq S \times T$. A pair (s, t) is said to be *in relation* R iff $(s, t) \in R$.

---

[13]The symbol itself is called a "perp".

- a *ternary relation* R over elements of type $S \times T \times U$ is a subset $R \subseteq S \times T \times U$. A triple (s, t, u) is said to be *in relation* R iff $(s, t, u) \in R$.[14]

Example 2.5 The quality relation and subset relation over sets of elements of type T as defined in Definition 2.2 are actually binary relations in the following sense:

$$\_ = \_ \subseteq Set(T) \times Set(T)$$
$$\_ \subseteq \_ \subseteq Set(T) \times Set(T)$$

We usually write $S_1 \subseteq S_2$ (so-called "infix notation") instead of $(S_1, S_2) \in \_ \subseteq \_$.

2.2.3 Functions

We intuitively understand what the addition or multiplication functions are, and maybe also the factorial function. Functions describe *maps* from objects of a certain type into objects of another[15] type. In functional programming languages, functions are first-class citizens. The programmer can define those functions syntactically. For us, however, functions are descriptive[16] and *not* programs or part thereof. We can describe functions as special relations which we will do next.

2.2.4 Partial Functions

Definition 2.10 (*Partial Functions*) Let A and B be sets of possibly different types of elements, e.g. $A \in Set(S)$ and $B \in Set(T)$. A *partial function* f from A to B is a subset of $A \times B$ (i.e. $f \subseteq A \times B$) satisfying the following uniqueness condition:

For all $a \in A$ there is *at most one* $b \in B$ such that $(a, b) \in f$.

To abbreviate that f is a partial function from A to B we briefly write $f : A \to B_\bot$, where we call A the argument type of f and B the result type. The reason for actually writing $B_\bot$ (defined in Definition 2.8) in this notation will become clear shortly when we define the following binary *application relation* $\_@\_ \subseteq (A \to B_\bot) \times A$ for a partial function $f : A \to B_\bot$ and an element $a \in A$:

$$f @ a = \begin{cases} b & \text{if } (a, b) \in f \\ \bot & \text{otherwise} \end{cases}$$

---

[14]At first glance, it is not obvious whether $S \times T \times U$ means $S \times (T \times U)$ or $(S \times T) \times U$. We assume cartesian products to be *associative*, identifying these two definitions, thus dropping the extra parentheses and writing (s, t, u) for triples, and similarly for n-tuples where $n > 3$.

[15]Which may possibly be the same type.

[16]Mathematical objects.

Instead of $f @ a$ we will be writing $f(a)$ which is the common notation for function application also widespread in programming languages. If $(a, b) \in f$ we therefore simply write $f(a) = b$ and say that "f *applied* to a equals b." We also call a the *argument* of the function (application) and b the *result* of the application. By the uniqueness condition we know that there can only be one result which always lies in $B_\perp$. It is possible that no such $b \in B$ exists, in which case $f(a) = \perp$ and we say that f is *undefined* for a and often use the short notation $f(a)\uparrow$ to express this. We also sometimes use $f(a)\downarrow$ to express that $f(a)$ is *defined* when we are not interested in the concrete result value.

Functions with more than one argument are simply described by using a cartesian product as argument type. In this case the function takes a tuple as input.

Example 2.6 Consider the integer division operator on natural numbers, *div*. This *partial* function takes two integers n and m and returns $\frac{n}{m}$ in case $m \neq 0$. Thus $div : \mathbb{N} \times \mathbb{N} \to \mathbb{N}_\perp$.

$$((n, m), r) \in div \ \text{ iff } \ \exists k < m . m \times r + k = n$$

where $m, n, r, k \in \mathbb{N}$. Note that we have that $div\,(n, 0)\uparrow$ as there is no natural number k that is smaller than 0.

2.2.5 Total Functions

Total functions are total in the sense that function application always returns a defined value. Therefore, total functions are just a very special case of partial functions.

Definition 2.11 (*Total Functions*) Let A and B be sets of possibly different types of elements, e.g. $A \in Set(S)$ and $B \in Set(T)$. A *total function* f from A to B is a subset of $A \times B$ (i.e. $f \subseteq A \times B$) satisfying the following two conditions (where the first is the uniqueness condition for partial functions):

1. For all $a \in A$ there is *at most one* $b \in B$ such that $(a, b) \in f$.
2. For all $a \in A$ there is *at least one* $b \in B$ such that $(a, b) \in f$.

To abbreviate that f is a total function from A to B we briefly write $f : A \to B$ where we again call A the argument type of f and B the result type. We can take the binary *application relation* defined in Definition 2.10 for partial functions and restrict its type to $\_@\_ \subseteq (A \to B) \times A$ for a total function $f : A \to B$ and argument $a \in A$. Since for total functions we know from the second condition that there must always be a $b \in B$ for every $a \in A$ (which is unique by the first condition) so we can never have $f @ a = \perp$. As for partial function application, we write $f(a)$ for $f @ a$ and if $(a, b) \in f$ we simply write $f(a) = b$ and say that "f *applied* to a equals b."

Example 2.7 Consider the factorial function on natural numbers, *fac*. Often the notation *n*! is used instead of application *fac*(*n*). This *total* function $fac : \mathbb{N} \rightarrow \mathbb{N}$ takes an integer *n* and returns the factorial of *n* defined as follows:

$$(n, r) \in fac \ \text{ iff } \ (n = 0 \wedge r = 1) \ \vee \ (n > 0 \wedge (n-1, s) \in fac \ \wedge \ r = n \times s$$

where $n, r, s \in \mathbb{N}$. This is a recursive (actually inductive) definition of *fac* as we use the function (functional relation) *fac* on the left and right hand side of the definition. The definition is, however, well defined as the argument for the application of *fac* on the right hand side uses a "smaller" argument $n - 1$ than the one on the left hand side (which uses *n*). When defining functions in this book we will normally not define the relation that defines the function but write the (equivalent) definition of function application, i.e. we define the result of $f(n)$ rather than defining the relation $(n, r) \in f$. For the factorial function we would typically write:

$$fac(n) \ = \ \begin{cases} 1 & \text{if } n = 0 \\ n \times fac(n-1) & \text{otherwise} \end{cases}$$

2.3 Problems

Our first quest is to find a problem that is not computable (or decidable by a computer program) to learn and understand that not everything is computable even with unlimited resources. In order to do this, we obviously need to define what we mean exactly by "*problem*" and what we mean exactly by "*computable*." Whereas the former is easy to do the latter is a bit more tricky.

We will allow ourselves to restrict the definition of *problem*. Since we are interested in the *Limits of Computation*, we are interested in *negative* results, i.e. what *cannot* be achieved in computing. If there is a problem of a restricted kind that is not computable then we still have found a problem that is not computable, so this restriction does not take anything away from our ambition.

A *problem* of the kind we are interested in is characterised by two features:

1. It is a uniform class of questions. *Uniform* refers to the domain of the problem, i.e. what data the problem is about. The type of domain must be precisely definable.
2. It can be given a definite and finite answer. The type of the answer must be also precisely definable.

The type in question can be any set, like for instance $\mathbb{N}$, $\mathbb{N}_\perp$, Σ^* and so on.

Definition 2.12 Let *S* and *T* some well defined (finite) types. A *function problem* is a uniform set of questions, the answers of which have a finite type. The solution of a function problem is given as a partial function $f : S \rightarrow_\perp T$ as described in Sect. 2.2.4. The uniform question of this problem is of the sort: "given an $x \in S$, what is a $y \in T$ such that a certain condition on *x* and *y* holds?"

A *decision problem* is a relation $R \subseteq S$. The uniform question of this problem is of the sort: "given an $x \in S$, does x belong to R, i.e. $x \in R$? The solution of a decision problem is given as a *total* function $\chi : S \to \mathbb{B}$, also called the *characteristic function of R*.

Example 2.8 Here are some examples of function and decision problems:

1. For a tree t, what is its height? Domain: trees (apparently with arbitrary number of children). Answer for any given tree t: a natural number describing the height of t (and we know what the meaning of "height of a tree" is). The answer type is the type of natural numbers.
2. For a list of integers l, what does l look like when sorted? In other words, what is the sorted permutation of l using the usual ordering on integers? Domain and answer type are here the type of integer lists.
3. For a natural number n, is it even? Domain: natural numbers. Answer for any given number n: a Boolean,[17] stating whether n is even or not (we understand what even and odd mean). The answer type is the type of boolean values.
4. For a given formula in number theory (arithmetic) ϕ, is it valid? Domain: formulae in arithmetic. Answer for a given formula ϕ: a Boolean, stating whether the formula ϕ is true (and we understand what it means for a formula to be true).

The first two examples above are function problems, the last two examples are decision problems.

Example 2.9 Here are some examples of problems that *do not qualify* as problems for us.

1. "What is the meaning of life?"[18] This is not a uniform family of questions. Moreover, we do not know what the answer type is. If we'd expect a string as answer then it would still not qualify as we don't know whether there *is* a definite answer.
2. "Is the number 5 even"? This is not a *uniform* class of questions, as this question only refers to the number 5.

---

[17]Boolean values are named after George Boole (2 November 1815–8 December 1864), an English mathematician and logician famous for his work on differential equations and algebraic logic. He is most famous for what is called Boolean algebra. Throughout this book, we will use the term "boolean" to indicate a truth value for which the corresponding algebra operations are available.

[18]This question is easily confused with the one famously asked in Douglas Adam's masterpiece: "The Hitchhiker's Guide to the Galaxy" [2] which actually is called: "Ultimate Question of Life, The Universe, and Everything". The computer in question, *Deep Thought*, after a considerable 7.5 million years answered famously: "42". Alas, nobody understood the question. So *Deep Thought* suggested to build an even more powerful super-computer to produce the question to the answer. This computer was later revealed to be planet "Earth" which was unfortunately destroyed 5 min before completion of the calculations.

2.3.1 Computing Solutions to Problems

According to the two types of problems introduced, we will consider two concrete kinds of "solving a problem": computing a function and deciding membership in a set.

The data type of Turing machines is the set of finite words over a finite alphabet. Recall that Σ^* denotes all finite words over the alphabet Σ, including the empty word. The comparison test for tape symbols is built into the construction set and from that it is possible to implement equality of words. In a general notion of effective procedure, the data type should be general enough to encode finite words and their equality test. The latter must be effective so it must be terminating. This means that equality of infinite objects is likely to be problematic and thus we do not cover computability over infinite objects in this book.

Definition 2.13 Provided a certain choice of effective procedures $\mathscr{P}$, a (function or decision) problem is called $\mathscr{P}$-computable if, and only if, its solution can be computed (calculated) by carrying out a specific such effective procedure in $\mathscr{P}$. A decision problem that is computable is also called $\mathscr{P}$-decidable.

If the kind of effective procedures is known by the context we also simply use the unqualified terms *computable* and *decidable*.

It is important to remember that programs are solutions to *computable problems*. The programs that solve computable decision problems are also called decision procedures.

Example 2.10 The solutions to the computable and decidable, resp., problems in Example 2.8 are given below as *programs*.

1. For a tree t what is its height? The solution is a function program that takes a tree as input and computes its height.
2. For a list of integers l what is l sorted? The solution is a program that takes a list of integers as input and returns a sorted copy of the list. The program can use various well known sorting algorithms, e.g. bubble-sort, merge-sort, or quicksort. They all perform the same task eventually, but use different methods to achieve this and also may take different time. This is an issue we will discuss in the complexity part.
3. For a natural number n, is it even? The solution in a program takes a natural number as the input and returns the boolean value true if the input is even and false if it is odd. We call such a program also a decision procedure for the property of "being even".
4. For a given formula in number theory (arithmetic) ϕ, is it valid? As discussed in the introduction, this is undecidable, so there can't be any program that takes as input as an arithmetic formula (suitably encoded) and returns true if the formula is satisfiable and false if it is not.

What Next?

Now that we know what we mean by "computable" and have seen that the historically first definition of computability via a machine involves tedious low level programming, we want to define a high-level language that can do the job as well. So in the next chapter we introduce the language WHILE and in the following chapter we show that WHILE-programs can be legitimately chosen for effective procedures.

Exercises

1. What is the "Entscheidungsproblem"? What is the type of its domain? Is it a decision problem?
2. Why did Alan Turing allow his (pencil and paper) computing device to use only finitely many symbols (on the tapes) and let the "state of mind" of the computer only glance at finitely many symbols at any given time?
3. Which of the following pairs of sets A and B are equal? Show either $A = B$ or $A \neq B$.

   a. $A = \mathbb{N} \times \mathbb{N}$ and $B = \mathbb{N}^2$
   b. $A = \{1, 3, 5\}$ and $B = \{1, 3, 5, 6\}$
   c. $A = \{1, 3, 3, 3\}$ and $B = \{1, 3\}$
   d. $A = \{x \in \mathbb{N} \mid x = x + 1\}$ and $B = \emptyset$
   e. $A = \{x \in \mathbb{N} \mid \text{even}(x) \wedge x < 11\}$ and $B = \{0, 2, 4, 6, 8, 10\}$

4. Describe the relation that one natural number can be divided by the second natural number without remainder as $R_{\text{divisible}} \subseteq \mathbb{N} \times \mathbb{N}$.
5. Give an example of a partial function of type $\mathbb{N} \to \mathbb{N}_\perp$ and an example of a total function $\mathbb{N} \to \mathbb{N}$, respectively.
6. What is the difference between a decision problem and a function problem?
7. Give an example of a problem that is neither a decision nor a function problem. Why is it acceptable that we consider only those specific kinds of problems?
8. Give two other examples of decision and function problems, respectively, that have not been mentioned in this chapter.
9. Assume that we have fixed the notion of effective procedures $\mathscr{P}$. When do we call a function problem $\mathscr{P}$-computable?
10. Assume that we have fixed the notion of effective procedures $\mathscr{P}$. When do we call a decision problem $\mathscr{P}$-decidable?

References

1. ACM Home Page, available via DIALOG, http://www.acm.org. Cited on 30 August 2015
2. Adams, D.: The Hitchhiker's Guide to the Galaxy. Pan Books (1979)
3. Church, A.: An unsolvable problem of elementary number theory. Am. J. Math. **58**(2), 345–363 (1936)
4. Copeland, J.: The Church-Turing Thesis. References on Alan Turing (2000). Available via DIALOG. http://www.alanturing.net/turing_archive/pages/Reference%20articles/The%20Turing-Church%20Thesis.html. Cited 2 June 2015

5. Goldin, D., Wegner, P.: The church-turing thesis: breaking the myth. In: Cooper, S.B., Löwe, B., Torenvliet, L. (eds.) New Computational Paradigms. Lecture Notes in Computer Science, vol. 3526, pp. 152–168. Springer, Heidelberg (2005)
6. Hilbert, D., Ackermann, W.: Grundzüge der theoretischen Logik. Springer, Berlin (1928). (Principles of Mathematical Logic.)
7. Jones, N.D.: Computability and Complexity: From a Programming Perspective. MIT Press, Cambridge (1997). (Also available online at http://www.diku.dk/neil/Comp2book.html.)
8. Makinson, D.: Sets, Logic and Maths for Computing, 2nd edn. Springer, UTiCS Series (2012)
9. Reneé Descartes. Entry in Encyclopædia Britannica, http://www.britannica.com/biography/ Rene-Descartes/Final-years-and-heritage. Available via DIALOG. Cited on 2 Sept 2015
10. Soare, R.I.: The history and concept of computability. In: Griffor, E.R. (ed.) Handbook of Computability Theory, pp. 3–36. North-Holland (1999)
11. The Joint Task Force for Computing Curricula 2005 (ACM, AIS, IEEE-CS): Computing Curricula 2005. Available via DIALOG, http://www.acm.org/education/curric_vols/CC2005-March06Final.pdf (2005)
12. Turing, A.: Systems of logic based on ordinals. Proc. London Math. Soc. **45**(1), 161–228 (1939)

Chapter 3
The WHILE-Language

What language do we use to write our "effective procedures"?

In the previous chapter, we have observed that Turing machines are tedious to program and thus not an ideal language to study computability and complexity issues. Moreover, we are used to program in high-level languages like Java, C, Haskell, Python, and so on. We thus prefer to program in a high-level language in order to write effective procedures in the sense of Turing. The notion of computable or decidable will then be based on the programmability in this language. We follow Neil Jones' idea [2, Preface, Page X] and use the language WHILE since "The WHILE language seems to have just the right mix of expressive power and simplicity."

In Chap. 7 it will become clear why this particular mix is important and desired for the purposes of investigating questions of computability. For now it just appears to be a sensible choice to be able to write programs (relatively) easily in the way we are used to and also have a language that is simple enough to understand the semantics of programs.

This chapter is a short introduction to this language. In the next chapter it will be argued that it is actually a good definition of "effective procedure" meeting Turing's criteria. Before we discuss the details, we pause to think about the data type the language with the "right mix" is supposed to provide. Since we want a simple language, it will not feature explicit types and a type system but use just one basic data type. But since we want an expressive language, the built-in datatype needs to be flexible enough to encode any data type the programmer would like to use, be it integers or natural numbers, be it lists or even abstract syntax trees. Traditionally, the simplest imperative language in use for this kind of purpose supports sequential composition of commands and while loops as well as assignments where all variables store natural numbers. It is however, extremely inconvenient to encode values of other data types as natural numbers.

© Springer International Publishing Switzerland 2016   29
B. Reus, *Limits of Computation*, Undergraduate Topics in Computer Science,
DOI 10.1007/978-3-319-27889-6_3

The idea followed in [2] is therefore to use a type that can encode others in a way similar to *JSON* (JavaScript Object Notation, see [6]) or *XML* (eXtensible Markup Language, see [8]). These are widely used document notations for exchanging data between software agents. It is important that such documents can be easily parsed and manipulated by machines but are at the same time readable by humans. All these formats rely on trees of attribute-value pairs. These trees are denoted linearly as nested lists. In Fig. 3.1 we exemplify this showing how a data record can be easily expressed in three different formats: XML, JSON, and (LISP style) s-expressions.

```
<scientist id="AMT">
  <firstName>"Alan"</firstName>
  <midInitial>"M"</midInitial>
  <lastName>"Turing"</lastName>
  <famousFor>
     <achievement>"crack Enigma code"</achievement>
     <achievement>"define computability"</achievement>
  </famousFor>
</scientist>
```

```
{
   "scientist": {
     "id": "ATM",
     "firstName": "Alan",
     "midInitial": "M",
     "lastName": "Turing",
     "famousFor": {
          { "achievement" : "crack Enigma code" },
          { "achievement": "define computability" }
          }
     }
   }
```

```
(scientist
  (id "ATM")
  (firstName "Alan")
  (midInitial "M")
  (lastName "Turing")
  (famousFor
     (achievement "crack Enigma code")
     (achievement "define computability")
  )
)
```

Fig. 3.1 Encoding data in XML, JSON and as an s-expression

The first elements in each list are called *atoms*. These *atoms* are indivisible and form the simplest form of trees consisting of just a leaf. It seems logical to choose a data type for WHILE that allows the encoding of such trees of attribute value-pairs, which can be expressed as nested lists. The type of nested lists, or s-expressions, can be represented by a very simple type of binary trees.

In the following we first discuss the data type of binary trees (Sect. 3.1) and introduce the (syntax of the) imperative untyped language WHILE that makes use of it (Sect. 3.2). Section 3.3 explains how various datatypes like natural numbers and lists can be encoded in WHILE. Finally, we present some sample programs in Sect. 3.4. Note that the semantics of WHILE will be discussed in detail in the next chapter.

3.1   The Data Type of Binary Trees

The values in our data type of trees are *binary trees* where every node has either two subtrees or is a leaf labelled with an atom. This datatype is similar to the LISP datatype of s-expressions. Here is an example of such a binary tree where all atoms are nil:

Definition 3.1 The set of binary trees is given inductively. It contains

1. the *empty tree*: nil
2. any tree constructed from two binary trees t_l and t_r:

   and which is written $\langle t_l.t_r \rangle$ in textual notation
3. and these are the only terms in the set of binary trees.

The set of binary trees is denoted $\mathbb{D}$ (short for "data").

Accordingly, the tree given above is written linearly as

$$\Big\langle \langle \text{nil}.\text{nil}\rangle.\big\langle \langle \text{nil}.\text{nil}\rangle.\langle\langle \text{nil}.\text{nil}\rangle.\text{nil}\rangle \big\rangle \Big\rangle$$

The size of a tree t, short $|t|$, is the number of leaves the tree has. Therefore the size of a tree can be described inductively as follows:

Definition 3.2 The function size : $\mathbb{D} \to \mathbb{N}$ is defined as follows

$$\text{size } t = \begin{cases} 1 & \text{if } t \text{ is an atom} \\ \text{size } l + \text{size } r & \text{if } t = \langle l.r \rangle \end{cases}$$

For example, the tree pictured above has size 7.

3.2 WHILE-Syntax

Let us now define the syntax of the programming language WHILE. This will be done in three steps: expression language, command language and programs. As common for programming language syntax, we will also give a BNF grammar for WHILE which is a formal and precise definition of the syntax.

3.2.1 Expressions

An expression of the WHILE-language denotes a binary tree as outlined in Sect. 3.1. We have two constructors:

1. denoting the *empty tree* nil, there is the expression nil
2. and to construct a nonempty tree there is the expression cons E F where E and F are also expressions (denoting binary trees).

We also have two destructors: hd, short for "head", and tl, short for "tail". They "disassemble" a tree (expression) into its left (head) and right (tail) subtree, respectively. They do not have any effect on the empty tree (as it does not have any subtrees). Furthermore, our expression language also contains variables.

The details of the evaluation of expressions will be defined formally in the semantics in Chap. 4.

Note that it is important to distinguish the *syntactic* language expressions that *denote* trees in WHILE and the actual *semantic* trees that are data values in $\mathbb{D}$. The expressions can be *evaluated* to values in $\mathbb{D}$ using the semantic interpretation for expressions explained in Chap. 4.

3.2.2 Commands

The simplicity of WHILE is also ensured by the fact that there are only four types of commands: assignment, conditional (if-then-else), a while loop that gives the language its name, and sequential composition to form statement blocks inside while loops and conditionals. These constructs should be all familiar to programmers and Computer Science students. Note that, unlike expressions, commands do not produce a value, but have side effects on the variables as they can change their values.

The conditional (if-then-else) and while loop both use a Boolean guard to determine control flow. We will discuss how we code boolean values in our datatype of tree in Sect. 3.3.1.

3.2.3 Programs

A program consists of a name, a fixed read statement for passing the input, a statement block, and a fixed write statement to return the output. So there is exactly one input value passed to a program and one output value produced. We will see later why having just one input and one output is not really a restriction. The passing of data in a read and write statement, respectively, is via a variable, which is called the input and output variable, respectively. So a program looks like this:

```
pname read X {
     B
}
write Y
```

where pname denotes any valid program name, and X and Y denote any valid variable (name). They can also denote the same variable, of course. The statement block B is the main body of the program. This block consists of a list of statements separated by semicolons denoting sequential composition. The exact syntax of programs is given in the next section and the semantics will be formally explained in the next chapter.

3.2.4 A Grammar for WHILE

From the above it should be obvious that WHILE is a really simple language which is, however, already quite expressive. We have basic imperative programming constructs as well as a rich data type.

Let us look now more formally at the syntax of WHILE programs. Figure 3.2 contains a complete BNF (Backus-Naur-Form) context-free grammar of the abstract syntax of WHILE. We denote non terminal symbols in angle brackets ⟨*nonterminal*⟩, and a rule is denoted as ⟨*nonterminal*⟩ :: = ... | ... where | denotes alternative productions for the same nonterminal. More on BNF grammars can be found e.g. in [1]. We assume special carriers for nonterminals ⟨*variable*⟩ and

⟨*expression*⟩	::=	⟨*variable*⟩	(variable expression)
	|	`nil`	(atom nil)
	|	`cons` ⟨*expression*⟩ ⟨*expression*⟩	(construct tree)
	|	`hd` ⟨*expression*⟩	(left subtree)
	|	`tl` ⟨*expression*⟩	(right subtree)
	|	(⟨*expression*⟩)	(parentheses)
⟨*block*⟩	::=	{ ⟨*statement-list*⟩ }	(block of commands)
	|	{ }	(empty block)
⟨*statement-list*⟩	::=	⟨*command*⟩	(single command list)
	|	⟨*command*⟩ ; ⟨*statement-list*⟩	(list of commands)
⟨*elseblock*⟩	::=	`else` ⟨*block*⟩	(else-case)
⟨*command*⟩	::=	⟨*variable*⟩ := ⟨*expression*⟩	(assignment)
	|	`while` ⟨*expression*⟩ ⟨*block*⟩	(while loop)
	|	`if` ⟨*expression*⟩ ⟨*block*⟩	(if-then)
	|	`if` ⟨*expression*⟩ ⟨*block*⟩ ⟨*elseblock*⟩	(if-then-else)
⟨*program*⟩	::=	⟨*name*⟩ `read` ⟨*variable*⟩	
		⟨*block*⟩	
		`write` ⟨*variable*⟩	

Fig. 3.2 BNF grammar for WHILE

⟨*name*⟩ which represent program variables and program names, respectively. Those names are supposed to be identifiers as used in other programming languages. The identifiers usually consist of characters (including digits and special symbols like $ or -), assuming that no identifier begins with a number. Terminal symbols in the grammar below are all given in this `font`.

The syntax presented here agrees largely with the one suggested in [3] which is also a variation of Neil Jones's WHILE as presented in [2].

3.2.5 Layout Conventions and Brackets

In order to make WHILE programs readable, and avoid too many brackets, we use the following conventions throughout:

- The command (body) of a program is indented with respect to `read` and `write`.
- The body of a while loop is often written below the `while E {` part and indented accordingly.
- Every assignment command is always written on a new line with the semicolon at the end and has the same indentation as the other commands in the same block.

Examples in Sect. 3.4 use this convention.

The expression language of WHILE uses prefix notation of operators and thus makes reading of expressions unique without the use of any brackets. For instance:

```
cons hd hd X cons Y nil
```

with no parenthesis can be uniquely parsed and denotes the same expression as

```
cons (hd (hd X)) (cons Y nil)
```

with explicit parentheses.

Comments

In WHILE-programs we will use comments of the form `(* ... *)`.

3.3 Encoding Data Types as Trees

Since our language is untyped we can, actually we must, use the binary tree type to encode anything we want to program with. This includes Boolean values, natural numbers and lists the encoding of which is described below. It is a relatively easy exercise to devise encodings for other types like integers or rational numbers. Throughout the book we use the notation $\ulcorner \_ \urcorner$ for the encoding into $\mathbb{D}$.

3.3.1 Boolean Values

The empty tree nil encodes false, any other tree, in particular $\langle \text{nil.nil} \rangle$ encodes true. In other words, we define an encoding operator on Booleans as follows:

Definition 3.3 We encode Boolean values as follows:

$$\ulcorner \text{false} \urcorner = \text{nil}$$
$$\ulcorner \text{true} \urcorner = \langle \text{nil.nil} \rangle$$

The while loop and the conditional interpret nil $\in \mathbb{D}$ as false, and *anything else* as true.[1] So one does not have to use $\ulcorner \text{true} \urcorner$ as guard to express true. In the next chapter on semantics it has to be ensured that the while loop and the conditional respect this interpretation of Boolean values.

---

[1] The language C similarly treats 0 as false and any non-zero value as true.

It is left as Exercise 5 how to implement programs for Boolean operators like conjunction ∧, disjunction ∨, and negation ¬.

3.3.2   Lists and Pairs

First note that lists can contain other lists as elements. They do not have to be homogenous in the sense that all the elements of a list have to be of the same type. This does not make sense since WHILE is untyped. As notation for lists we use here square brackets like in Haskell [5] and not round brackets as in Lisp [4]. So we write the empty list as [ ] and the list containing numbers 2, 4 and 8 for instance as [2, 4, 8]. Lists can be encoded using the inductively defined operator ⌜_⌝:

Definition 3.4 The empty list is encoded by the empty tree nil and appending an element at the front of the list is modelled by ⟨_._⟩. More formally we define:

$$\ulcorner [\,] \urcorner = \text{nil} \tag{3.1}$$

$$\ulcorner [a_1, a_2, \ldots, a_n] \urcorner = \langle \ulcorner a_1 \urcorner . \langle \ulcorner a_2 \urcorner . \langle \cdots \langle \ulcorner a_n \urcorner . \text{nil} \rangle \rangle \cdots \rangle \rangle \tag{3.2}$$

Let _::_ be the operation that appends an element at the front of a list. From the encoding it follows that $a::l$ can be encoded as ⟨ $a.l$ ⟩. Moreover, for a list encoded as ⟨ $e.f$ ⟩ the element e is the head of the list and f is the tail of the list. Accordingly, the WHILE expressions hd and tl compute the first element, and the rest of a list, respectively, if they are applied to an expression that actually encodes a list in the sense above.

Example 3.1 Here are a few more examples of encoded lists:

1. The list [[ ], [ ]] that contains two empty lists can be encoded as:

$$\ulcorner [[\,], [\,]] \urcorner = \langle \text{nil}. \langle \text{nil.nil} \rangle \rangle$$

   The two-dimensional representation of this tree looks as follows:

2. The list containing two trees ⟨ nil.nil ⟩ and ⟨ nil.⟨ nil.nil ⟩ ⟩ can be encoded as:

$$\ulcorner [\langle \text{nil.nil} \rangle, \langle \text{nil.} \langle \text{nil.nil} \rangle \rangle] \urcorner = \langle \langle \text{nil.nil} \rangle . \langle \langle \text{nil.} \langle \text{nil.nil} \rangle \rangle . \text{nil} \rangle \rangle$$

3. The list consisting of three lists, namely a list of two nil, a list of one nil and an empty list, can be encoded as:

$$\ulcorner [[\text{nil, nil}], [\text{nil}], [\,]] \urcorner = \langle \langle \text{nil.} \langle \text{nil.nil} \rangle \rangle . \langle \langle \text{nil.nil} \rangle . \langle \text{nil.nil} \rangle \rangle \rangle$$

From the definition of the encoding and the $\mathbb{D}$, it follows that we can "listify" every tree:

Proposition 3.1 *Every tree encodes a list via the* $\ulcorner\_\urcorner$ *operator. In other words, for every* $t \in \mathbb{D}$ *there is a* list *l such that* $t = \ulcorner l \urcorner$.

Proof The list encoded by a given tree can be defined simply by "running down the right spine" of the tree until a nil is found, the "list terminator". Then anything that dangles off the spine to the left is a list element. Details are discussed in Exercise 6.

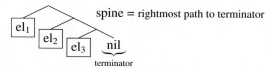

With lists we can also easily encode *pairs*, i.e. tuples (x, y) as we encode them simply as lists of two elements:

$$\ulcorner (e,f) \urcorner = \ulcorner [e,f] \urcorner \quad .$$

We could also have encoded a pair (e, f) as binary tree $\langle \ulcorner e \urcorner . \ulcorner f \urcorner \rangle$ but we avoid this to have a uniform parameter passing technique for all arities (not just 2) via lists.[2]

3.3.3 Natural Numbers

For natural numbers one can use various encodings. Most common are unary and decimal representation.

3.3.3.1 Unary Representation

Unary representation or *"unary numeral system"* known from *tally marks* is used e.g. when keeping track of scores. A (tally) stick | indicates "+1" such that for instance three sticks | | | represent the number three.[3] Instead of a stroke in our type of trees we can use nil. Then we can encode numbers simply as lists of nil atoms where the encoding of 0 is the empty list.

We define the encoding based on our underlying type $\mathbb{D}$ of binary trees and then use our list encoding to view those as lists.

---

[2]For beginners it turned out to be quite confusing to have the choice between encodings as lists or trees.

[3]Unlike tally marks though we don't group in fives.

Definition 3.5 We encode numbers inductively as follows:

$$\ulcorner 0 \urcorner = \text{nil} \qquad\qquad (3.3)$$

$$\ulcorner n + 1 \urcorner = \langle \text{nil}.\ulcorner n \urcorner \rangle \qquad\qquad (3.4)$$

Example 3.2 We give two examples of how numbers are encoded as binary trees:

1. $\ulcorner 1 \urcorner = \langle \text{nil}.\ulcorner 0 \urcorner \rangle = \langle \text{nil}.\text{nil} \rangle$,
2. $\ulcorner 3 \urcorner = \langle \text{nil}.\ulcorner 2 \urcorner \rangle = \langle \text{nil}.\langle \text{nil}.\ulcorner 1 \urcorner \rangle \rangle = \langle \text{nil}.\langle \text{nil}.\langle \text{nil}.\ulcorner 0 \urcorner \rangle \rangle \rangle = \langle \text{nil}.\langle \text{nil}.\langle \text{nil}.\text{nil} \rangle \rangle \rangle$,
    depicted two-dimensionally:

```
      nil
          nil
             nil    nil
```

Note that not every tree has the required form to encode a natural number. If we interpret the encoding trees as lists according to Definition 3.4, the number n will be encoded by a list containing n nil atoms:

$$
\begin{aligned}
\ulcorner 0 \urcorner &= [\,] \\
\ulcorner 1 \urcorner &= [\text{nil}] = [\,[\,]\,] \\
\ulcorner 2 \urcorner &= [\text{nil}, \text{nil}] = [\,[\,], [\,]\,] \\
\ulcorner 3 \urcorner &= [\text{nil}, \text{nil}, \text{nil}] = [\,[\,], [\,], [\,]\,] \\
\ulcorner 4 \urcorner &= [\text{nil}, \text{nil}, \text{nil}, \text{nil}] = [\,[\,], [\,], [\,], [\,]\,]
\end{aligned}
$$

and so on

In WHILE we can thus simply express the successor function on unary encodings by mapping n to cons nil n. And similarly the predecessor function $n - 1$ by mapping n to tl n.

We can now also encode lists of, say, natural numbers like the list that contains 1 and 2, [1, 2], as follows:

```
      ⌜1⌝
         ⌜2⌝  nil
```

Unfolding the encodings of natural numbers gives rise to the list:

$$[\,[\text{nil}], [\text{nil}, \text{nil}]\,]$$

If we unfold the list encodings too, we obtain the following tree in two-dimensional representation:

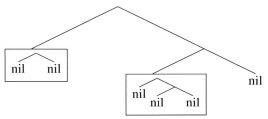

3.3.3.2 Decimal Representation

In daily life numbers are most commonly expressed in the decimal system with base 10. For instance, we know that we can represent the number 742 as $7 \times 10^2 + 4 \times 10^1 + 2 \times 1$.

Definition 3.6 We encode numbers as decimal numbers inductively as lists of digits 1–9 encoded again as natural numbers as discussed in Sect. 3.3.3. To indicate that and decimal representation we write $\ulcorner\_\urcorner^{10}$.

$$\ulcorner 0 \urcorner^{10} = [\,0\,]$$
$$\ulcorner n \urcorner = \ulcorner d_m 10^m + d_{m-1} 10^{m-1} + \ldots + d_1 10^1 + d_0 10^0 \urcorner^{10} = [\,d_0, d_1, \ldots, d_{m-1}, d_m\,]$$

where $n > 0$, $m = \log_{10} n$, $r_i = n \div 10^i$ and $d_i = r_i \mod 10$. To avoid problems with leading zeros, the encoding list begins with the least significant bit for $10^0 = 1$. So the encoding list contains the figures of decimal representation in reverse order.

Example 3.3 According to the decimal representation of Definition 3.6 we get decimal encodings of numbers as follows:

$$\ulcorner 3 \urcorner^{10} = [\,3\,]$$
$$\ulcorner 56 \urcorner^{10} = [\,6, 5\,]$$
$$\ulcorner 1332 \urcorner^{10} = [\,2, 3, 3, 1\,]$$

3.3.3.3 Binary Representation

Computer Science students will have already learned that integers can be represented in various numeral systems. There is not just the decimal system with base 10 as seen above, but e.g. the binary system with base 2. The binary system is, after all, most important for the representation on real hardware. For instance we know how to represent the following numbers: $3 = 2^1 + 2^0$, $7 = 2^2 + 2^1 + 2^0$ and $16 = 2^4$. Any natural number n can thus be represented in the dual system as $a_m a_{m-1} \ldots a_0$ where $m = \log_2 n$ such that

$$n = \sum_{i=0}^{m} a_m \times 2^i$$

Thus $3 =_2 11$, $7 =_2 111$ and $16 =_2 10000$ where the subscript in $=_2$ indicates that we use binary representation. Obviously, the binary representation, just like the decimal representation, is logarithmically shorter than the unary representation. This is a convenient fact. Imagine having to write large numbers in unary form. Binary representation is also used in all modern computer hardware. In our type of binary tree $\mathbb{D}$ we can encode binary numbers as lists of 0 and 1's (again encoded as seen above).

Definition 3.7 We encode numbers as binaries inductively as lists of 1s and 0s as follows:

$$\ulcorner 0 \urcorner^2 = \ulcorner [\, 0\,]\urcorner$$
$$\ulcorner n \urcorner = \ulcorner k + 1 \urcorner^2 = \ulcorner [\, d_0, d_1, \ldots, d_{m-1}, d_m \,]\urcorner$$

where $m = \log_2 n$, $r_i = n \div 2^i$ and $d_i = r_i \mod 2$. To avoid problems with leading 0s, the encoding list begins with the least significant bit for $2^0 = 1$. So the encoding list contains the figures of dual representation in reverse order.

Example 3.4 According to the binary representation of Definition 3.7 we get binary encodings of numbers as follows:

$$\ulcorner 1 \urcorner^2 = \ulcorner [\, 1\,]\urcorner = \langle\,\langle\, \mathrm{nil.nil}\,\rangle.\mathrm{nil}\,\rangle$$
$$\ulcorner 3 \urcorner^2 = \ulcorner [\, 1, 1\,]\urcorner = \langle\,\langle\, \mathrm{nil.nil}\,\rangle.\langle\,\langle\, \mathrm{nil.nil}\,\rangle.\mathrm{nil}\,\rangle\,\rangle$$
$$\ulcorner 6 \urcorner^2 = \ulcorner [\, 0, 1, 1\,]\urcorner = \langle\, \mathrm{nil}.\langle\,\langle\, \mathrm{nil.nil}\,\rangle.\langle\,\langle\, \mathrm{nil.nil}\,\rangle.\mathrm{nil}\,\rangle\,\rangle\,\rangle$$
$$\ulcorner 16 \urcorner^2 = \ulcorner [\, 0, 0, 0, 0, 1\,]\urcorner = \langle\, \mathrm{nil}.\langle\, \mathrm{nil}.\langle\, \mathrm{nil}.\langle\, \mathrm{nil}.\langle\,\langle\, \mathrm{nil.nil}\,\rangle.\mathrm{nil}\,\rangle\,\rangle\,\rangle\,\rangle\,\rangle$$

3.3.4 Finite Words

Turing machines use the datatype of finite words over a finite alphabet. We can encode those in $\mathbb{D}$ as well. First of all, to encode the (finite set of) symbols Σ we can either add Σ to the set of atoms or encode them by natural numbers via Definition 3.5. Then we can simply encode a word $w = s_1 s_2 \cdots s_n$ as a list of symbols $[\ulcorner s_1 \urcorner, \ulcorner s_2 \urcorner \cdots, \ulcorner s_n \urcorner]$ with the help of Definition 3.4.

3.4 Sample Programs

This section presents two concrete WHILE programs. The first one is addition and it makes use of the *unary* encoding of natural numbers. Since addition needs two inputs but WHILE programs only allow one, one needs to encode two arguments as one. In general, one often needs to encode n arguments. The easy (uniform) solution is to use one input variable that represents a *list of n data values*. The WHILE program that is passed such a list of length n will first assign the individual n components to n variables that then represent the n inputs.

3.4.1 Addition

The following program in Fig. 3.3 takes two numbers as input– encoded as a list with two elements $[\, m, n\,]$ where m and n are encoded as unary numbers themselves—and outputs $m + n$ as a (unary) natural number. As addition requires two arguments, the

```
add read L {
    X := hd L;                  (* X is first argument  m *)
    Y := hd tl L;               (* Y is second argument n *)
    while X {                   (* run through X *)
          Y := cons nil Y;      (* Y: = Y+1 *)
          X: = tl X             (* X := X-1 *)
          }
    }
    write Y                     (* finally, Y is m+n *)
```

Fig. 3.3 Adding two unary numbers in WHILE

program has to first unwrap these arguments X and Y from its input list in variable
L. It then uses the unary interpretation of numbers (as lists of nils) to run through
X. For every unit (i.e. +1) in X, variable Y is incremented by one. In other words,
the above program implements the following algorithm for addition:

$$0 + Y = Y \quad (X + 1) + Y = (X + Y) + 1$$

Replacing the operation $\_ + 1$ by *Succ* for successor (of a natural number) to improve
readability, we obtain the recursive definition:

$$0 + Y = Y \quad Succ(X) + Y = Succ(X + Y) \tag{3.5}$$

If we run the above program on input [4,4] then at each run of the loop body
X and Y contain the values listed in Table 3.1. We can read off the final result in Y.

3.4.2 List Reversal

The program in Fig. 3.4 interprets its input as a list (of something, what exactly does
not matter, the program is "polymorphic" in that respect) and the output as a list and

Table 3.1 Variable values
during execution of add

X	Y
4	4
3	5
2	6
1	7
0	8

```
reverse read X {
   Y := nil;                    (* initialise accumulator Y *)
   while X {                    (* run through input list X *)
      Y := cons hd X Y;         (* append first element of X to Y *)
      X := tl X                 (* remove first element from X *)
   }
}
write Y
```

Fig. 3.4 Reversing a list in WHILE

Table 3.2 Variable values during execution of reverse

X	Y
[1, 2, 3, 4, 5]	[]
[2, 3, 4, 5]	[1]
[3, 4, 5]	[2, 1]
[4, 5]	[3, 2, 1]
[5]	[4, 3, 2, 1]
[]	[5, 4, 3, 2, 1]

implements the reversal of its input list X. In this program, Y is an auxiliary variable in which we build the result list initialised with nil.[4]

If we run the above program on input [1, 2, 3, 4, 5] then at each run of the loop body we get values for X and Y as outlined in Table 3.2.

3.4.3   Tail Recursion

It has been observed already in [7] that recursive calls at the end of a procedure in tail position (hence the name) can be implemented easily by a "goto" without the need of a stack as it would be necessary to implement a general recursive procedure. Indeed, tail elimination is built into many optimising compilers. Instead of a goto-jump, in a high level language one can also use a while loop. Since WHILE has loops but no recursive procedures, tail elimination is a useful tool for programming in WHILE.

Definition 3.8 A recursive program is called *tail recursive* if the result of the program is produced by *one top level* recursive call. The result of the one recursive call can not be used in any further computation.

---

[4]Technically this initialisation is not needed because we will see soon that variables are automatically initialised with nil but the explicit assignment makes the program more readable.

Example 3.5 The standard definition of factorial function:

$$\text{fac}(0) = 1 \quad \text{fac}(n+1) = (n+1) \times \text{fac}(n)$$

is obviously *not* tail recursive but the following version is:

$$\text{fac2}(x, y) = \begin{cases} 1 & \text{if } x = 0 \\ \text{fac2}(x - 1, x \times y) & \text{otherwise} \end{cases}$$

One obtains the unary factorial function by initialising the accumulator correctly as 1, i.e. setting $\text{fac}(n) = \text{fac2}(n, 1)$.

3.4.4 Analysis of Algorithms

Let us revisit the algorithms implemented in Figs. 3.3 and 3.4. The addition algorithm for "+" as given in Eq. 3.5 uses a recursive call that is not yet tail recursive. However, with a bit of rewriting one can turn it into a tail recursive program:

$$\text{add}(0, acc) = acc \quad \text{add}(Succ(X), acc) = \text{add}(X, Succ(acc)).$$

Instantiating the right initial accumulator value acc as n, i.e. setting $m + n = \text{add}(m, n)$, one obtains the desired result. It follows that the WHILE program for addition given in Fig. 3.3 *does actually encode* this tail recursive addition algorithm with the help of a while loop (after tail elimination so to speak).

Consider the following algorithm for reversing a list that uses an accumulator and a tail recursive call:

$$\text{reverse}([\,], Y) = Y \quad \text{reverse}([\,a_1, a_2, \ldots, a_n\,], Y) = \text{reverse}([\,a_2, \ldots, a_n\,], a_1 :: Y))$$

If we initialise the accumulator Y as nil—setting $\text{rev}(Y) = \text{reverse}(Y, \text{nil})$—we get the desired result. The WHILE program in Fig. 3.4 implements this tail recursive algorithm rev with the help of a while loop. The _::_ operator in WHILE is realised simply by cons due to the list encoding as specified in Sect. 3.3.

The fact that both algorithms above use tail recursive calls allows us to simply use a while loop to implement recursion. If we use recursive calls that are not tail recursive, matters get significantly more complicated. This will be discussed in more detail in Chap. 7.

What Next?

Now that we know the syntax of the WHILE language and its data type, we need to make sure that programs of this language actually qualify as notion of computation. In other words, are WHILE programs good enough to be used to describe "effective procedures"? What Turing once did for his machines, we need to do for WHILE, i.e. we have to describe precisely and formally, in any case very carefully, how WHILE programs are executed. We will do this without reference to any compiler or interpreter directly and, like Turing did, on paper in the next chapter.

Exercises

1. Lists in WHILE-programs:

   a. Draw the following elements of $\mathbb{D}$:

   $$\langle\,\langle\, \text{nil}.\text{nil}\,\rangle.\text{nil}\,\rangle \quad \text{and} \quad \langle\, \text{nil}.\langle\, \text{nil}.\text{nil}\,\rangle\,\rangle$$

   as trees.
   b. Are the given two trees equal?
   c. Do the trees represent numbers or lists? If they do, which number or list, respectively, do they denote? If they denote lists, are they lists of (encoded) numbers?

2. Is there a difference between cons _ _ and $\langle\,\_.\_\,\rangle$? Both appear to be binary operations on trees.
3. What are the encodings of numbers 0 and 3 in the WHILE-datatype $\mathbb{D}$ of binary trees?
4. What are the encodings of lists [nil, nil] and [nil, $\langle$ nil.nil $\rangle$, nil] in the WHILE datatype $\mathbb{D}$ of binary trees?
5. Write WHILE-programs that implement the Boolean algebra operators $\vee$, $\wedge$, and $\neg$ as programs or, and, and neg via the encoding of Boolean values. The former two take two arguments encoded as a pair.
6. Program the "listify" operation from Proposition 3.1.

   a. Define the algorithm listify that takes as input a tree $t \in \mathbb{D}$ and outputs a list l such that $t = \ulcorner l \urcorner$.
   b. Code the data type $\mathbb{D}$, lists, and the algorithm listify in your favourite programming language. Functional languages will be particularly appropriate for that purpose.

7. Write a (short) WHILE-program nonterm that does not terminate for any input.
8. Write a WHILE-program length that takes a list as input and returns its length. Since the length is a natural number, the result must be a binary tree encoding a natural number. Use the *unary* number encoding.
9. Write a WHILE-program last that takes a list as input and returns the last element of this list.
10. Write a WHILE-program elem that takes a list [n, l] as input where n encodes a natural number and l encodes a list and returns the $n + 1$th element of this list. If n is greater than or equal the length of the list l then there is no $n + 1$th element and the program shall return nil in this instance. Use the *unary* number encoding.
11. Using the *unary* encoding of natural numbers, write a WHILE-program subtract that takes input of the form [m, n] and returns $m - n$ if $n \leq m$ and 0 otherwise.

12. Develop a `WHILE`-program `count` that adds all the numbers in the input list which is supposed to be a list of numbers (in unary encoding). If the input list is not a list of numbers we don't care what the program does. The function `count` that is to be implemented can be defined recursively as follows:

$$\text{count}([\,]) = 0$$
$$\text{count}([\,a_1, a_2, \ldots, a_n\,]) = a_1 + \text{count}([\,a_2, \ldots, a_n\,])$$

Do we need a stack to implement this function? In other words, is the function `count` tail recursive?

13. Develop a `WHILE`-program that implements function *size* from Definition 3.2. The given definition is recursive but not tail recursive.

   a. Develop first a tail recursive version *size2* that takes two arguments: a stack of trees (given as a list) and a natural number as accumulator, such that

   $$size2([e_1, e_2, \ldots, e_n], acc) = acc + size(e_1) + size(e_2) + \cdots + size(e_n)$$

   b. Implement *size2* in `WHILE`.
   c. Implement *size* by adapting the program for *size2* accordingly. Recall that the input of *size* is just a single tree.

References

1. Hopcroft, J.E., Motwani, R., Ullman, J.D.: Introduction to Automata Theory, Languages and Computation, 3rd edn. Addison-Wesley (2007)
2. Jones, N.D.: Computability and Complexity: From a Programming Perspective. MIT Press, Cambridge (1997). (Also available online at http://www.diku.dk/neil/Comp2book.html.)
3. Karp, A.: A Programming Language Oriented Approach to Computability. Bachelor Thesis. University of Bern. Available via DIALOG, https://github.com/zombiecalypse/Bachelor-Thesis/wiki. Cited on 7 Sept 2015 (2013)
4. McCarthy, J.: Recursive functions of symbolic expressions and their computation by machine. Part I. Commun. ACM **3**(4), 184–195 (1960)
5. Official Haskell Website. Available via DIALOG. https://www.haskell.org (2015). Accessed 9 June 2015
6. Standard Ecma 404: The JSON Data Interchange Format. Available via DIALOG http://www.ecma-international.org/publications/files/ECMA-ST/ECMA-404.pdf. Cited on 9 June 2015
7. Steele Jr, G.L.: Debunking the "expensive procedure call" myth or, procedure call implementations considered harmful or, LAMBDA: the Ultimate GOTO. In: Ketchel, J.S., et al. (eds.) Proceedings of the 1977 annual conference ACM'77, pp. 153–162. ACM, Seattle, Washington (1977)
8. W3C: Extensible Markup Language (XML). Available via DIALOG http://www.w3.org/XML/. Cited on 9 June 2015

Chapter 4
Semantics of WHILE

What is the exact meaning of our programming language? Do its programs qualify as effective procedures?

If we want to take WHILE-programs as effective procedures we better make sure we understand exactly how to execute each command and ensure it can be executed using finite resources. Normally programs in high-level languages like WHILE are not directly executed on a machine like Turing machine programs are. Usually, they are interpreted by a special program, an *interpreter*, that can then be executed on any given computer's hardware. Sometimes the program is also *compiled* (translated) into code that can be directly run on the hardware.

However, we want to abstract away from all the details of hardware, but still ensure that WHILE-programs qualify as effective procedures. In order to do that we have to give WHILE a formal semantics that withstands any scrutiny. It must not be vague and it must be described by finite means only. The area of *semantics of programming languages* is an established one and there are many different kinds of semantics one can apply (see e.g. [2, 4] or [3] for functional programming or [1] for a subset of Java). The type of semantics that is most accessible is usually the *operational semantics*.

From the description of WHILE in the previous chapter we might already have a good idea what the semantics is, as we are familiar with while loops and assignments and we have discussed the data type in the previous chapter in detail. But looking at the details, there may still be questions like:

- What is the result of hd nil?
- How do we evaluate the guard of a while loop?
- What is the value of variables if we have not explicitly initialised them?

The semantics of WHILE will require us to describe the *behaviour* of any program. The behaviour of a program specifies the result of the program (what output does it

© Springer International Publishing Switzerland 2016

B. Reus, *Limits of Computation*, Undergraduate Topics in Computer Science, DOI 10.1007/978-3-319-27889-6_4

write at the end?) given any input. WHILE is an *imperative* programming language.[1]
Therefore its computation is driven forward by commands that are assignments to
variables. Many such assignments happen (sequentially) during execution. In order
to model the behaviour, the semantics, of WHILE-programs we must be able to record
the effects of such assignments. In real machines the variables' values are kept and
modified in memory. In our more abstract, machine-independent, semantics we will
introduce an abstract memory that we call *store*, and they are explained first in
Sect. 4.1. After that, the semantics of programs (Sect. 4.2), commands (Sect. 4.3),
and expressions (Sect. 4.4) is defined formally.

4.1 Stores

First of all, we need to define the state of WHILE-programs during execution. Since
we have an imperative language the state is a type of memory that assigns values
to each variable of the program. We call those *stores* and usually denote them with
Greek symbol σ decorated with various subscripts and superscripts.

Definition 4.1 Let *name* $\in$ *VariableName* be a variable name and *val* be an element
in $\mathbb{D}$, then pair $(name, val) \in$ *VariableName* $\times \mathbb{D}$ is called a *variable binding* (for
variable *name*).

Definition 4.2 A *store* for a WHILE-program is a finite set of variable bindings (as
defined in Definition 4.1) and thus an element in *Set*(*VariableName* $\times \mathbb{D}$). A store,
sometimes also called *environment*, thus contains the values of variables produced by
a program. It changes during the program's execution due to the assignments made to
variables. We use the abbreviation *Store* $=$ *Set*(*VariableName* $\times \mathbb{D}$), so *Store* denotes
the set of all possible stores for a WHILE-program.

Since a store is a finite set of pairs we can view it as a set of key-value pairs (also
called *dictionary* or *associative array* in programming languages) where the keys
are the variable names.[2]

Definition 4.3 We define the following operations and notations on stores:

Notation   Since a store is essentially a set of key-value pairs we can write concrete
     stores accordingly as set of pairs "(variable name,value)" as follows:

$$\{X_1 : d_1, X_2 : d_2, \ldots, X_n : d_n\}$$

where $d_1, \ldots, d_n$ are values in $\mathbb{D}$.

---

[1]"Imperative" here is to be understood literally, referring to "command" (as in "imperative mood").
This means that programs of an imperative language consist of *commands*.

[2]In PHP, for instance, one writes `array(key1=>value1,...,keyN=>valueN)` or in
Javascript `array={'key1': 'value1',...,'keyN':'valueN'}` for such associative
arrays.

Lookup   We write $\sigma(\text{X})$ for the lookup operation in a store σ for variable X. This returns a value in $\mathbb{D}$. If X does not occur in σ it returns value nil. This will be in line with the use of variables that are assumed to be initialised with value nil.

Update   We write $\sigma[\text{X} := d]$ for the update operation of X in σ which is defined as follows:

$$\sigma[\text{X} := d] = \sigma \setminus \{(\text{X}, val)\} \cup \{(\text{X}, d)\} \tag{4.1}$$

Example 4.1   The following expression:

$$\{\text{X} : \langle \text{nil.nil} \rangle, \text{Y} : \text{nil}\}$$

denotes a store which maps X to $\langle \text{nil.nil} \rangle$ and Y to nil.

For the execution of a WHILE-program we need a store to start off with, the *initial store*.

Definition 4.4   We denote the *initial store* for a program p with input d by $\sigma_0^{\text{p}}(d)$. Note the extra argument d which denotes the input of the program that will be initially assigned to the input variable. We thus get the following definition of the initial store $\sigma_0^{\text{p}}(d)$ for a program p read X {S} write Y with input d:

$$\sigma_0^{\text{p}}(d) = \{\text{X} : d\}$$

We now have the ingredients to define the semantics of commands and expressions.

4.2   Semantics of Programs

The semantic map $[\![\_]\!]^{\text{WHILE}}$ takes a WHILE-program and maps it into its semantics which is a description of its input/output behaviour and thus a function from binary trees to binary trees. The function however may not always return a result as programs may not terminate for certain (or even all) input. We therefore need to model that a program may not terminate. Semantically, we express this by a special outcome called "undefined", which is abbreviated $\perp$. The semantics of a program is a partial function from binary trees to binary trees, i.e. a function of type $\mathbb{D} \to \mathbb{D}_\perp$ where $\mathbb{D}_\perp$ is the set of trees $\mathbb{D} \cup \{\perp\}$ as described in Definition 2.9. The "undefined" is needed in lieu of a missing result for the semantics of a program in cases where it does not terminate.

Now $[\![p]\!]^{\text{WHILE}}(d) = e$ means that running program p with input d terminates and the output is e and $[\![p]\!]^{\text{WHILE}}(d) = \perp$ means that running program p with input d does *not* terminate. The semantics of commands is going to be defined as a relation between commands and initial and final store. We write this as a *judgment*

$$\text{S} \vdash \sigma_1 \to \sigma_2$$

which is a concise description of the fact that execution of a list of commands S in state σ_1 terminates and produces state σ_2 as result. The stores represent the memory (and its change) used in the imperative computation model. For our simple language, that does not use an implicit heap or stack, we use *Store* as introduced above. However, the meaning of the judgment is formally a relation of a command and two stores. This relation will be defined formally further below in Definition 4.6. For the moment let us assume we already have this definition.

Definition 4.5 Let p read X {S} write Y be a WHILE-program where S is a statement list. The semantics of p is defined as follows:

$$[\![p]\!]^{\text{WHILE}} (d) = \begin{cases} e & \text{if } S \vdash \sigma_0^p(d) \rightarrow \sigma \text{ and } \sigma(\text{Y}) = e \\ \bot & \text{otherwise} \end{cases} \tag{4.2}$$

So the semantics of p (which is a partial function) applied to argument d equals e, if running the body of p (S) in the initial state $\sigma_0^p(d)$ for p (which sets the input variable X to d and all other variables to nil) terminates in a state σ in which the output variable Y has the value e. It equals $\bot$ (undefined) if the body of p in the initial state $\sigma_0^p(d)$ does not terminate.

4.3 Semantics of Commands

It remains to define the meaning of the judgment $S \vdash \sigma_1 \rightarrow \sigma_2$ that describes the execution of commands. For a program p this is a ternary relation on cartesian product *StatementList* × *Store* × *Store* where *StatementList* denotes the non-empty lists of commands as described syntactically by nonterminal $\langle statement\text{-}list \rangle$. In other words, elements in this relation are triples, and we write $S \vdash \sigma_1 \rightarrow \sigma_2$ instead of $(S, \sigma_1, \sigma_2) \in R_{\text{SemanticsStmtList}}$ in the same way as we write $e_1 = e_2$ instead of $(e_1, e_2) \in R_{\text{Equality}}$.

Definition 4.6 The relation $\bullet \vdash \_ \rightarrow \_$ is defined as the smallest relation satisfying six rules some of which use the relation (recursively) again in their definition. Statement lists are either just one command or a command followed by a statement list. In the first case we distinguish the three different commands: assignment, conditional and while loop. In the second case we have a command followed by another statement list.

The first rule is for assignment commands and describes the effect of assigning E to variable X. The effect is that the value for X in the initial store σ is updated to be the result of evaluating E. Thus, we will need to say what expressions mean. For now let us simply assume we already know how to evaluate expressions and let $\mathcal{E}[\![\text{E}]\!]\sigma$ denote the value of the expression E in store σ. This store σ is actually needed since the expression E may contain variables the values of which are obviously kept in the store. With this we get the first rule more formally as

$$X := E \vdash \sigma \to \sigma[X := e] \quad \text{if} \quad \mathscr{E}[\![E]\!]\sigma = e \tag{4.3}$$

where the notation $\sigma[X := e]$ (as explained in Definition 4.3) denotes a state that is identical to σ with the exception that the value for X is set to be e.

The next pair of rules is for the conditional. With this we get the first rule more formally as

$$\text{if } E \{S_T\} \text{ else } \{S_E\} \vdash \sigma \to \sigma' \quad \text{if} \quad \begin{cases} \mathscr{E}[\![E]\!]\sigma \neq \text{false and } S_T \vdash \sigma \to \sigma' \\ \mathscr{E}[\![E]\!]\sigma = \text{false and } S_E \vdash \sigma \to \sigma' \end{cases} \tag{4.4}$$

For the case where there is no "else" branch we obtain the simpler rule:

$$\text{if } E \{S_T\} \vdash \sigma \to \sigma' \quad \text{if} \quad \begin{cases} \mathscr{E}[\![E]\!]\sigma \neq \text{false and } S_T \vdash \sigma \to \sigma' \\ \mathscr{E}[\![E]\!]\sigma = \text{false and } \sigma = \sigma' \end{cases} \tag{4.5}$$

The semantics prescribes that one first has to evaluate E. If E evaluates to false (nil) we continue executing the else branch, if it evaluates to true, i.e. something different from nil we continue executing the "then" branch.

The next command is the while loop. We need two rules describing the two possible cases: whether the loop terminates or whether the body is executed (at least once). Let us deal with the more complicated case first. If we have while $E \{S\}$ and E evaluates to something that is not representing false (i.e. nil) then we need to execute the statement list S (the loop body) and then execute the entire loop again, hoping that the changed values for the variables appearing in E have brought the loop nearer to its termination. Again, we need an intermediate state σ' which is the result of having executed the body once. This can be more formally phrased as follows:

$$\begin{aligned} \text{while } E \{S\} \vdash \sigma \to \sigma'' \quad &\text{if} \quad \mathscr{E}[\![E]\!]\sigma \neq \text{nil and } S \vdash \sigma \to \sigma' \text{ and} \\ &\quad \text{while } E \{S\} \vdash \sigma' \to \sigma'' \end{aligned} \tag{4.6}$$

The termination case is easy as here we don't do anything if the value of the guard E equals nil (which represents false). Formally, we get:

$$\text{while } E \{S\} \vdash \sigma \to \sigma \quad \text{if} \quad \mathscr{E}[\![E]\!]\sigma = \text{nil} \tag{4.7}$$

Note that in the definition of $\vdash$ it says that the meaning is the *smallest* relation on *StatementList* $\times$ *Store* $\times$ *Store* that satisfies the five rules mentioned above. Why is this? This is to make sure that in cases where a block (statement list) S does *not* terminate in state σ there is actually no state σ' such that $S \vdash \sigma \to \sigma'$. For a non-terminating loop while $E \{S\}$ the condition E will always evaluate to something different from nil so Rule 4.6 will always be applicable and so the semantics then boils down to

$$\text{while } E \{S\} \vdash \sigma \to \sigma'' \text{ if } S \vdash \sigma \to \sigma' \text{ and } \text{while } E \{S\} \vdash \sigma' \to \sigma''$$

which we can simply fulfil using any σ''. But this would clearly violate our intuition of the semantics. The condition that the semantics of $\vdash$ is the *smallest* relation means that in this (non-termination) case *no* σ'' should be in relation with S and σ, in other words for no σ'' we have $S \vdash \sigma \rightarrow \sigma''$ which is what we want (as we want $S \vdash \sigma \rightarrow \sigma''$ to express termination).[3]

It remains to look at the case where the command to be executed is actually not a single command but a proper statement list C; S, i.e. a command followed by a statement list. The semantics simply states that one first executes C and then (inductively) the statement list S. Using the judgments we will need to use an intermediate store which acts as the initial store for the second statement list S. It should be clear that the result store of the first command is the right intermediate store here. We thus obtain more formally:

$$C; S \vdash \sigma \rightarrow \sigma'' \quad \text{if} \quad C \vdash \sigma \rightarrow \sigma' \quad \text{and} \quad S \vdash \sigma' \rightarrow \sigma'' \tag{4.8}$$

Since we can have empty blocks we also need a rule to deal with empty statement lists although they are not part of the grammar:

$$\vdash \sigma \rightarrow \sigma$$

4.4   Semantics of Expressions

Finally, it only remains to give the expression semantics. So for each kind of expression we need to inductively define its value. Since expression can contain variables we need to use the store as argument of the semantics to look up the values. An important feature of `WHILE` is that expressions have *no side effects* so they don't change the store at all (and their evaluation always terminates) and thus the result of the interpretation of an expression is simply the value of the expression.

Definition 4.7 We define the interpretation function for expressions $\mathscr{E}[\![E]\!] : Store \rightarrow \mathbb{D}$ inductively by case analysis. If the expression is a variable we get the value from the store giving us the rule:

$$\mathscr{E}[\![X]\!]\sigma = \sigma(X) \tag{4.9}$$

explaining why we need to evaluate expressions *in a store*. If the expression is atom `nil` we simply interpet it as the empty tree nil:

$$\mathscr{E}[\![\texttt{nil}]\!]\sigma = \text{nil} \tag{4.10}$$

The fonts of the "nil" on the left and on the right hand side of the definition are intentionally different, because the left hand side "nil" is a syntactic expression of

---

[3]See also Exercise 5.

WHILE whereas the "nil" on the right hand side denotes the empty tree in $\mathbb{D}$. If we enriched the language to contain more atoms their evaluation would be defined analogously.

For a constructor expression we inductively first evaluate the (smaller) left and right argument of the cons expressions and then build the tree:

$$\mathscr{E}[\![\text{cons E F}]\!]\sigma = \langle\, \mathscr{E}[\![\text{E}]\!]\sigma.\mathscr{E}[\![\text{F}]\!]\sigma\, \rangle \tag{4.11}$$

For the head hd and tail tl expressions, respectively, we first evaluate their argument and then take the left, or right subtree, respectively:

$$\mathscr{E}[\![\text{hd E}]\!]\sigma = \begin{cases} e & \text{if } \mathscr{E}[\![\text{E}]\!]\sigma = \langle\, e.f\, \rangle \\ \text{nil} & \text{otherwise} \end{cases} \tag{4.12}$$

$$\mathscr{E}[\![\text{tl E}]\!]\sigma = \begin{cases} f & \text{if } \mathscr{E}[\![\text{E}]\!]\sigma = \langle\, e.f\, \rangle \\ \text{nil} & \text{otherwise} \end{cases} \tag{4.13}$$

It is worth pointing out that the tree operations hd and tl have no obvious meaning when applied to an atom nil, which represents the empty tree, as the empty tree does not have a left and right subtree. In a more sophisticated language we might want to raise a runtime error or throw an exception but we do not have the possibility in WHILE so we pick as result the most sensible tree, namely the empty one.

This completes the presentation of the semantics of the language WHILE. We will see some extensions soon but the language presented here is our core language.

What Next?

In this chapter we have defined the semantics of WHILE-programs as partial functions that describe the programs' input/output behaviour. The semantics explains how to execute programs and proves that WHILE-programs can be used as effective procedures. Before we write more WHILE-programs we look into some "syntax sugar" (like the inclusion of equality discussed earlier) and extensions of the WHILE-language that will make writing programs even easier.

Exercises

1. Let p be a WHILE-program with input variable X.

   a. What does *Store* denote?
   b. Why, or for what, do we need *Store* when executing p?
   c. Why, does the initial store for p only initialise variable X?

2. What does the judgment $C \vdash \sigma \to \sigma'$ mean for a WHILE-statement C and stores σ and σ'?

3. For which of the following extensions of the expression syntax would it be possible to extend the interpretation function of type $\mathscr{E}[\![\text{E}]\!] : Store \to \mathbb{D}$ accordingly? Why?

   a. Tree literals of the form `"d"`.

   b. Macro calls of the form `<`⟨*name*⟩`>` ⟨*expression*⟩, where ⟨*name*⟩ refers to the names of programs.

4. Evaluate the following expressions in a store $\sigma = \{X : \langle \text{nil.nil} \rangle\}$. Note that there is intentionally no binding for Y (why is this not a problem?)

   a. `cons nil cons nil nil`
   b. `tl tl X`
   c. `hd cons X cons Y Y`
   d. `cons Y cons hd cons X Y X`

5. Show that if one did *not* require the semantics of commands $\bullet \vdash \_ \to \_ \subseteq$ *StatementList* × *Store* × *Store* to be the *smallest relation* satisfying the rules given in Definition 4.6, we could give any result to a non-terminating program like program `nonterm` from Exercise 7 in Chap 3, i.e. we could define

$$\text{nonterm} \vdash \{X : \text{nil}\} \to \sigma$$

for any σ we like.

6. Let us do once (and only once) what Turing did for his machines, namely compute a WHILE-program on paper. After that, we will use an interpreter for that. So consider the program `add` from Fig. 3.3 in Chap. 3.

   Going through the formal definition of the semantics of WHILE, determine the result of this program when run on input [2, 1] subject to the encoding for lists and numbers.

References

1. Cenciarelli, P., Knapp, A., Reus, B., Wirsing, M.: A structural operational semantics for multi-threaded Java. In Alves-Foss, J. (ed.) Formal Syntax and Semantics of Java. LNCS, vol. 1523, pp. 157–200. Springer (1999)
2. Fernández, M.: Programming Languages and Operational Semantics—A Concise Overview, UTiCS Series. Springer, Heidelberg (2014)
3. Streicher, T.: Domain-Theoretic Foundations of Functional Programming. World Scientific, Singapore (2006)
4. Winskel, G.: Formal Semantics of Programming Languages Paperback. MIT Press, Cambridge (1993)

Chapter 5
Extensions of WHILE

How can we make our programming language more expressible without changing the semantics?

The WHILE-language was simple enough to give it a precise semantics in the previous chapter. Accordingly, we adopt it as our language to write "effective procedures". Nevertheless, it turns out that some language features not included in WHILE make the programmer's life so much easier and WHILE-programs so much more readable. Clearly, in an introductory textbook aimed at Computer Science students used to program in high-level languages such language features are highly desirable. Examples of such features are built-in equality (Sect. 5.1), literals for number and Boolean data values (Sect. 5.2), additional atoms (Sect. 5.3), explicit syntactic support for writing lists (Sect. 5.4), macro calls (Sect. 5.5), and a switch statement (Sect. 5.6). These extensions are the topic of this chapter.

To our delight we will see that all these extensions are not real extensions in the sense that we need to extend our semantics. All the extensions of this chapter are features that one can already express in WHILE-programs anyway, however with some tedious efforts. The added syntax just makes programming easier and the resulting programs more readable. We will use these extensions throughout the book knowing that we could always "translate them away".

5.1 Equality

Equality on trees is an arbitrary complex operation as it may involve arbitrarily large trees. It does not seem to be in line with Turing's idea of every computation step being atomic and thus was not included in our core WHILE-language. However, one can easily *program* equality with the other given features in WHILE, so in a way the equality is already syntax sugar for a macro call. This will be done in Exercise 7 and the code will reveal the detail of the complexity of such a general equality operator.

© Springer International Publishing Switzerland 2016
B. Reus, *Limits of Computation*, Undergraduate Topics in Computer Science,
DOI 10.1007/978-3-319-27889-6_5

We can therefore consider equality as a special built-in macro call of WHILE and extend the expression language with explicit equality on trees:

$$\langle expression \rangle ::= \ldots$$
$$\qquad | \quad \langle expression \rangle = \langle expression \rangle \qquad \text{(equality)}$$
$$\qquad \vdots$$

Example 5.1 Example uses of equality are given below

```
1. if X=Y { Z := X } else { Z := Y }
2. Z := hd X = cons nil nil;
3. while X = hd Z { X := tl X }
```

We do not have to write parentheses around the equality test in the `if` and `while` statements.

5.2 Literals

Literals as expressions are very common in programming languages, most of all string and number literals. Double (or single) quotes are pretty common to identify string literals. To make programming in WHILE easier we also introduce some literals as extensions.

5.2.1 Number Literals

We have seen in Sect. 3.3.3 how to encode natural numbers in WHILE as lists of nil atoms and as such they can be perfectly viewed as expressions. With the help of literals one can write e.g. the number 2 as con nil cons nil nil. But this becomes tedious for larger numbers. So we introduce syntax sugar for numbers. We therefore add to the BNF syntax of Fig. 3.2 the clause:

$$\langle expression \rangle ::= \ldots$$
$$\qquad | \quad \langle number \rangle \qquad \text{(number literals)}$$
$$\qquad \vdots$$

where nonterminal $\langle number \rangle$ contains the natural numbers $\mathbb{N}$. Note that in this case we do not require any double quotes as numbers are otherwise not part of the language syntax and thus cannot be misread. Despite having those number literals we still do not have any built-in arithmetic for numbers. Arithmetic operators have to be implemented by programs.

Example 5.2 Here are some examples for expressions that use number literals according to the extension described above:

1. `2`
2. `hd 99`
3. `cons 3 cons 1 nil`

Since the numbers can be unfolded to lists of nil atoms we know the semantics of these expressions. For (1) it is [nil, nil]; for (2) it is nil; and for (3) it is [[nil, nil, nil], [nil]].

One might wish to have number literals that are translated into decimal or binary representation. This would require a more complicated translation of programs.

5.2.2 Boolean Literals

We have seen in Sect. 3.3.1 how to encode Booleans in WHILE. We thus add to the BNF syntax of Fig. 3.2 the clause:

$$\langle expression \rangle ::= \ldots$$
$$| \quad \texttt{true} \qquad \text{(true literal)}$$
$$| \quad \texttt{false} \qquad \text{(false literal)}$$
$$\vdots$$

5.3 Adding Atoms

The data type $\mathbb{D}$ of our core WHILE-language only has one atom: nil and the expression language accordingly knows only one atom constructor, `nil`.

In order to do some encodings, it is often advantageous to have several atoms. We always restrict to finitely many atoms in order to ensure we have an effective way to interpret the basic WHILE-expressions and commands.[1]

Definition 5.1 Let *Atoms* denote a finite set of atoms that always includes the atom nil. The extended set $\mathbb{D}$ of data values is defined inductively as follows:

1. any atom $a \in Atoms$ is a tree consisting of just a leaf labelled a.
2. for any two trees l and r, such that $l \in \mathbb{D}$ and $r \in \mathbb{D}$ the tree with left subtree l and right subtree r, written linearly $\langle a.b \rangle$ is also in $\mathbb{D}$.

We must extend the syntax accordingly and thus adapt the BNF-syntax of Fig. 3.2 replacing the clause:

---

[1]This will also be important in the Complexity part when we want to measure time usage.

$$\langle expression \rangle \ :::= \ldots$$
$$| \quad \texttt{nil} \qquad (\text{atom nil})$$
$$\vdots$$

by a new clause allowing more atoms:

$$\langle expression \rangle \ :::= \ldots$$
$$| \quad \texttt{a} \qquad (\texttt{a} \in Atoms)$$
$$\vdots$$

We will use some additional atoms already in the next chapter for the purpose of encoding programs as data.

5.4   List Constructor

We have already seen how to encode lists in $\mathbb{D}$ in Sect. 3.3.2. To construct a list of natural numbers [1, 2, 3] in WHILE one writes expression cons 1 cons 2 cons 3 nil. But often the list elements will be generated on the fly rather than being (number) literals. So syntax sugar for building lists (and not just list literals like [1, 2, 3]) would be useful. Therefore, we introduce another extension to the WHILE-language, *list expressions*. We introduce list expressions using the square brackets [ ] (known from Haskell and other languages) already used to denote list *values*. A list expression will always evaluate to a list value without side effects. The elements of a list expression are themselves expressions separated by commas.

We add to the BNF-syntax of Fig. 3.2 the clause

$\langle expression \rangle \qquad ::= \ldots$
$\qquad | \quad [] \qquad\qquad\qquad\qquad\qquad\qquad\qquad$ (empty list constructor)
$\qquad | \quad [\langle expression\text{-}list \rangle] \qquad\qquad\qquad$ (nonempty list constructor)
$\qquad \vdots$
$\langle expression\text{-}list \rangle ::= \langle expression \rangle \qquad\qquad\qquad\qquad$ (single expression list)
$\qquad | \quad \langle expression \rangle , \langle expression\text{-}list \rangle \quad$ (multiple expression list)

Example 5.3 Here are some examples for expressions that use the list constructor. We assume variables X and Y that have values nil and 5, respectively.

1. [X]
   is a list expression that in the given store will evaluate to [nil] (or [0] if the element is considered a number).
2. [tl Y, cons X X, X]
   is a list expression that in the given store will evaluate to [4, 1, 0].

3. `[ Y , [ tl Y, tl tl Y, tl tl tl Y ]   ]`
   is a list expression that in the given store will evaluate to $[5, [4, 3, 2]]$.
4. If we combine this with number and tree literals we can also write:
   `[1, 2, 3]` or `[ nil, [1,2,3], 1 ]`.

5.5 Macro Calls

The main reason of the success of high-level programming languages is the level of abstraction from the machine level they provide. Procedural abstraction allows one to implement procedures with parameters that can be used and re-used several times in a larger program but have to be defined only once. Every use of the procedure may instantiate the parameters differently.

When writing larger WHILE-programs it makes a significant difference if one can use procedures as it reduces the amount of code one has to write while at the same time improving the readability of the code.

We won't extend WHILE by proper (recursive) procedures or functions as this would need a proper language extension that cannot be simply achieved by a compilation. Yet, we will allow WHILE-programs to call *other* WHILE-programs by means of macro expansion. The emphasis here is on *other* programs since recursive self-calls will not be permitted as they would need a stack to keep intermediate results and thus a semantics extension too. We already have given names (using nonterminal ⟨*name*⟩) to programs which can be used to call other programs. Of course it is important that no two programs have the same name. These names are used as handles in the macro call when they are enclosed in angular brackets.

We accordingly add to the BNF-syntax of Fig. 3.2 the clause

⟨*command*⟩ ::= ...
$$| \quad \langle variable \rangle := < \langle name \rangle > \langle expression \rangle \qquad \text{(macro call)}$$
⋮

The meaning of the macro calls is intuitive. First, the expression argument is to be evaluated[2] and then the called program with the given name is to be run with the obtained value as input. The resulting output is then to be assigned to the variable in the assignment. We limit macro calls to assignments as we do not wish to change an important property of expressions, namely that they do not have side effects. If we allowed macros to be called in general expressions and the program in question does not terminate then we would have undefined expressions to deal with. The details of how such a macro could be translated into a pure WHILE-program are discussed in Exercise 2.

---

[2] As for call-by-value parameter passing.

```
succ read X {
   X := cons nil X
}
write X

pred read X {
   X := tl X
}
write X

add read L {
  X:= hd L;
  Y:= hd tl L;
  while X {
      X := <pred> X;
      Y := <succ> Y
      }
}
write Y
```

Fig. 5.1 A WHILE-program with macro calls

Example 5.4 The program add that implements addition on unary numbers, which we encountered already in Chap. 3, Fig. 3.3, can be written differently to highlight the meaning of hd and tl on numbers as outlined in Fig. 5.1.

5.6 Switch Statement

In WHILE-programs one often has to analyse the structure of a given tree, then break it down into parts and process those parts. This can be done, of course, with the help of a conditional, equality and a sequence of tl and hd applications. A more concise and elegant way of achieving the same is with the help of *pattern matching*, known from functional languages like Haskell [3], Scala [4] and F# [1]. This is also used in [2]. In order to avoid the complications of *pattern matching* we introduce just a useful form of case analysis in order to avoid cascading conditionals, called a switch statement known from most programming languages.

We add to the BNF-syntax of Fig. 3.2 the clauses in Fig. 5.2.

The nonterminal ⟨*expression-list*⟩ is as defined in the list constructor extension from Sect. 5.4. The meaning of the switch clause is as follows: the value of the expression argument is compared to the values in the rules in the case expressions. The rules are inspected in the order of appearance in the switch statement. As soon as a match is found, the corresponding statement list of the matching rule is executed.

$$\langle command \rangle ::= \dots$$
$$| \quad \texttt{switch} \ \langle expression \rangle \ \{ \ \langle rule\text{-}list \rangle \ \} \quad \text{case analysis}$$
$$| \quad \texttt{switch} \ \langle expression \rangle \ \{ \ \langle rule\text{-}list \rangle$$
$$\quad \texttt{default} : \langle statement\text{-}list \rangle \ \} \quad \text{with default case}$$
$$\vdots$$

$$\langle rule \rangle \quad ::= \texttt{case} \ \langle expression\text{-}list \rangle : \langle statement\text{-}list \rangle \quad \text{case rule}$$

$$\langle rule\text{-}list \rangle \quad ::= \langle rule \rangle$$
$$| \quad \langle rule \rangle \ \langle rule\text{-}list \rangle$$

Fig. 5.2 BNF grammar for switch statement

```
switch X   {
  case 0            :   Y := 0
  case 1, 3         :   Y := 1
  case cons 2 nil   :   Y := 2
  }
```

Fig. 5.3 An example for a `switch` statement

Unlike in some other languages, we do not allow that more than one rule is ever triggered.[3]

Example 5.5 For instance the switch statement in Fig. 5.3 can be rewritten to the WHILE-program in Fig. 5.4 that still uses number literals, which could also be translated away of course.

```
if X = 0
   { Y := 0 }
else { if X = 1
          { Y := 1 }
       else
          { if X = 3
               { Y := 1 }
            else
               {  if X = cons 2 nil
                    { Y := 2 }
               }
          }
     }
```

Fig. 5.4 A WHILE-program with if-cascade instead of `switch` command

---

[3]We thus do not require any `break` command.

What Next?

Now that we are familiar with the WHILE-language, its semantics, and syntax sugar
for writing readable WHILE-programs, we can look into using WHILE in order to
write some interesting programs. For instance, we may wish to code an interpreter,
that is a program that computes the semantics of another program when run with
some additional input. We will look at this in detail in Chap. 7, but in order to write
such a program we first need to consider how we can represent WHILE-programs as
data objects in the WHILE-datatype of binary trees, $\mathbb{D}$. This is the topic of the next
chapter.

Exercises

1. Give a compilation scheme for natural number literals that translates any expressions with natural numbers into a correct pure WHILE-expression.
2. Give a compilation scheme for macro calls that translates any program with macro
   calls into a correct pure WHILE-program.
3. Give a compilation scheme for switch-statements that translates any switch statement into a correct pure WHILE-block.
4. Write a program equalsNat in core WHILE, i.e. without using the equality
   extension presented in this chapter, that takes a list of two natural numbers (in the
   unary encoding presented in Sect. 3.3.3.1) [m, n] as input and returns $m = n$ as
   Boolean value (using the encoding presented in Sect. 3.3.1). Use the conditional
   if that allows one to test whether a tree equals nil.
5. Implement the equality of natural number in the extended WHILE-language presented in this chapter, but use the *decimal* representation of numbers. In other
   words, write a program equalsNatD that takes as input a suitably encoded list
   of two natural numbers (in the decimal encoding presented in Sect. 3.3.3.2) and
   returns $m = n$ as Boolean (using the encoding presented in Sect. 3.3.1).
6. The WHILE-language includes a conditional, if then else, statement.
   Explain how this conditional could be considered an extension so that it could
   have been omitted from the core language.
7. Let us generalise Exercise 4 to general trees. Write a program equals in "core"
   WHILE, i.e. without using the equality extension presented in this chapter, that
   takes a (suitably encoded) list of two trees [X, Y] as input and returns a Boolean
   stating whether trees X and Y are equal. Use the conditional if that allows one
   to test whether a tree equals nil.
   *Hint: Instead of recursion, which is not directly available, use a while loop and
   a stack to store intermediate subtrees that still require testing for equality.*
8. Using extended WHILE as described in this chapter, implement multiplication on
   unary numbers, i.e. write a program mult that takes an encoded list of two unary
   numbers [m, n] and returns an encoding of $m \times n$ as unary number.
   Hint: Use add from Fig. 5.1 as macro.

References

1. F# Software Foundation: Official F# Website. Available via DIALOG. http://fsharp.org (2015). Accessed 9 June 2015
2. Jones, N.D.: Computability and Complexity: From a Programming Perspective. MIT Press, Cambridge (1997)
3. Official Haskell Website. Available via DIALOG. https://www.haskell.org (2015). Accessed 9 June 2015
4. Official Scala Website. Available via DIALOG. http://www.scala-lang.org/index.html (2015). Accessed 9 June 2015

Chapter 6
Programs as Data Objects

> *How can we express and manipulate programs by another program?*

Many interesting (types of) programs take input as program or a program and some other data. In order to be able to write such programs in WHILE, we need to be able to treat other WHILE-programs (or Turing machine programs for instance) as data.

Three types of programs in particular use programs as input: compilers, interpreters and program specialisers.

A *compiler* takes a program as input and translates it into another program (most likely in another language). The resulting program must have the same semantics as the original program, otherwise the compiler would not be correct. A compiler may involve three languages: the source language, the target language and the implementation language of the compiler.

An *interpreter* takes a program and a data value from the program's datatype and returns as output the value that the corresponding program would return when run with the data value as input. An interpreter may involve two languages: the interpreted language and the implementation language of the interpreter.

A *program specialiser* takes a program with two (more generally n) inputs and one data value from the programs datatype as input and *partially evaluates* the program with the given data value as partial input, thus generating a new program with $2 - 1 = 1$ (or more generally $n - m$) inputs preserving the program's behaviour. Or as Neil Jones puts it "*A program specialiser is given an S-program p together with part of its input data s. Its effect is to construct a T-program which, when given p's remaining input d, will yield the same result that p would have produced given both inputs*" [1, p. 58]. A specialiser may involve two languages: the language of the source and target program (S and T in the quote above) and the implementation language of the specialiser. Often, for practical purposes, source and target languages are identical. The main purpose of a specialiser is to achieve faster programs in certain situations where some input is known.

© Springer International Publishing Switzerland 2016
B. Reus, *Limits of Computation*, Undergraduate Topics in Computer Science,
DOI 10.1007/978-3-319-27889-6_6

In this chapter we introduce the notion of interpreter more formally (Sect. 6.1), motivating programs as data. In the next chapter we will then write a special kind of interpreter. Section 6.2 reviews the concept of *abstract syntax trees* (ASTs). Finally, we explain how one can encode ASTs of WHILE-programs in the WHILE-datatype of binary trees (Sect. 6.3).

6.1 Interpreters Formally

To be able to be more precise (and formal) about programs that take other programs (in various languages) as input, we need to define the semantics of a programming language in general. We already said what the semantics of WHILE (programs) is. This can be generalised now:

Definition 6.1 A *programming language* L consists of

1. two sets: L-programs (the set of L-programs) and L-data (the set of data values described by the datatype used by this language).[1]
2. A function $[\![\_]\!]^{L}$: L-programs $\rightarrow$ (L-data $\rightarrow$ L-data$_{\perp}$) which maps L-programs into their semantic behaviour, namely a partial function mapping inputs to outputs, which are both in L-data.

Definition 6.2 A programming language L defined as above *has pairing* if its data type, L-data, permits the encoding of pairs. For a general (unknown) language that has pairing we denote pairs (a, b), i.e. using parenthesis and a comma.

From Sect. 3.3.2 we recall that WHILE has pairing.

Definition 6.3 A programming language L defined as above *has programs as data* if its data type, L-data, permits the encoding of L-programs. For a general (unknown) language that has programs as data the encoding of a program p is denoted $\ulcorner p \urcorner$.

The rest of the chapter will be devoted to proving that WHILE has programs as data. With this concept one can define exactly what an interpreter int for a language S written in L is:

Definition 6.4 Assume S has programs as data, S-data $\subseteq$ L-data and L has pairing. An interpreter int for a language S written in L must fulfil the following equation for any given S-program p and $d \in$ S-data the following equation:

$$[\![\texttt{int}]\!]^{L} (\ulcorner p \urcorner, d) = [\![p]\!]^{S} (d) \qquad\qquad (6.1)$$

---

[1] Again, we make some simplifying assumptions here in the sense that we only have one datatype. We talk about untyped languages so it makes sense to have just one type.

6.2   Abstract Syntax Trees

In order to be able to write compilers, interpreters, and specialisers in WHILE, we need a representation of WHILE-programs as data objects ("programs as data"). Since the WHILE-datatype of binary trees is not ideal for representing strings, it makes sense to represent programs directly as *abstract syntax trees* (a concept known from compilation). Such trees are usually obtained by the so-called *lexical analysis phase* that takes the string of ASCII symbols and represent the structure of the program as abstract syntax tree (for compilation or interpretation purposes). In a way, by directly using abstract syntax trees as encodings, we dodge all issues of lexical analysis which in effect makes it even easier to use programs as data.

As binary trees in $\mathbb{D}$ have no labels for (inner) nodes we represent abstract syntax trees (ASTs) as lists. Traditionally, maybe in a compiler textbook, an abstract syntax tree would look like the one given in Fig. 6.1.

A tree of the form $op(t_1, \ldots, t_n)$ where t_i are the children (subtrees) of the labelled node op is represented as a list the first element of which is a "label" representing op followed by the encoding of the n subtrees, which, in turn, have to be encoded the same way. Leaves for this abstract syntax tree are operations without arguments and the only ones of this kind we have are variables and literals.

6.3   Encoding of WHILE-ASTs in $\mathbb{D}$

In this section we describe how "core" WHILE-programs, i.e. WHILE-programs without extensions, can be represented as data values in $\mathbb{D}$. As any WHILE-program using extensions can be translated (compiled) into one written in "core" WHILE, this does not pose any restrictions on us.

Fig. 6.1 A "traditional" abstract syntax tree for a while program

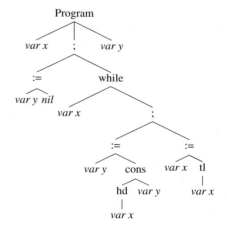

$$\ulcorner \texttt{progname read X \{S\} write Y} \urcorner = \quad \bigl[varnum_X, \ulcorner \texttt{S} \urcorner, varnum_Y \bigr]$$

$$
\begin{aligned}
\ulcorner \texttt{while E B} \urcorner \quad &= \quad [\texttt{while}, \ulcorner \texttt{E} \urcorner, \ulcorner \texttt{B} \urcorner] \\
\ulcorner \texttt{X} := \texttt{E} \urcorner \quad &= \quad [:=, varnum_X, \ulcorner \texttt{E} \urcorner] \\
\ulcorner \texttt{if E B}_T \texttt{ else B}_E \urcorner \quad &= \quad [\texttt{if}, \ulcorner \texttt{E} \urcorner, \ulcorner \texttt{B}_T \urcorner, \ulcorner \texttt{B}_E \urcorner] \\
\ulcorner \texttt{if E B} \urcorner \quad &= \quad [\texttt{if}, \ulcorner \texttt{E} \urcorner, \ulcorner \texttt{B} \urcorner, []] \\[4pt]
\ulcorner \{ \texttt{C}_1; \texttt{C}_2; \ldots; \texttt{C}_n \} \urcorner \quad &= \quad [\ulcorner \texttt{C}_1 \urcorner, \ulcorner \texttt{C}_2 \urcorner, \ldots, \ulcorner \texttt{C}_n \urcorner] \\[4pt]
\ulcorner \texttt{nil} \urcorner \quad &= \quad [\texttt{quote}, \texttt{nil}] \\
\ulcorner \texttt{X} \urcorner \quad &= \quad [\texttt{var}, varnum_X] \\
\ulcorner \texttt{cons E F} \urcorner \quad &= \quad [\texttt{cons}, \ulcorner \texttt{E} \urcorner, \ulcorner \texttt{F} \urcorner] \\
\ulcorner \texttt{hd E} \urcorner \quad &= \quad [\texttt{hd}, \ulcorner \texttt{E} \urcorner] \\
\ulcorner \texttt{tl E} \urcorner \quad &= \quad [\texttt{tl}, \ulcorner \texttt{E} \urcorner]
\end{aligned}
$$

Fig. 6.2 Encoding of WHILE-programs as data

First we need to encode labels in order to represent the type of operation in ASTs. We use special additional atoms.

Definition 6.5 (*Extra atoms*) We add several special atoms $\mathbb{D}$ that so far only used one atom: nil. These extra atoms are var, cons, :=, while, if, tl, hd, quote.

This extension can be done according to our discussion in Sect. 5.3.

There is one more issue we need to sort out yet for the representation of programs as data objects, namely the encoding of variables. We cannot use a finite number of atoms as there are an infinite number of possible variables. We use numbers to encode variable names. This encoding is produced by the map *varnum* : *VariableName* $\to \mathbb{N}$ for which it holds that $varnum_X = varnum_Y$ implies that X = Y. In other words, *varnum* uniquely encode variable names. The concrete numbers are actually not important, it is only important that the same variable is encoded by the same number.[2]

The representation $\ulcorner p \urcorner$ of a WHILE-program p as data can now be defined by the map $\ulcorner \_ \urcorner$: WHILE-programs $\to$ WHILE-data as outlined in Fig. 6.2. On the right hand side of each equation definition we define lists and not trees as those are more readable. To obtain proper elements in $\mathbb{D}$ one just needs to apply encoding operator $\ulcorner \_ \urcorner$. It has been discussed already in Sect. 3.4 how this is defined (as well as encodings of natural numbers i due to Definition 3.5). Some comments are in order. The program name is *not* included in the AST because names are only needed for macro calls in the extended language, but we define the encoding only for "pure" WHILE-programs.

Example 6.1 In Fig. 6.3 the WHILE-program p on the left and $\ulcorner p \urcorner$ as list representation of the corresponding AST on the right. Indentation is used to highlight the structure of the AST in list representation on the right hand side.

---

[2]One can use unary or binary representation of numbers actually, and in the following we may use one or the other, according to the task at hand.

```
reverse read X {        [0,
  Y:= nil;                [[:=,1,[quote,nil]],
  while X {               [while,[var,0],
    Y:= cons hd X Y;        [ [:=,1,[cons,[hd,[var,0]],[var,1]]],
    X:= tl X                  [:=,0,[tl,[var,0]]]
  }                         ]
}                         ]],
write Y                   1]
```

Fig. 6.3 A WHILE-program in concrete and abstract syntax "as data"

The "tags" for commands and expression operators, like : = and var, are the extra atoms introduced earlier. They could also be (in a more tedious fashion) encoded via natural numbers to avoid extra atoms.

What Next?

We have seen how one can encode WHILE-programs as data in the form of abstract syntax trees. Since we can also encode pairing, we are in a position to write a WHILE-interpreter in WHILE, a so-called self-interpreter. We will do this in detail in the next chapter.

Exercises

1. Assuming that we start counting variables from 0, give the programs as data representation of the WHILE-program given in Fig. 6.4.
2. Consider the tree t depicted in Fig. 6.5.

   a. Why is t a correct tree in $\mathbb{D}$ although items like 1, 0, cons, : =, quote, var appear at its leaves rather than just nil?
   b. Write tree t in list notation.
   c. Does t correctly encode a WHILE-program in abstract syntax? If this is the case, write the corresponding WHILE-program p for which $\ulcorner p \urcorner = t$ holds in concrete syntax. If this is not the case, apply minimal corrections to t such that the resulting new tree t' encodes a WHILE-program p. Write p in concrete syntax.

```
test read X {
    while hd X {
        X := tl X }
  }
  write X
```

Fig. 6.4 Sample program for Exercise 1

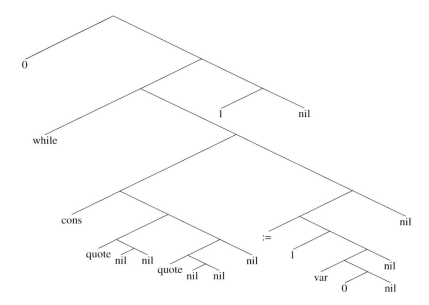

Fig. 6.5 A binary tree

3. In the style of Definition 6.4, formally explain what a compiler from language
   S to language T written in in language L is. For simplicity, you can assume that
   S-data = T-data. Assume that L has (S-)programs as data.
4. Discuss what changes in the answer to Exercise 3 are required if we give up
   condition S-data = T-data.
5. In the style of Definition 6.4, formally explain what a program specialiser from
   language S to language T written in language L is that partially evaluates an
   S-program with two inputs (encoded as pair) and a value in S-data and returns
   a T-program that only has one input. For simplicity, you can assume that S-data
   = T-data = L-data. Assume that S has programs as data and admits pairing.
6. Assume language L has pairing and programs as data. Assume further that you
   have an S-interpreter int written in L and a specialiser spec for L, i.e. a specialiser from L to L, written in another language, say T. Explain how you can
   compile S-programs into L programs with the help of int and spec and what
   conditions you have to place on the data of S, L, T (if any).
   Hint: This is the first Futamura projection.

Reference

1. Jones, N.D.: Computability and Complexity: From a Programming Perspective. MIT Press,
   Cambridge (1997) (Also available online at http://www.diku.dk/~neil/Comp2book.html)

Chapter 7
A Self-interpreter for WHILE

How can we write a WHILE interpreter in WHILE? How do we traverse abstract syntax trees?

We already know what an interpreter is. In this chapter we will focus on an interpreter for WHILE written in WHILE itself. Why is this interesting? From a pragmatic point of view it is certainly desirable to have an interpreter for WHILE. That we can write one in WHILE itself is evidence that our WHILE -programs were an acceptable choice for effective procedures. Or as *Neil Jones* in his book put it: *"The WHILE language seems to have just the right mix of expressive power and simplicity."* [1, Preface, p. X].

But there is further good use in computability and complexity for such a self-interpreter (and for "efficient" self-interpreters), which we will use in future chapters. It is important to recall that a self-interpreter is an interpreter written in the language it interprets. It is therefore *not just* an interpreter that interprets itself as it can interpret *any* program written in the same language. Formally, we can define a *self-interpreter* as follows:

Definition 7.1 A *self-interpreter* or *universal program* for a programming language: L is an interpreter for L, written in L, i.e. an L-program s such that

$$[\![s]\!]^{\mathrm{L}}(p, d) = [\![p]\!]^{\mathrm{L}}(d)$$

for all L-programs p and L-data d.

Historically, the first person to define a self-interpreter was *Alan Turing*. His famous paper [2] contained a (Turing) machine that can take the description of another (Turing) machine and an input word, and simulate the run of the given machine on the given input. Turing called this machine *universal* for obvious reasons. Self-interpreters are therefore often called *universal programs*.

© Springer International Publishing Switzerland 2016
B. Reus, *Limits of Computation*, Undergraduate Topics in Computer Science,
DOI 10.1007/978-3-319-27889-6_7

In the following a possible implementation of the self-interpreter for WHILE is presented and discussed in detail. We mainly follow Neil Jones' definitions in [1, Chap. 4] and first write an interpreter for a slightly simpler language than WHILE, called WH^{1}LE(Sect. 7.1). This language is like WHILE but programs must only use one variable. Section 7.2 then presents the actual self-intepreter for WHILE.

7.1   A Self-interpreter for **WHILE** -Programs with One Variable

In this section we build a self-interpreter for WHILE with just one program variable.

Definition 7.2 (*WH^{1}LE*) This language WH^{1}LEis like WHILE but programs must only use *one* program variable. The number 1 alludes to this fact.

This means for WH^{1}LE-programs that input and output variable must also be this one variable. Note that one can compile any WHILE-program into a WH^{1}LE-program by storing the values of all variables in a list which is stored in the single variable one is allowed to use in a WH^{1}LE-program. But this is not our concern now and is covered in Exercise 7. The reason for using WH^{1}LE first is that the memory handling is much easier so one can see much better how the self-interpreter works.

7.1.1   General Tree Traversal for ASTs

The interpreter needs to traverse the abstract syntax tree and produce intermediate results "on the fly". Those intermediate results will be stored on the value stack (St). The template of a traversal algorithm for an AST is as described in Fig. 7.1 in pseudo-code:

This template is implemented in the outline of the interpreter in Fig. 7.2. The while loops in both algorithms correspond. However, in Fig. 7.2 we have left the details of the tree traversal to a macro STEP that uses the "state" of the traversal, i.e. the code stack (which corresponds to the tree stack in Fig. 7.1), the data stack (previously called value stack or intermediate result stack), and a value which represents the execution state of a WH^{1}LE-program.

Since interpretation of commands and evaluation of expressions has to be broken down into its constituent sub-expression and sub-commands we need a stack to keep the parts of the AST that still need to be visited (evaluated) and, at the same time, store the intermediate results of what we have already computed. We will use tags on the command stack to indicate that some bits of evaluation or execution still need to be done. These tags are similar to the ones we used to encode abstract syntax trees (e.g. := or tl). We thus need more atoms, namely the following auxiliary atoms: doHd, doTl, doCons, doAsgn, doIf, and doWhile. With their help one can now define the STEP macro.

```
initialise tree and value stack to be empty
push tree (to be traversed) on tree stack
while tree stack not empty do
 pop a tree t from tree stack
 if t is just an opcode with arity n
 then                          (* auxiliary marker encountered *)
   pop n results r1,...,rn from value stack
   r := o(r1,...,rn)           (* compute intermediate result *)
   push r on value stack       (* push intermediate result *)
 else                          (* t is proper tree *)
   if t's top level operation has n>0 arguments
   then                        (* break down AST *)
     push t's opcode as auxiliary marker on tree stack
     push n subtrees of t on tree stack
     (* these pushed subtrees are still to be evaluated *)
   else                        (* t is a leaf *)
     evaluate t and push result on value stack
   endif
 endif
endwhile
```

Fig. 7.1   A generic algorithm for traversing an AST

7.1.2   *The STEP Macro*

We already know that the STEP macro manipulates the state of the traversal, i.e. the code stack, the data stack and the value of our one program variable. For the sake of readability,[1] let us present the behaviour of STEP as a set of rewrite rules of the form

$$[\text{CSt}, \text{DSt}, \text{val}] \Rightarrow [\text{CSt}_{\text{new}}, \text{DSt}_{\text{new}}, \text{val}_{\text{new}}]$$

that describe how the data structures of the traversal, i.e. DSt, the value stack, CSt, the code stack, and val, the store for execution that consists of just one variable, change according to their current values. Diagrammatically, such a rule is depicted as:

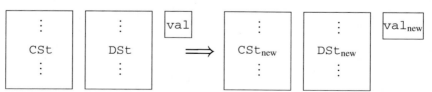

Note that stacks grow upwards here, pushing and popping happens from the top.

   Such a rule states that if the current code stack is CSt, the current value stack is DSt, and the value of our one variable is val, then the STEP macro has to change the code stack to CSt_{new} the value stack to CSt_{new} and the variable value to val_{new}.

---

[1] Also following the pedagogical route taken by [1].

```
    read PD  {              (* input is a list [P,D] *)
      P := hd PD ;          (* P = [X,B,X] *)
      D := hd tl PD;        (* D input data *)
      B := hd tl P;         (* B is program block *)
      CSt := B;             (* CSt is code stack  *)
                            (* initially  commands of B *)
      DSt := nil;           (* DSt is computation stack for *)
                            (* intermediate results *)
      val := D;             (* D is initial value of variable *)
      state := [ CSt, DSt, val ];
                            (* wrap up state for STEP macro *)
      while CSt {           (* main loop for interpretation *)
        state := <STEP> state; (* loop body macro *)
        CSt := hd state     (* get command stack *)
        }
      val := hd tl tl state (* get final value of variable *)
    }
    write val               (* return value of the one variable *)
```

Fig. 7.2 The WH[1]LE-interpreter in WHILE

Below we will give those rules in diagrammatic form, showing more clearly how the stacks (and the value of the variable) change according to their top elements.

A set of those rules then describes, in a way, a big case-statement that tells us what happens depending on what values can be found on top of the code stack and value stack, respectively.

Now according to the interpreter of Fig. 7.2 the initial state sets the code stack to the body of the program B, the data stack to nil and the value of the one variable used to val, which in diagrammatic form is depicted as follows:

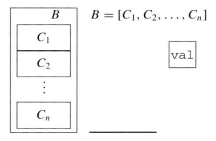

The STEP macro needs to behave as prescribed by the following rewrite rules in diagrammatic form:

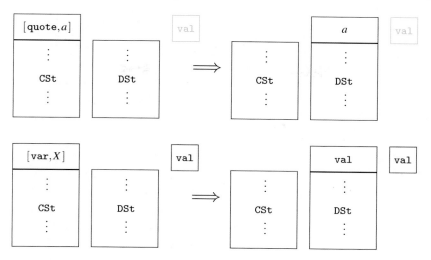

Fig. 7.3 STEP macro for evaluating quote and var

7.1.2.1   Nullary Expressions

We consider the leaves of ASTs, i.e. atoms of the form [quote, a] and variables of the form [var, X] where the number encoding the variable name does not matter as it is always the same variable that will be used in a WH^1LE-program. The diagrammatic form of the rules for AST leaves is depicted in in Fig. 7.3.

The evaluation of a quoted literal returns the literal as value which is therefore pushed on the data stack. The quote expression is popped from the code stack (as it has been dealt with).

The evaluation of a variable (recall we have only one in language WH^1LE) returns the value of this one variable which is pushed onto the result stack. The variable expression is popped from the code stack.

7.1.2.2   Compound Expressions

Next, we need to handle single argument compound expressions hd and tl. The diagrammatic form of the rules for those is depicted in Fig. 7.4.

To evaluate an AST expression of the form [hd, E] we first need to evaluate E. So a marker doHd is pushed onto the command stack, followed by the still to be evaluated expression E. Nothing happens on the data stack as no value is produced nor consumed at this stage.

Once a marker doHd is found on top of the code stack we know that we can finish the computation of a hd expression. We also know that the evaluated argument has been pushed on top of the data stack. So we pop this value v off the data stack,

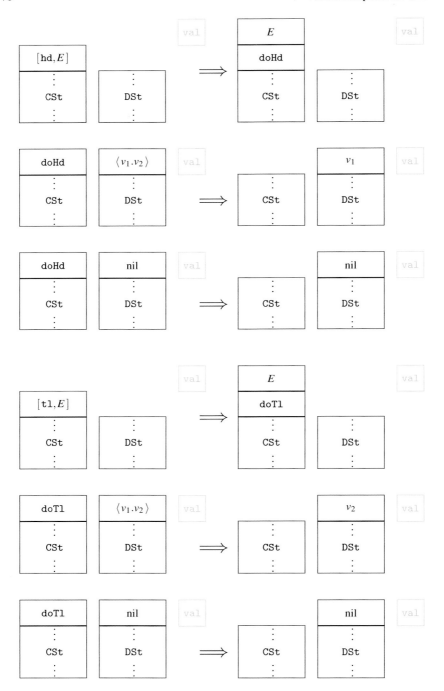

Fig. 7.4 STEP macro for evaluating hd and tl

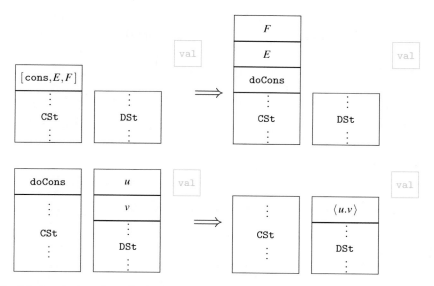

Fig. 7.5 STEP macro for evaluating cons

compute hd v and push the result on the data stack. Since we have now dealt with the hd expression we pop the marker off the code stack.

For an AST of the form [tl, E] we do exactly the same as for hd just replacing doHd by doTl and instead of the result of hd v we push the result of tl v on the data stack.

Next, we discuss the binary operator cons. The rules are shown in diagrammatic form in Fig. 7.5.

To evaluate an AST expression of the form [cons, E, F] we first need to evaluate expressions E and F. So a marker doCons is pushed onto the code stack, followed by the still to be evaluated expression trees E and F. Note that the order in which we push both arguments is arbitrary in principle, but it is important that it is *fixed* such that the rule for doCons knows in which order to take values from the data stack to build the resulting tree correctly. Nothing happens on the value stack as no value is produced nor consumed at this stage. Once a marker doCons is found on top of the command stack, we know that we can finish the computation of a cons expression. We also know that two evaluated arguments have been pushed on top of the result stack. So we pop both values u and v off the data stack, construct the corresponding tree $\langle u.v \rangle$, which denotes the result of the cons operation according to the semantics, and push its result on the data stack. Here is where we need to know that the first argument was pushed first in order to produce the right result. We have now dealt with the cons expression, hence we can pop the marker off the code stack.

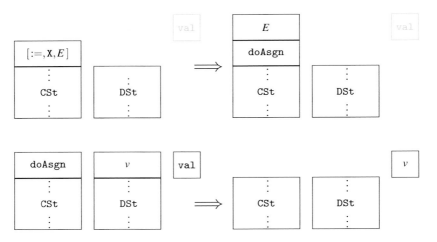

Fig. 7.6 STEP macro for evaluating assignment

7.1.2.3   Commands

Finally, we have to consider how to execute the various commands: assignment, conditional, and while loop. First we look at the assignment operator. The rules for that are depicted diagrammatically in Fig. 7.6.

To interpret an AST encoding an assignment of the form [:=, X, E] we first need to evaluate the expression E. So we push first a marker doAsgn on the code stack followed by the still to be evaluated E. Nothing happens on the value stack as no value is produced nor consumed at this stage. Once a marker doAsgn is found on top of the command stack, we know that we can finish the interpretation of an assignment command. We also know that the value to be assigned has been pushed onto the data stack. So we pop this value v off the data stack and update the "memory", in this case the value for the one variable of the program. Since we have now dealt with the assignment we pop the marker off the code stack.

For the conditional [if, E, B_T, B_E] we first need to evaluate the expressions E like for the assignment. We thus push it onto the command stack, but not before we pushed a marker to recall that we still have to finish the conditional, doIf and the entire conditional itself. In fact, it would suffice to just push both blocks for the then-case B_T, and the else-case, B_E, respectively. This is necessary to decide what branch to execute once we know the value of E. Once the expression E has been evaluated, it has disappeared from the command stack where we will find the marker doIf. In this case we inspect the top element of the data stack which is supposed to be the value of E and act accordingly. If the top element represents true, i.e. is a non-nil value, we pop the marker and the conditional and push the commands of the then-block B_T onto the command stack. We also pop the top element, the used value of E, from the data stack. In case the top element of the data stack is nil, representing false, we analogously pop marker and conditional and push the commands of the

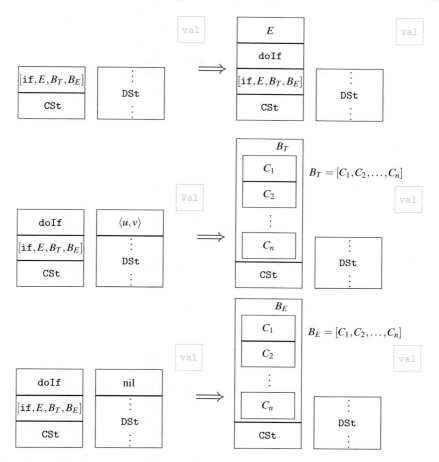

Fig. 7.7 STEP macro for evaluating if-then-else

else-block B_E instead. Again we pop the top element off the data stack. It is important to notice that one does not push the blocks B_T and B_E, respectively, in their entirety as a list but rather pushes all commands of the corresponding list in the correct order. This means that we push the last command first and the first command last, such that the first command of the block will now be executed first. The rules for the conditional are depicted in Fig. 7.7.

The final command, and the most complicated to execute, is the while loop [while, E, B].

To interpret an AST representing a while loop [while, E, B], we first need to evaluate the guard expression E. Since we might need to evaluate the body of this loop for a yet unknown number of times we first push the entire while command on the code stack, followed by a marker doWhile, followed by the expression E still to be evaluated. Nothing happens on the value stack as no value is produced nor

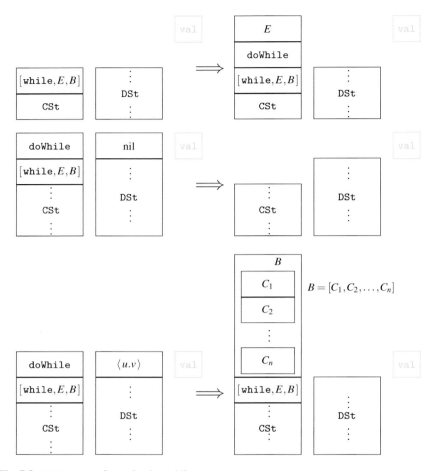

Fig. 7.8 STEP macro for evaluating while

consumed at this stage. Once a marker doWhile is found on top of the code stack we know that we have finished the evaluation of the guard expression and can decide what to do next accordingly. We also know that the value of the guard expression has been pushed onto the data stack. We consider two cases (Fig. 7.8):

- if the top element of the data stack is nil we pop it off. Since this means we terminate the loop, we also pop the marker as well as the pushed while loop off the code stack as the interpretation of this loop is now finished.
- if the top value of the data stack is different from nil, i.e. of the form $\langle u.v \rangle$, we pop it off the data stack. Since this means we need to execute the loop body another time, we pop the marker doWhile and push in exchange the body B onto the code stack. Since B is a block, it is of the form $[C_1, C_2, \ldots, C_n]$, and we push the individual elements C_i of this list accordingly, making sure that the first one ends

up on the top of the code stack as it is the first one to be executed. We leave the entire while loop on the code stack "below the loop body B", so that the process of executing the loop can begin from the top, once the execution of block B has finished and it has disappeared from the top of the code stack.

7.2 A Self-interpreter for WHILE

From the self-interpreter for WH^{1}LE it is relatively straightforward to derive a self-interpreter for WHILE. It remains to explain how we can generalise from one program variable to arbitrarily many. We need to implement a store, i.e. a set of variable bindings which can be represented as a *list* of variable bindings where bindings are themselves pairs (lists with two elements): first the variable name (encoded as number) and second the value of the variable. Choosing this representation, the order in which the variable bindings are stored is not relevant for the result. We find the right element by looking up the first entry in the binding list.

The program from Fig. 7.2 has to be adapted accordingly. Instead of a single value val, the store now will be kept in variable St, a store encoded as a list, and macro functions update and lookup are in use to deal with variable update and lookup. They will be explained further below. The resulting new main program can be found in Fig. 7.9.

```
read PD {                        (* input is a list [P, D] *)
   P  := hd PD ;                  (*   P = [X,B,Y] *)
   D  := hd tl PD;                (* D input data *)
   X  := hd P;                    (* X is input var name *)
   Y  := hd tl tl P;             (* Y is output var name *)
   B  := hd tl P;                 (* B is program code block *)
   CSt := B;                      (* CSt is code stack   *)
                                  (* initially contains only B *)
   DSt := nil;                    (* DSt is data stack for *)
                                  (* intermediate results *)
   bind := [ X, D ];
   St := [ bind ];                (* initialise store *)
   state := [ CSt, DSt, St ];     (* wrap state for STEP macro *)
   while CSt {                    (* main loop for interpretation *)
     state := <STEPn> state;      (* loop body macro   *)
     CSt := hd state              (* get command stack *)
     };
   St := hd tl tl state;          (* get final store *)
   arg := [ Y, St ];              (* wrap argument for lookup *)
   Out := <lookup> arg            (* lookup output variable value *)
   }
   write Out                      (* return value of result variable *)
```

Fig. 7.9 A WHILE-interpreter in WHILE

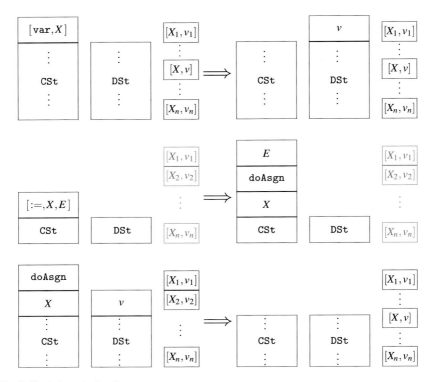

Fig. 7.10 Adapted rules for STEPn macro

The new STEPn macro is very similar to the STEP macro described in Sect. 7.1.2 and needs to be changed only at those places where variables are concerned: the evaluation of variables and the assignment command. The corresponding rules must now contain a store list. This will be depicted as an array starting at the top with the first binding. Although the order of bindings is irrelevant, we usually write the input variable's binding as first element and the output variable's binding as last element of the store list (Fig. 7.10).

Let us have a look at the changes we needed to make because variables now come with a number in the form [var, X]. For the evaluation of such a variable we now need to find the right element in the list [[X_1, v_1], [X_2, v_2], ..., [X_n, v_n]] which will be stored in variable St. To obtain the value of variable X we must find the binding for X in the store list and then use the corresponding value v from binding [X, v]. To achieve this we will use macro lookup described in the following Sect. 7.2.1. As before, the corresponding value v is then pushed on the data stack and the variable expression popped off the command stack.

For an AST representing an assignment [:=, X, E] we again push the marker and the still to be evaluated expression E, but we also need to remember which variable this assignment is for so we push the variable number X onto the command stack before we also push the marker. Once we then hit the doAsgn marker on the command stack, we know the value of the assignment expression has been evaluated and resides on the top of the data stack.

We pop the marker doAsgn and the following variable number X off the command stack, pop the result value v off the data stack, and update the element that corresponds to variable X in the list $[[X_1, v_1], [X_2, v_2], \dots, [X_n, v_n]]$ to contain v in the binding for X. This will be done in code using the auxiliary macro update described in the following Sect. 7.2.1.

The code for the lookup and update, respectively, is based on simple list manipulation checking for the right binding using the atom encoding the variable name. Their code is briefly discussed in the following subsection which can be skipped if one is not interested in coding the self-interpreter.

Finally, the concrete code for the STEP and STEPn macros is left as Exercise 1.

7.2.1 Store Manipulation Macros

7.2.1.1 Lookup

The program lookup from Fig. 7.11 takes as input a list [X, St] where X is a unary number encoding a variable name and St is a list containing variable bindings, i.e. two-element lists of the form [X_i, v_i] where each X_i is a number encoding a variable name and $v_i \in \mathbb{D}$ is the current value of this variable. There is one exceptional case: if X does not appear as variable in the list St, i.e. if one looks up a variable that

```
lookup read XS  {          (* argument XS is list [X, Store] *)
   X := hd XS;              (* X is number encoding variable *)
   St := hd tl XS;          (* Store contains list of bindings *)
   (* do the lookup; this destroys the Store *)
   while St  {              (* run through store list *)
      bind := hd St;        (* bind is binding *)
      N := hd bind;         (* N is var number of bind *)
      if N=X {              (* found right variable *)
       { Res := hd tl bind; (* Res is value of X *)
         St := nil          (* reset St to exit loop *)
       }
      }
      else {                (* variable X not found *)
      St := tl St           (* drop first binding off store *)
      }                     (* end of if-the-else  *)
   }                        (* end of while loop *)
}
write Res
```

Fig. 7.11 Variable lookup in store

84

7 A Self-interpreter for WHILE

```
update read AL   {            (* argument AL is list [X,V,St] *)
   X := hd AL;                (* X is variable code *)
   V := hd tl AL;             (* V is value used for update *)
   nb := [ X, V ];            (* nb = new binding for X *)
   St := hd tl tl AL;         (* St is store to be updated *)
   miss := true;
   NSt := nil;                (* new store to be produced *)
   while St     {             (* run through St destructively *)
      bind := hd St;          (* bind is first binding of store *)
      N := hd bind;           (* N is variable code of bind *)
      if N=X {                (* variable X found   *)
         NSt := cons nb NSt;  (* update new store *)
         miss:= false         (* set flag to remember this *)
         }
      else {
        NSt := cons bind NSt  (* copy binding in new store *)
      };                      (* end of if-the-else *)
      St := tl St             (* move to next binding in St *)
      };                      (* end of while loop *)
      if miss                 (* if var X not found add binding *)
      { NSt := cons nb NSt
      }
   }
   write NSt                  (* return updated store *)
```

Fig. 7.12 Variable update in store

has not been initialised, then the result Res will be nil (as in this case it will not have been assigned anything). The consequence is that uninitialised variables are implicitly looked up with value nil which is exactly what we need.

7.2.1.2 Update

The program update from Fig. 7.12 takes as argument a list [X, V, St] where X is the number encoding of the name of the variable to be updated, V is the new value of the variable to be used for the update, and St is the list of variable bindings in which the update takes place.

The program copies St into the new result store NSt, and in case the encoded variable name of the found binding equals X, a new binding nb is produced that uses the new value V and is inserted into NSt instead of the old binding for X. If there is no binding for X in St then the newly produced NSt will just be a copy of St in reverse order and we need to add the new binding. As the order of appearance of bindings in a store list is not important, there is no need to reverse the store list at the end.

7.2.1.3 A More Efficient Store Implementation

Given that we decided to encode variable names as unary numbers, the implementation of the store as a list of bindings is not the most efficient one. One can avoid the comparisons $N = X$ in the programs of Figs. 7.11 and 7.12 by using the variable number as a kind of "pointer" into the list of values in the store. Therefore, a more efficient variable management can be obtained by implementing the store as a list of values $[v_1, v_2, \ldots, v_n]$ where the value v_i is supposed to be the value of the ith variable X_i (assuming we start counting variables from 1).[2]

The macro lookup now only has to travel along the spine of the store list, as many times as the variable number of X states minus one, to find the right position in the list where the value of X is stored. The macro update needs to perform the same action, keeping track however of the values of the variables it sees along the way. Once it found the right value for X, the first element of the remaining list is updated to the new value and the kept prefix of the travelled store list needs to be put back at the front of the store list (in the right order). The details for this version of the macros are delegated to Exercise 3. One other change has to be made to the interpreter itself in Fig. 7.9. The initialisation of the store St has to be changed as we don't keep explicit bindings any longer.

What next?

We have seen that WHILE is expressive enough to write a self-interpreter in it. This is further proof that WHILE is an appropriate language to be used for the investigation of computability and complexity. In the next chapter we can therefore use WHILE-programs as notion of effective procedure to define WHILE-decidability and encounter a first problem that is *not* WHILE-decidable. To prove the latter we will discuss a specific technique called diagonalisation.

Exercises

1. Implement the macros STEP and STEPn in WHILE.
2. Discuss the correctness of the auxiliary store manipulation programs in Figs. 7.11 and 7.12.
3. Implement the macros lookup and update for the store implementation as list of values in the order of variable numbers discussed in Sect. 7.2.1.3.
4. Check that the following invariant for expressions is guaranteed by the WH[1]LE interpreter discussed in Sect. 7.1.

$$[\langle E.Cd \rangle, St, d] \Rightarrow^* [Cd, \langle e.St \rangle, d] \quad \text{if and only if} \quad \mathscr{E}[\![E]\!]\{X : d\} = e$$

where $\Rightarrow^*$ refers to a finite number of applications of the rewrite rules explained in this chapter.

---

[2]We can also start from 0.

5. Check that the following invariant for statement lists (blocks) is guaranteed by the WH^1LE interpreter discussed in Sect. 7.1.

$$\langle C_1.\langle C_2.\ldots.\langle C_{n-1}.\langle C_n.Cd \rangle\rangle\ldots\rangle, St, d \,] \Rightarrow^* [\,Cd, St, e\,]$$
$$\text{if and only if}$$
$$C_1; C_2; \ldots; C_n \vdash \{X : d\} \to \{X : e\}$$

where $\Rightarrow^*$ refers to a finite number of applications of rewrite rules explained in this chapter.

6. Extend the self-interpreter with tree literals.
7. Program a *compiler* while2one from WHILE to WH^1LE. This program takes a WHILE program encoded as a list according to our AST encoding and returns a WH^1LE-program encoded as a list that only uses one variable.
   Hint: The one variable in the constructed WH^1LE-program should be a list containing the values of all the variables of the original While-program. Where necessary, the compiler needs to generate the correct access code for the intended variable in this list.

References

1. Jones, N.D.: Computability and Complexity: from a Programming Perspective (Also available online at http://www.diku.dk/neil/Comp2book.html). MIT Press, Cambridge (1997)
2. Turing, A.M.: On computable numbers, with an application to the Entscheidungsproblem. Proc. London Math. Soc. Ser. 2 **42** (Parts 3 and 4), 230–265 (1936)

Chapter 8
An Undecidable (Non-computable) Problem

Is there a problem that is undecidable by any WHILE program?
How can one show this?

After having decided to use WHILE-programs as our notion of effective procedures and after having discussed WHILE at great length, it is now time to investigate our first major result in computability theory, and our first interesting problem. More precisely, we find a problem that is undecidable, i.e. that no computer program can solve. That means we cannot write a (WHILE) program that computes the decision procedure for this problem.

The chosen undecidable problem is the one that Alan Turing used in his seminal paper [7]. It refers to checking the termination of a given program with some given input and is therefore called the "Halting problem". We will actually show that this problem cannot be solved by a program and in order to do that we will use a famous technique called "diagonalisation".

Before we do this we recall the definition of decidability (Sect. 8.1) and instantiate it now for the concrete "effective procedures" chosen, i.e. for WHILE programs. Then we define the Halting problem (Sect. 8.2), carefully introduce the concept of diagonalisation (Sect. 8.3), discussing some interesting paradoxes along the way, and finally prove in Sect. 8.4 that the Halting problem is WHILE-undecidable.

8.1 WHILE-Computability and Decidability

First of all, we formally define what computable means in WHILE. We do this for function and decision problems, by basically instantiating WHILE for the class of effective procedures $\mathscr{P}$ in Definition 2.14. Consequently, we will use the data type of WHILE to encode the domain of any problem we want to solve. We have already seen in Sect. 3.3 that we can encode many data types in $\mathbb{D}$.

© Springer International Publishing Switzerland 2016
B. Reus, *Limits of Computation*, Undergraduate Topics in Computer Science,
DOI 10.1007/978-3-319-27889-6_8

Definition 8.1 A partial function $f : \mathbb{D} \to \mathbb{D}_\perp$ is WHILE-*computable* if there is a WHILE-program p such that $f = [\![p]\!]^{\text{WHILE}}$, in other words if f is equal to the semantics of p (we can also say "if p implements f").

The slogan here is:

A WHILE-computable (partial) function on trees is one that can be implemented in WHILE.

The above definition of f being WHILE-computable means in more detail that for any $d, e \in \mathbb{D}$ (for all trees d and e):

1. If $f(d)\uparrow$ then $[\![p]\!]^{\text{WHILE}}(d)\uparrow$
   Whenever f applied to d is undefined, then so is the semantics of p, in other words program p does not terminate when run on input d.
2. If $f(d) = e$ then $[\![p]\!]^{\text{WHILE}}(d) = e$
   Whenever f applied to d equals e, then so does semantics of p applied to d, in other words program p terminates when run on input d and produces e as output.

As we can encode numbers in WHILE's data type, we know that a WHILE-computable partial function on natural numbers is one that can be implemented in WHILE. This is important when comparing WHILE-computability with other notions of computation that use the natural numbers as data type (this is quite common).

A decision problem A with domain $\mathbb{D}$ corresponds to a set $A \subseteq \mathbb{D}$ of trees. The uniform question problem for A is the following:

Is a given element $d \in \mathbb{D}$ in A?

Or in other words: Given $d \in \mathbb{D}$, does $d \in A$ hold? Note how this is a problem in our sense: it is uniform, the input type is well defined ($\mathbb{D}$) as well as the output type which has to allow for answers 'yes' and 'no' ('true' and 'false') and is thus the type of Boolean values. Elements of each of both types are finite and precisely describable.

We can now instantiate the generic definition of a problem computable or decidable by an effective procedure, as we know we use WHILE as effective procedures: A decidable set (or a decidable problem) is one for which the membership test can be computed by a WHILE-program that always terminates.

These definitions can also be given more precisely and formally:

Definition 8.2 A set $A \subseteq \mathbb{D}$ is WHILE-decidable if, and only if, there is a WHILE-program p such that $[\![p]\!]^{\text{WHILE}}(d)\downarrow$ (meaning $[\![p]\!]^{\text{WHILE}}(d)$ is defined) for all d in $\mathbb{D}$, and, moreover, $d \in A$ if, and only if, $[\![p]\!]^{\text{WHILE}}(d) = \text{true}$.

Note that in the (older) literature, when computability is defined in terms of natural number computation, a decidable set is often called a *recursive set*.

8.2   The Halting Problem for WHILE

Let us look at the following problem which is about WHILE-programs (represented as elements of $\mathbb{D}$, we already know how to encode programs as data so this is not a problem).

Definition 8.3 The *Halting problem*—as set HALT $\subseteq \mathbb{D}$—is defined as follows:

$$\text{HALT} = \{\, [p, d] \in \mathbb{D} \mid \llbracket p \rrbracket^{\text{WHILE}} (d) \downarrow \,\}$$

What does the problem say informally speaking, given as a class of (uniform) questions? Well, the question is this:

> Given a WHILE-program p (as data) and a value $d \in \mathbb{D}$, does program p terminate if we run it on input d?

Note again the uniformity, the answer must be given for an arbitrary program and data. We also have a problem about two pieces of information here, the program and its input, but we can pair them together into one input tree in the way we already encountered for other WHILE-programs using the encoding of pairing $\ulcorner(p, d)\urcorner$ as list $[p, d]$ which can be itself encoded in $\mathbb{D}$. We drop the encoding brackets from the list to improve readability. We will drop those encoding brackets most of the time in the following chapters too.

We will show in this section that the set HALT is not WHILE-decidable in the sense explained earlier (Definition 8.2). This is equivalent to showing that the function *halt* defined below is not computable by a WHILE-program. In other words, the (total) function

$$halt(a) = \begin{cases} \text{true} & \text{if } a = [p, d] \text{ and } \llbracket p \rrbracket^{\text{WHILE}} (d) \downarrow \\ \text{false} & \text{otherwise} \end{cases}$$

(which is the membership test for the Halting set HALT) is not computable by a WHILE-program.

It should be clear that it would be really useful if we could compute/solve this problem. A compiler could then check if we had inadvertently written a non-terminating program for some specific input and warn us like a type checker warns us about incompatible types when we have used expressions in an ill-typed way.

It is desirable to be able to recognise undecidable problems. Nobody wants to start working on a project that seeks to develop a program that computes the solution for a non-computable problem. One would like to avoid wasting time on that. It is of course not always easy to find out whether a problem is undecidable and, maybe quite surprisingly, there is an abundance of undecidable problems. We will soon present more of them in Chap. 9. It turns out that some of them can be easily recognised.

It should be pointed out that there is still a considerable amount of research going on how to find out whether a program might not terminate for some input. Termination checkers have been recognised as a useful tool even if they cannot work for all input due to the undecidability of the Halting problem. The proof that a program terminates might be particularly important for systems code (operating system). For an overview over this exciting topic see [3].

8.3   Diagonalisation and the Barber "Paradox"

In the proof of the undecidability of the Halting problem we will use an old idea first discovered by Georg Cantor[1] in 1891, called *diagonalisation*, which was used in [2] to show that there are uncountable sets of real numbers. Let us first look at the diagonalisation in another context. The quite famous "Barber paradox" is derived from a paradox found by Bertrand Russell.[2] However, it is *not* a paradox in the Quinean definition: "a paradox is an argument whose conclusion contradicts a widely shared opinion" [5]. The "Barber paradox" has "only an incidental resemblance to Russell's Paradox" [5]. Actually, it is just a set of contradicting statements, but this is good enough for our purposes. The Barber of Seville says:

> In my town Seville, I shave all men who do not shave themselves. Those who shave them-selves I do not shave.

> Now we can ask:

> Does the barber shave himself?

Let us first remind ourselves that the Barber of Seville is a man and lives in Seville. Then let us look at this question and assume for the time being the answer was "yes". That means the barber shaves himself, but he stated that he does not shave the men in Seville who shave themselves. So since he is a man from Seville he does not shave himself. But then he stated that he shaves the men in Seville who do not shave themselves, so he must shave himself, and so on and so on.

Similarly, if we assume the answer is "no" we know that the barber does not shave himself but he stated that he shaves all men in Seville who do not shave themselves and he is a man living in Seville, so he must shave himself but then he said he does not shave those who shave themselves, so he must shave himself and so on and so on.

We see quickly that there is no answer to this question. The fact that there is no such (male) barber (in Seville) resolves any paradoxical situation and explains again why this is not a paradox. A paradox could only arise if there was a postulate that such a barber must exist. This is not the case, however, hence this is only a "pseudo-paradox". Any now slightly confused reader can study the philosophical subtleties in e.g. [5]. In any case, the so-called "Barber paradox" is good enough to

---

[1] Georg Ferdinand Ludwig Philipp Cantor (March 3, 1845–January 6, 1918) was a German mathe-matician. He is considered the inventor of set theory. For his revolutionary ideas he was as heavily critiqued during his lifetime as he is applauded now.

[2] Bertrand Arthur William Russell (May 18, 1872–February 2, 1970) was a famous Welsh logician and philosopher.

Table 8.1 Diagonalisation table for the Barber paradox

number of man in Seville	number of man in Seville									
	1	2	3	4	5	6	...	3466	3467	...
1	yes	yes	no	yes	no	no	...	no	no	...
2	no	no	yes	no	no	yes	...	no	yes	...
3	no	yes	no	no	no	no	...	no	yes	...
4	yes	yes	no	no	yes	yes	...	yes	yes	...
5	no	no	no	no	no	no	...	no	no	...
6	yes	yes	no	no	no	yes	...	yes	yes	...
⋮					⋮					
3466	yes	yes	no	no	yes	yes	...	yes	yes	...
3467	no	yes	yes	no	yes	yes	...	yes	no	...
⋮	⋱									

show the non-existence of the barber, so we can follow this line of argumentation when showing that there is a no program that can decide a certain problem.[3]

We can visualise the non-existence of the barber by a two-dimensional table where on the x-axis and y-axis we list the men of Seville. We can clearly do this. We use numbers to represent them. In each cell at row x and column y we then write whether person number x shaves person number y. So, for instance for row 2 and column 3 we have 'yes' meaning that person number 2 shaves person number 3. Table 8.1 contains (part of) such a sample table. Note that the entries are arbitrarily chosen in this example.

If there seems to be an "awful lot of shaving going on" in Seville, then this is by chance. Please note that it does not matter for demonstrating the argument what the concrete entries are.

Now the question is: what does the barber's row look like? What entries does it have for each column? Assume we look at column y. So we ask whether the barber shaves person number y. Well he does shave y in case that y does *not* shave y, otherwise he does not shave y. So we have to look into the diagonal of the table at (y, y) and negate the entry. If the entry was 'yes' then the entry in column y for the barber must be 'no' and vice versa. Therefore the barber's row is the negated diagonal so to speak. For the above table the barber's row then looks as follows:

no	yes	yes	yes	yes	no	...	no	yes	...

---

[3]Raising a (pseudo-)paradoxical question is a typical Sci-Fi plot to overcome some computer, robot, or probe that is about to destroy the spaceship/earth/galaxy. Trying to answer the question it runs into an endless cycle of questions indicated above and eventually explodes. That trick has been used at least several times in the classic Star Trek TV series (see also Exercise 1).

But this means the barber's row cannot be one of the rows in the table. The reason is that the row is different from any other row in the table that we assumed listed all men in Seville. Why is the barber's row different from all rows in the table? Well, it is so by construction. Assume without loss of generality, the barber's row was the one for person number 2. In this case we look at the entry in the 2nd column. The entry for the barber will be 'no' if the entry for person number two was 'yes' and will be 'yes' if the entry of person number 2 was 'no'. Remember this is because the barber's row is the negated diagonal. We can play the same "spiel" for any number, not just 2. Thus for every row k we can find a column (namely the kth column) where the barber's row and the kth row are different. Consequently, the barber cannot be already in the table. This is a problem though as he must be from Seville (by assumption). Thus we get a contradiction (unless we remove the assumption that the barber is from Seville or that he is a man).

Now, the "trick" here is the self reference of the barber's statement. When he talks about *all men in Seville* he implicitly includes himself. So the statement becomes self-referential. We try to learn from this and apply the same technique to show the undecidability of the Halting Problem.

8.4 Proof of the Undecidability of the Halting Problem

We now show that the Halting problem HALT is not decidable in WHILE, in other words, it is (WHILE-) undecidable. The original proof for Turing Machines (they will be defined precisely in Sect. 11.3) has been carried out in Turing's seminal paper [7]. The plan of the proof is as follows: assume that HALT is decidable, construct another program from this and then show using diagonalisation that this leads to a contradiction. Thus, the assumption that the Halting problem is decidable must have been wrong. This is called a proof by contradiction. "The proof is a beautiful example of self-reference. It formalizes an old argument about why you can never have perfect introspection: because if you could, then you could determine what you were going to do ten seconds from now, and then do something else" [1].

Let us assume more concretely that there is a WHILE-program h that decides the Halting problem and, without loss of generality, looks like described in Fig. 8.1 with input variable A, main block B and output variable C. Here B is the body of the program where all the "action happens" (we don't need to know what it is exactly).

```
h read A {
   B;
   }
write C
```

Fig. 8.1 WHILE-program h that is (wrongly) assumed to decide HALT

```
r read X {
    A := [ X, X ];    (* diagonalisation of argument   *)
    B;                (* run h on input [X, X] *)
    Y := C;           (* store result of running h in Y *)
    while Y {
            Y := Y
            }
    }
    write Y
```

Fig. 8.2 WHILE-program *r* used in the proof of the undecidability of HALT

Let us now construct a WHILE-program *r* that will correspond to the barber of the previous example. This program *r* is defined in Fig. 8.2 and uses A, B and C from the (assumed) program *h* above which is important as we will see shortly.

We now ask a question similar to "Does the barber shave himself?", namely

Does program r terminate when run on itself as input?

In other words, more concisely, "Does $[\![r]\!]^{\text{WHILE}} (r) \downarrow$ hold?" To answer this, let us go through *r* and interpret it with *r* as input. In the following, we drop the encoding brackets from the program arguments, so we do not distinguish between program *r* and its data representation $\ulcorner r \urcorner$ for the sake of readability.

1. Initially, we read in the input and afterwards X contains (the AST of) *r*.
2. After the first assignment of the program, A := [X, X], has been executed, A has value [*r*, *r*].
3. Then block B is executed which decides the Halting problem (this was our assumption). Thereafter, variable C contains the result of the decision procedure, i.e. $C = [\![h]\!]^{\text{WHILE}} [r, r]$.
4. Next, Y := C is executed and afterwards $Y = [\![h]\!]^{\text{WHILE}} [r, r]$.
5. For the while loop that follows, we have to distinguish two cases: whether Y is true or not.

   First assume Y is indeed true, then we are obviously caught in a non-terminating loop, so we know that *r* does not terminate when run on input *r*, in other words, $[\![r]\!]^{\text{WHILE}} (r) = \bot$. But, hang on a minute, was Y not supposed to be true? Indeed it was. In this case $Y = [\![h]\!]^{\text{WHILE}} [r, r]$ must be true as well. This in turn means by assumption (*h* decides the Halting problem) that *r* *does* terminate when run with input *r*, in other words $[\![r]\!]^{\text{WHILE}} (r) \downarrow$. This contradicts $[\![r]\!]^{\text{WHILE}} (r) = \bot$.
   Ok, maybe this means Y is false then. In this case *r* will not go into the non-terminating loop but instead terminates with result false. Thus, $[\![r]\!]^{\text{WHILE}} (r) =$ false or, ignoring the result, $[\![r]\!]^{\text{WHILE}} (r) \downarrow$. But, hang on a minute, was Y not supposed to be false? Indeed it was, and therefore $Y = [\![h]\!]^{\text{WHILE}} [r, r] = \text{false}$ which means by assumption (*h* decides the Halting problem) that *r* *does not* terminate when run with input *r*, in other words $[\![r]\!] (r) = \bot$. This contradicts $[\![r]\!]^{\text{WHILE}} (r) \downarrow$.
   So whichever case we consider, we always get a contradiction.

The diagonalisation in this argument can be shown by writing a table similar to the one for the Barber's paradox. Instead of the men in Seville, we list all WHILE-programs on both axes. We can do that as we can enumerate (generate) all programs and give each a number.[4] After all, programs are just finite *strings* (syntax).

Now the entry in the table for row x and column y describes whether program x terminates when run with input being program y. We recall that r plays the role of the barber here. Its row of entries is again the negated diagonal of the table by construction. So r cannot appear in the table itself and thus it cannot be a WHILE-program. Yet, we have seen that r *is* a WHILE-program *provided that the WHILE-program h exists*. Since the existence of h was the only assumption in our argument, we know that this assumption must be wrong. Thus we know for sure that the WHILE-program h cannot exist and thus the Halting problem is not decidable (by a WHILE-program).

What Next?

We have seen a first problem that is not computable by a WHILE-program and due to [7] not by a Turing machine either. In Chap. 11 we will discuss that the actual choice of programming language (as long as it is expressive enough) does not matter and we could have formulated the Halting problem also for Java or Haskell programs. In the next chapter we will look at more undecidable problems. We will see, among other things, that any interesting semantical property of a program is actually undecidable. And we will discuss a finer distinction between undecidable problems. For so-called semi-decidable problems we can only write "half of a decision procedure" where answers are guaranteed to be computed only if the answer is yes.

Exercises

1. (Liar Paradox) Consider the statement

    This sentence is false.

   sometimes also phrased as the "Liar paradox"[5]:

    Everything I say is false.

   Explain why this is a self referential statement and why it is a paradox.
2. One might think a way to resolve the paradox is to add to 'true' and 'false' a third logical value 'meaningless'. Thus, a statement can be 'meaningless' so it does not have to be either 'true' nor 'false'. Explain why using such a logic with 'meaningless' does not mean there can't be any paradox like the one of Exercise 1. *Hint: Combine 'meaningless' and 'false' logically in a self-referential statement that will still give rise to a new paradox, whether this sentence is meaningless, true or false.*

---

[4]The enumerability of programs follows from the fact that for any alphabet Σ the set of all finite strings over Σ, typically denoted Σ^*, is enumerable. The latter is a left as Exercise 6.

[5]A variation of this paradox is used in the *Star Trek* (Original TV Series) episode "I, Mudd", by Captain Kirk and Harry Mudd to confuse and eventually shut down an android called Norman, which was holding them prisoners [6].

3. (Russell's paradox) Let R be the set of all sets that do not contain themselves. Formally, $R = \{x \mid x \notin x\}$. Explain where the paradox lies.

4. (Crocodile Dilemma)[6] A crocodile that has stolen a child promises the mother that her child will be returned if, and only if, she can correctly predict whether or not the crocodile will return the child. Consider the following two cases and discuss whether a paradox can arise.

   a. The mother guesses the child will be returned.
   b. The mother guesses the child will not be returned.

5. Explain how every undecidable problem gives rise to a non-computable function as well.

6. In the diagonalisation for the proof of undecidability of HALT we used the fact that all WHILE-programs are enumerable. Discuss how this enumeration can be defined.

7. Show by a diagonalisation argument similar to the one given in this chapter that the "Uniform Halting problem" (UNIHALT) is also WHILE-undecidable. In other words, show that it is WHILE-undecidable whether a given WHILE-program terminates for *all input*.

8. Georg Cantor proved 1891 that the real numbers are not countable, i.e. there are more real numbers $\mathbb{R}$ than there are natural numbers $\mathbb{N}$ by what is known as the first proof by diagonalisation. Let us try to understand his argument. Assume that there are exactly as many real numbers in the interval $[0, 1[$ (all real numbers greater or equal 0 and smaller than 1) as there are natural numbers (note that both sets of numbers are infinite). Derive a contradiction with the help of a diagonalisation table. On both axes of this table we list the natural numbers. The entry for cell (i, j) shall be the jth digit after the decimal point of the ith real number. So, for instance, if the 5th real number starts $0.123456666\ldots$ then cell $(5, 3)$ contains 3. Now finish the argument why there must be more real numbers in the interval $[0, 1[$ than there are natural numbers.

9. Prove that the problem "is $d \in \mathbb{D}$ encoding a WHILE-program as data" is WHILE-decidable by writing the decision procedure as a WHILE-program.

References

1. Aaronson, S.: Who can name the bigger number? The blog of Scott Aaronson. Available via DIALOG. http://www.scottaaronson.com/writings/bignumbers.pdf (1999). Accessed 21 July 2015
2. Cantor, G.: Über eine elementare Frage der Mannigfaltigkeitslehre. Jahresbericht der Deutschen Mathematiker-Vereinigung, Band 1, 75–78, Teubner Stuttgart (1892)
3. Cook, B., Podelski, A., Rybalchenko, A.: Proving program termination. Commun. ACM **54**(5), 88–98 (2011)
4. Fargo Episode List. IMDb. Available via DIALOG. http://www.imdb.com/title/tt3097534/ (2015). Accessed 13 Oct 2015

---

[6]This is also the title of an episode of 2014 TV series Fargo [4].

5. Raclavský, J.: The barber paradox: On its paradoxicality and its relationship to Russell's paradox. Prolegomena J. Philos. **13**(2), 269–278 (2014)
6. Star Trek Episode List. IMDb. Available via DIALOG. http://www.imdb.com/title/tt0708432/?ref_=ttep_ep8 (2015). Accessed 13 Oct 2015
7. Turing, A.M.: On computable numbers, with an application to the Entscheidungsproblem. In: Proceedings of the London Mathematical Society, series 2, vol. 42 (Parts 3 and 4), pp. 230–265 (1936)

Chapter 9
More Undecidable Problems

In this chapter we will look at more undecidable problems. There are quite a number of them, infinitely many actually, and they appear in disguises more often than one might think. For instance, it turns out that any non-trivial purely semantical property of programs is undecidable.

But before we prove this result, known as Rice's Theorem (Sect. 9.2), let us make good use of the self-interpreter for WHILE from Chap. 7. We use it to show that HALT (our version of the Halting problem), which we already know from Chap. 8 is WHILE-undecidable, is at least WHILE-*semi-decidable* (Sect. 9.1). The notion of semi-decidability is another important concept of computability theory. For a semi-decidable property we cannot program a decision procedure but, so-to-speak "half" (semi) of such a decision procedure. More precisely, the "half" that answers yes in case the argument is in the set in question. A semi-decision procedure may not necessarily return false in case the argument is not in the set. A semi-decision procedure may therefore diverge for such arguments.

A few more famous undecidable problems are presented (Sect. 9.5) including the Tiling Problem (Sect. 9.3) which is introduced in more detail.

The very important and useful concept of *reduction* is explained in Sect. 9.4. It turns out we have already used it to show Rice's theorem. We will look into the reasoning principles reduction gives rise to.

The chapter concludes with a short discussion about how to deal with undecidable problems (Sect. 9.6) and introduces a function $\mathbb{N} \to \mathbb{N}_\perp$ that is not computable (Sect. 9.7) because it grows more quickly than any computable function.

9.1 Semi-decidability of the Halting Problem

We already know what a WHILE-decidable set or problem is. For sets (or problems) that are not decidable, we often can still show a weaker property that allows us to make at least statements about the elements that are in the set.

© Springer International Publishing Switzerland 2016
B. Reus, *Limits of Computation*, Undergraduate Topics in Computer Science,
DOI 10.1007/978-3-319-27889-6_9

A *WHILE-semi-decidable* set (or a semi-decidable problem) is one for which an approximation of the membership test can be computed by a WHILE-program that may not terminate if the answer is supposed to be false. Strictly speaking, the WHILE-program does not actually compute the membership test completely, only half of it (thus the name "semi"-decidable).

More formally we can define:

Definition 9.1 A set $A \subseteq \mathbb{D}$ is WHILE-semi-decidable if, and only if, there exists a WHILE-program p such that for all $d \in \mathbb{D}$ the following holds: $d \in A$ if, and only if, $\llbracket p \rrbracket^{\text{WHILE}}(d) = \text{true}$.

Note that in the (older) literature, when computability is defined in terms of natural number computation, a semi-decidable set is usually called a recursively enumerable set.

Recall the WHILE-interpreter written in WHILE from Chap. 8 and call this self-interpreter u for universal program. With the help of u one can show:

Theorem 9.1 *The Halting problem (for WHILE-programs),* HALT, *is WHILE-semi-decidable.*

Proof To show that HALT, the Halting set or Halting problem, is semi-decidable it suffices to come up with a WHILE-program *sd* such that for all input values $d \in \mathbb{D}$ and all WHILE-programs p we have that $[p, d] \in \text{HALT}$ if, and only if, $\llbracket sd \rrbracket^{\text{WHILE}}[p, d] = \text{true}$. In other words (expanding the definition of HALT), HALT is semi-decidable if, and only, if $\llbracket p \rrbracket^{\text{WHILE}}(d) \downarrow$ iff $\llbracket sd \rrbracket^{\text{WHILE}}[p, d] = \text{true}$. We can write such a program *sd* as outlined in Fig. 9.1. This program clearly returns output true if, and only if, Res := <u> PD terminates, which by the fact that u is the self-interpreter is exactly the case if PD $\in$ HALT.

In this subsection we state some important facts about (WHILE) decidability, semi-decidability, and their relationship. We can actually prove these results relatively easily by writing some programs.

Theorem 9.2   *1. Any finite set $A \subseteq \mathbb{D}$ is WHILE-decidable.*
2. If $A \subseteq \mathbb{D}$ is WHILE-decidable then so is $\mathbb{D} \backslash A$, its complement in $\mathbb{D}$.
3. Any WHILE-decidable set is WHILE-semi-decidable.
4. A problem (set) $A \subseteq \mathbb{D}$ is WHILE-decidable if, and only if, both A and its complement, $\mathbb{D} \backslash A$, are WHILE-semi-decidable.

```
sd read PD {
  Res := <u> PD;   (* call  self-interpreter *)
  X := true        (* X is true *)
}
write X
```

Fig. 9.1 Semi-decision procedure for HALT

Proof Exercise 1.

For later usage, we define the class of semi-decidable problems and decide problems using as basis for their abbreviations the old names found in the literature. Historically, i.e. in the 1930s, (partial) recursive functions were used to define computability and so-called *recursively enumerable sets* which are empty or the range of a *recursive function* as defined by *Kurt Gödel* who first used the term *general recursive function*. Accordingly, the field of *computability theory* was originally called *recursion theory*. A detailed exposition about the history and the terminology can be found, for instance, in [22]. Accordingly, **RE** stands for *recursively enumerable sets* which are the *semi-decidable problems* and **R** stands for *recursive set* which are the *decidable problems*.

Definition 9.2 Let **RE** denote the class of semi-decidable (often also called recursively enumerable) problems and **R** denote the class of decidable problems.

9.2 Rice's Theorem

Next, we generalise the result of the undecidability of the Halting problem. The Halting problem is a problem of (WHILE-) programs, namely whether such a program terminated on a specific input. It describes a property of WHILE-programs. We can actually generalise this and show that more properties of programs are undecidable, actually all "*interesting*" properties of programs are undecidable if they exclusively rely on the behaviour (semantics) of the program. We need to say what we mean by that.

A property of a program is *interesting* if it is a non-trivial property, i.e. one that not every program has but that at least one program has, and *extensional*. An extensional property is one that depends exclusively on the semantics, i.e. the input/output behaviour of the program. So interesting program properties are semantical program properties.

More precisely we define:

Definition 9.3 A *program property* A is a subset of WHILE-programs.

A program property A is *non-trivial* if $\{\} \neq A \neq$ WHILE-programs.

A program property is *extensional* if for all $p, q \in$ WHILE-programs such that $[\![p]\!]^{\text{WHILE}} = [\![q]\!]^{\text{WHILE}}$ it holds that $p \in A$ if and only if $q \in A$.

The last property states that an extensional property is one that cannot distinguish between semantically equal programs. If we have two different programs that have the same semantics either both must be in the set or none of them are in the set. It cannot be that one is in the set and the other one is not.

With the help of this definition we can now formulate Rice's theorem, named after Henry Gordon Rice[1]:

---

[1] Henry Gordon Rice (July 18, 1920–April 14, 2003) was an American logician and mathematician. He proved "his" theorem in his doctoral dissertation of 1951.

Theorem 9.3 (Rice [20]) *If A is an extensional and non-trivial program property then A is undecidable.*

Theorem 9.3 is a remarkable result as from this we get an entire family of undecidable problems. Informally speaking, this theorem states that "*any interesting property about the semantics of computation itself is undecidable*". We also obtain a "recipe" how to show that a problem about programs (in other words a program property) is undecidable, namely by applying Rice's theorem.

But let us first prove this theorem.

Proof Once again we do a proof by contradiction. We are given a property of a program (i.e. a problem about programs, namely whether they have this property) A that is extensional and non-trivial. We now assume A is decidable. Then we show that the Halting problem is decidable. The latter is clearly false so our assumption that A is decidable must have been false and therefore A must be *undecidable*.

How can we derive the decidability of the Halting problem from the assumption that A is decidable? We will show that any given program p on any given input e terminates if, and only if, another program (that we construct) is in A. As a consequence A is decidable if, and only if, the Halting problem is decidable.

In order to do carry out this plan, we need two auxiliary programs called *diverge* and *comp*. Let us define *diverge*, as the program that never terminates like e.g. in Fig. 9.2.

We have that $[\![diverge]\!]^{\text{WHILE}}(d) = \bot$ for any $d \in \mathbb{D}$. By non-triviality of A we know that there is a program that is not in A. Let us call this program *comp*.

Now assume A contains *diverge* (if this is not the case just swap the roles of A and its complement and perform the analogous proof). By extensionality of A we know that A contains all programs that never terminate for any input.

To show that the Halting problem is decidable assume we are given a program p and its input e. We wish to decide whether $[\![p]\!]^{\text{WHILE}}(e) \neq \bot$.

Without loss of generality, assume that the above program p and the program *comp* (that is not in A) have no variables in common. If they have, just consistently rename them without changing the semantics. We now construct a program q as outlined in Fig. 9.3.

We now consider the behaviour of q and whether it is in A or not. If $[\![p]\!]^{\text{WHILE}}(e) = \bot$ then clearly $[\![q]\!]^{\text{WHILE}}(d) = \bot$ for *all* $d \in \mathbb{D}$. On the other hand, if $[\![p]\!]^{\text{WHILE}}(e) \downarrow$

```
diverge read X {
  while true {
  }
 }
write Y
```

Fig. 9.2 WHILE-program that never terminates

```
q read X {
    Y := <p> "e";        (* run p on value e *)
    Res := <comp> X      (* run comp on input X *)
}
write Res
```

Fig. 9.3 Auxiliary program q to show Rice's Theorem

then $[\![q]\!]^{\text{WHILE}}(d) = [\![comp]\!]^{\text{WHILE}}(d)$ for *all* $d \in \mathbb{D}$. Therefore we get that

$$[\![q]\!]^{\text{WHILE}} = \begin{cases} [\![diverge]\!]^{\text{WHILE}} & \text{if } [\![p]\!]^{\text{WHILE}}(e) = \bot \\ [\![comp]\!]^{\text{WHILE}} & \text{if } [\![p]\!]^{\text{WHILE}}(e) \neq \bot \end{cases}$$

So if p does not halt on e we have that $[\![q]\!]^{\text{WHILE}} = [\![diverge]\!]^{\text{WHILE}}$ and so by extensionality of A and the fact that $diverge \in A$ we get that $q \in A$. On the other hand, if p does halt on e then $[\![q]\!]^{\text{WHILE}} = [\![comp]\!]^{\text{WHILE}}$ and so by extensionality of A and the fact that $comp \notin A$ we get that $q \notin A$. Therefore we get that $q \notin A$ if, and only if, p halts on e so the latter would be decidable if the former was. This completes the proof.

The implications of Rice's Theorem can hardly be underestimated. It implies that there can be no general program verification tool, that can check for any program and any specification that the program meets the specification. So, for instance, there can be no program that can check whether any other given program is a virus. This follows immediately from the fact the the properties that have to be checked in both cases are *extensional* (and clearly nontrivial) if defined naturally. It turns out that programmers actually run into these issues quite regularly. They sidestep them using means discussed in Sect. 9.6.

Of course, one can try and define what verification or being a virus means in terms of the syntax (e.g. the byte code) of a given program. But even then one runs into problems as arithmetic is already undecidable (see *Entscheidungsproblem*). All this, however, is not contradicting the fact that verification[2] and anti-virus tools *do actually exist*. Rice's Theorem just tells us that those tools can't possibly work correctly for *all inputs*.

9.3 The Tiling Problem

This is a classic problem (also called *Domino problem*) of which there are many variations. It dates back to the 1960s when mathematician, logician, and philosopher

---

[2]Program verification is actually a thriving discipline and quite considerable progress has been made in the last few decades [21] and has even reached giants like Facebook [4].

Fig. 9.4 Example of correct tiling

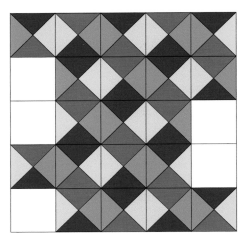

Hao Wang[3] proposed Tiling systems as an interesting class of formal system. A tile (now also called Wang tile) has a quadratic shape and is divided up diagonally into four sections (triangles), each of which can have a different colour (but the colours do not have to be necessarily different). The four sections are obtained by the cuts of the two diagonals of the square as follows:

where N, S, W, E stand for the four directions North, South, West, and East. When laying out the tiles we say that tiles put next to each other *match* if the colours of their adjacent edges are identical in all four directions: N, S, W, E. Note that it is not allowed to *rotate the tiles*. If the colours of each edge match, one obtains nice *diagonal* patterns since matching tiles next to each other give rise to a diagonal square in one colour (made up of the two halves of the matching tiles) as explained in Fig. 9.4. Therefore we can *tile* a quadratic floor (of size $n \times n$) with a given set of tile types if we can put tiles of the given types next to each other (without rotating them) such that their adjacent edges have the same colour. In particular, about such a tiling system Wang asked the following question:

> Given a certain (finite) number of tile types (colourings) and an infinite supply of tiles for each type, can one tile an arbitrary large quadratic floor (obeying the rules mentioned above)?

---

[3]Wang (20 May 1921–13 May 1995) was born in China but later taught at Harvard where he was also the thesis advisor of Stephen Cook (of whom we will hear more later).

Wang proposed this originally for a specific set of four tile types. He conjectured in 1961 that this question was decidable. In his proof argument he assumed that any set that could tile the plane would be able to do so periodically, i.e. with a repeating pattern like a wallpaper. Alas, he was wrong. His student Robert Berger managed to prove this wrong in 1966 [2]. He presented a case where the tiles would only tile the plane without repeating pattern, originally using 20,426 distinct tile shapes, later reducing this to 104. Even later, other researchers managed to complete the same proof with a much smaller number of tiles. Karel Culik only needed 13 in 1996 [7]. The idea of Berger's proof was to show that if the tiling problem was decidable then the Halting problem would be decidable too. This is the same proof technique that we used for proving Rice's theorem. So the trick here is to come up with tile patterns (you can use arbitrarily many as long as they are finitely many) that can encode the configurations a specific Turing machine goes through during execution for a specific input. Doing this, one must ensure that one can tile an arbitrary large quadratic floor with those tile types if, and only if, the encoded Turing machine run does *not* terminate (and runs for an arbitrarily large amount of time). Since the run is only determined by the Turing machine (program) and its input we know that we can tile a quadratic floor of any size if, and only if, the Halting problem for Turing machines was decidable.

9.4 Problem Reduction

What we have seen just now with the proof of the undecidability of the Tiling problem or the proof of Rice's theorem seems to be a good recipe for obtaining new results of undecidability from old ones. Indeed, what we have done there is *reducing* the Halting problem to the problem we want to show to be undecidable. If this reduction works, it means that we could decide the Halting problem if the problem in question was decidable too. But we already know that the Halting problem is undecidable so we can conclude that the problem in question must also be undecidable. A quite useful principle arises with the help of which one can show that a problem is undecidable.

So let us pin down this principle a bit more formally. In particular, we need to define more precisely what "reduction" means exactly.[4] In our case, dealing with computability, we will need to require that the reduction itself is computable (otherwise our argument above would not stand).

Assume A and B are problems (or sets) about items in X and Y, respectively. (For WHILE-decidability X and Y will be $\mathbb{D}$, the type of binary trees.) Informally, we have a reduction from problem A on X (or the question "is a given $x \in X$ in A") to problem B on Y (or the question "is a given $y \in Y$ in B") if we can find:

---

[4]Problem reduction is a strategy that human beings or companies use all the time in an informal manner. For instance to "solve" the problem of getting from home to work, one could reduce it for instance to the problem of getting to the train station and on the right train, provided one already has a ticket.

Fig. 9.5 Solving A in terms
of B via f: $A \leq_{\text{rec}} B$

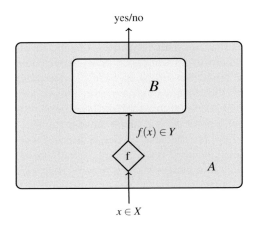

1. a computable total function $f : X \rightarrow Y$, mapping every element of X to an element of Y
2. such that we know that $x \in A$ if, and only, if $f(x) \in B$. This condition is important to guarantee that the reduction is correct.

What do these two conditions mean? The first condition states that any input $x \in X$, of which one can ask the question whether it is in A, must be translatable into an element $y \in Y$, of which one can ask whether it is in B; and, moreover, this translation cannot be magic, it must be computable. The second condition states that the question whether x is in A, or whether the translated x, namely $f(x)$, is in B, are equivalent. If you know the answer to one, you know the answer to the other.

The reduction and its ingredients are depicted in Fig. 9.5 where the decision procedure's input is at the bottom such that the simpler (or at least not harder than B) problem A is below B.

Formally the above explanations can be stated as follows:

Definition 9.4 (*Reduction*) Suppose one is given $A \subseteq X$ and $B \subseteq Y$. Define A to be *effectively reducible* to B if there is a total computable function $f : X \rightarrow Y$ such that for all $x \in X$, we have $x \in A$ if, and only if, $f(x) \in B$.
Symbolically we write this relation as $A \leq_{\text{rec}} B$ ("A is effectively reducible to B").

The above definition specifies just one particular type of reduction. There are many others. The one we use here is also often called "recursive many-one reduction". This is because *many* elements in B can correspond to one and the same element[5] in A via f and because, historically[6], *recursive* is another word for effective or computable.

---

[5]That means that for different $x \neq y \in A$ we can still have $f(x) = f(y)$.

[6]More precisely, "recursive" refers to the class of "partial recursive functions" on numbers (as introduced by Kleene in the late 1930s) which one can show are also a reasonable notion of computation equivalent to all the one we have presented.

The following theorem now explains the importance of this concept of reduction. We have implicitly used it already to prove Rice's theorem and argue for the undecidability of the Tiling problem:

Theorem 9.4 *If $A \leq_{rec} B$ and B is decidable then A is also decidable. Contrapositively, if $A \leq_{rec} B$ and A is undecidable then B is also undecidable.*

Proof The proof of this little theorem is simple and shows how both conditions of a reduction, $A \leq_{rec} B$, put together ensure that we can find a decision procedure for A if we already have one for B: just plug the implementation of the translation function f together with the decision procedure you already have for B and, hey presto, you have a decision procedure for A. Why? Well, the resulting "plugged together" procedure always terminates (since the decision procedure for B and the implementation of f always terminate). It remains to check that the new procedure run on x returns true if, and only if, $x \in A$. But the new procedure when run on x will return whatever the decision procedure for B returns on $f(x)$ and by assumption (the second condition in the definition of reduction) this is equivalent to $x \in A$.

The notion of reduction is a very important tool in the toolbox of computability and complexity researchers and we will meet another version that is most useful in the complexity part later.

9.5 Other (Famous) Undecidable Problems

We have met a number of famous undecidable problems already and we have been given some recipes how to obtain new ones. Below is a selective list of other famous undecidable problems to demonstrate that undecidable problems are not that rare in computer science as well as mathematics, that some were quite hard to prove undecidable and that new undecidable problems can be found, are still being found, and will be found in the future. Note however, that all undecidable problems, whatever their topic, are in a way computer science related as undecidability for a problem means that we cannot write a computer program that produces the answer to the given (uniform) problem for all input.[7]

The first two problems in the list are about context free grammars. A *context free grammar* (CFG) is a formal system similar to Backus–Naur-Form (BNF) description of syntax that is used to describe (generate) a language that is context free. Although a context free language can contain words with correctly bracketed expressions it

---

[7] Again, note that this does not mean we cannot write a program that might produce the answer for a specific input value. If this input is finite then a brute force search will always work. For instance, we can clearly write a program that decides whether a floor of a given fixed size (say 5×5) can be tiled with given, say k, tile types since there are k^{25} possible ways to tile such a floor. Note that there is a combinatorial explosion of the number of possible tilings but this does not concern us in computability. It will concern us in the part about complexity.

cannot "remember declarations (context)". For instance the language of words $a^n b^n c^n$ is not context free. The theory of CFGs is highly relevant for *parsing* programs.

- Do languages generated by two different context free grammars overlap? Undecidability of this problem appears to be first shown in [1].
- Is a context free grammar ambiguous? In other words: Are there two different derivations of the same word (string) using the rules of this grammar? According to the textbook by Hopcraft and Ullman [9], undecidability has been independently proved by D.C.Cantor [5], Floyd [8], and Chomsky and Schützenberger [6].
- Does a context free grammar generate all words over a given alphabet?
- Given two strings (over a finite alphabet) and a set of rewrite rules, does the first string rewrite into the second one using the given rules? Undecidability has been proved independently by Emil Post [17] and A.A. Markov [11].
- Is a given program typable in a functional programming language with polymorphic types[8] where the set of all types is itself a type? [24].
- Given a set of polynomial equations with integer coefficients, does it have an integer solution? This was a long-standing open problem posed by David Hilbert known as Hilbert's Tenth Problem (from the list of problems he presented at a conference in Paris 1900), until *Yuri Matiyasevich* famously proved undecidability in 1977 [12].
- Word problem for groups: given a finitely generated group, are any two of the generators representing the same element? This has been shown to be undecidable by Pyotr Novikov in 1955 [15].
- The "Entscheidungsproblem" (German for *decision problem*): Is a given arithmetic formula valid (or satisfiable)? This has been famously proved to be undecidable by Alan Turing in 1936 [23].

9.6   Dealing with Undecidable Problems

We have seen that many problems, in particular in the context of program development, are undecidable. This is bad news and one might wonder what impact this has on the life of software engineers and programmers.

As usual in life, one has to find ways around fundamental problems like this. There are a few coping strategies:

1. Approximate the decision procedure, accepting that the approximated decision procedure may return false negatives. This occurs, for instance, in certain cases of type checking. Semantically, type soundness is undecidable by Rice's theorem (Theorem 9.3). From this follows immediately that any nontrivial, sound and decidable type system must necessarily be incomplete. It will reject some pro-

---

[8]Haskell and ML are such programming languages but they exclude the type of all types.

grams that are actually type sound.[9] Instead of working with semantic types, e.g. all trees in $\mathbb{D}$ that denote natural numbers, one does work with syntactic types which are an approximation of the semantic types. For instance, one introduces type expressions into the language and either requires the programmer to annotate declarations with types or provides a type inference algorithm that may sometimes still need typing hints from the programmer.[10] Of course, to be useful, the type system must be expressive enough to type most "common" programs.

2. Give up on uniformity, and just test a few input values. This method is used to replace undecidable program verification (undecidable again by Theorem 9.3). Instead of verifying that a program is correct for all input values, just test whether it is correct for some, ideally the most likely, input values. One does not obtain a guarantee of correctness and thus not a decision procedure. This may be combined with probabilistic methods, if one has an idea how likely the inputs actually are.
3. Give up on the automation provided by the decision procedure. This strategy is very popular in program verification. Instead of a decision procedure that simply takes a program and a specification and returns a Boolean values stating whether the program meets the specification, builds an interactive tool that will be guided by the user (the human verifier) to find the right proof.
4. Simplify the problem. Instead of solving the undecidable problem, decide a simpler one. An example is arithmetic. Validity of arithmetic sentences is undecidable, a fact we have already discussed in the previous section. But if we take away multiplication, we obtain a decidable theory, called *Presburger* arithmetic [18]. The latter will feature again in the complexity part in Sect. 18.4. In program verification again, this strategy is also used, by restricting, for instance, the complexity of properties that a program can be shown to have (e.g. memory safety) and/or by restricting the range of programs itself.

9.7 A Fast-Growing Non-computable Function

We have seen that for undecidable problems the characteristic function[11] of the problem is non-computable. But a characteristic function is a function into the Boolean values (as the test result is 'yes' or 'no'). The example discussed in this section actually computes arbitrarily large natural numbers.

So, for a change, we now present a *non-computable function that computes natural numbers*, i.e. that is of type $\mathbb{N} \to \mathbb{N}_\perp$.

---

[9]It may also return false positives, i.e. accept programs that are not type sound, this happens e.g. in the case of division-by-zero.

[10]For a good textbook on types see [16].

[11]The characteristic function χ of a set A of elements of type T is of type $T \to \mathbb{B}$ and defined by $\chi(d) = d \in A$.

Tibor Radó[12] in his 1962 paper [19] introduced the *"Busy Beaver Game"* which corresponds to the following question for Turing machine programs (we can adapt this to WHILE-programs and will do so further below):

> for every input number n return the greatest number that can be outputted by any Turing machine that has at most n states when run on the empty input.

To produce an output, the Turing machine in question needs to terminate. Observe also that in Radó's original formulation the input/output alphabet of the Turing machine consists of just the symbol 1, so on the tape one can write only blanks and symbol 1 and so the output number is represented in unary form (three 1 s denote number 3 just like our encoding of numbers in WHILE is a list of three nils). Therefore, the *Busy Beaver* function for Turing machines is also often described as the highest number of non-blank symbols that a Turing machine with n states can write on the tape. It is relevant that the input is empty, so that one cannot "inject" simply big numbers via the input and then return them again as output.

For WHILE-programs we need to translate "n states" into something comparable for WHILE-programs:

> Can we compute the function that for every input number n returns the greatest number that can be outputted by any WHILE-program that is (as syntactic string) at most n characters long when run on input nil.

Definition 9.5 The (total) function $BB : \mathbb{N} \to \mathbb{N}$ is defined as follows

$$BB(n) = max \left\{ [\![p]\!]^{\text{WHILE}} (\text{nil}) \; \middle| \; \begin{array}{l} p \text{ is a WHILE-program with} \\ |p| \leq n, \text{ and } [\![p]\!]^{\text{WHILE}} (\text{nil}) \downarrow \end{array} \right\}$$

where $|p|$ denotes the length of the textual program p (*not* the size of the abstract syntax tree).

It should be obvious why this is called Busy Beaver, since the implementation of such a function needs to be "busy as a beaver" to compute the largest of all possible outputs for any program meeting the size constraint.

Maybe surprisingly, the Busy Beaver function attracted some interest in the research community (in Theoretical Computer Science) and still does so. The reason is the following:

Theorem 9.5 ([19]) *$BB(x)$ grows more quickly than any computable function f : $\mathbb{N} \to \mathbb{N}$. That means that for any computable f there is an $x_0 \in \mathbb{N}$ such that for all $x > x_0$ we have that $BB(x) > f(x)$.*

From this it follows quickly that:

---

[12]Tibor Radó (June 2, 1895–December 29, 1965) was a Hungarian mathematician who moved to the USA and worked in Computer Science later in his life. The publication on the *Busy Beaver* function is maybe his most famous one [19].

Corollary 9.1 *The Busy Beaver function is not computable.*

Proof Assume *BB* was computable. From the above Theorem 9.5, setting $f := BB$, it would follow that there is an x such that $BB(x) > BB(x)$ which is impossible.

Although the function is not computable, it is clearly the case that for a fixed argument n the value $BB(n)$ *is* computable, as finitely many programs need to be taken into consideration. This again stresses the importance of the difference between a function being computable and a specific result of a function being computable. Of course, the number of programs having to be taken into consideration increases rapidly with the size of the argument of the Busy Beaver function. And the sheer number of programs makes it difficult to compute the value of *BB* for even mildly large input values (such as 5). Traditionally, the values of *BB* have been investigated for Turing machines, as they are simple, and one has only to count the size of states (i.e. instructions) and not entire programs.

Lin and Radó himself could easily compute the result for Turing machines of size 1 to 3 as the number of Turing machines with 3 states is still reasonably small [10]. The results are as follows: $BB(1) = 1$, $BB(2) = 6$, $BB(3) = 21$. In [3] Brady has shown that $BB(4) = 107$. Then, interestingly, at size $n = 5$ the numbers go up quite dramatically, and for any number larger than 4 we do not have exact results but only lower approximations. By the work of Marxen and Bundtrock [13] we know that $BB(5) \geq 47{,}176{,}870$. They used computers to look through as many Turing machine programs of size 5 as possible, simulating their runs as efficiently as possible (the problem always is what do with non-terminating Turing machines).

As recent as 2010, Pavel Kropitz found the latest record lower bound for $BB(6) \geq 3.5 \times 10^{18267}$, a truly freakishly large number. Details can be found in [14] which provides also a good overview of all historic *Busy Beaver* results.

What Next?

We have seen some undecidable problems where one key trick was to use diagonalisation and thus self reference. This leads us to another question. What about self-referencing programs? Do they also lead to a paradox? Are programs that refer to themselves well-defined? And what can we do with such programs? This is the topic of the next chapter.

Exercises

1. Let us prove Theorem 9.2 which means in detail:

   a. Show that any finite set $A \subseteq \mathbb{D}$ is WHILE-decidable.
      Hint: assume without loss of generality that the finite set has n elements $A = \{d_1, d_2, \ldots, d_n\}$ and write the decision procedure for A as a WHILE-program.
   b. Show that, if $A \subseteq \mathbb{D}$ is WHILE-decidable then so is the complement of A, that is $\mathbb{D} \backslash A$.
      Hint: assume A is WHILE-decidable and thus we have a WHILE-program p *that decides A. Now write a WHILE-program* q *that decides the complement of A. Of course, you can and should use* p.

   c. Show that any WHILE-decidable set is WHILE-semi-decidable.
     Hint: do we actually have to program anything here?
   d. A problem (set) $A \subseteq \mathbb{D}$ is WHILE-decidable if, and only if, both A and its
     complement, $\mathbb{D} \backslash A$, are WHILE-semi-decidable.
     Hint: one has to prove two directions of the "iff" here. One follows easily from
     what we have already proved above. For the $\Leftarrow$ direction (if), one needs to
     write a decision procedure in WHILE using the semi-decision procedures for
     A and its complement. This is, however, not so easy (Why?) and it is suggested
     that one uses an extension of the self-interpreter that interprets two programs
     at the same time.

2. Below you find a list of properties of *(encodings of) WHILE-programs*. Which of
   these properties (or problems) are WHILE-decidable and which are not WHILE-
   decidable? In other words, for which properties A can we write a WHILE-program
   that takes d as input and decides whether d has property A (returning true or false
   accordingly)? Explain your answer *briefly* and look for possibilities to apply
   Rice's Theorem 9.3. If it is applicable explain carefully why that is the case.

   a. WHILE-program p terminates for *all* inputs (i.e. $[\![p]\!]^{\text{WHILE}}$ (d) $\downarrow$ for *all* $d \in \mathbb{D}$).
   b. WHILE-program p is either
```
read X { while true { } } write X
```
     or
```
read Z { Z:= cons X nil } write Z.
```
   c. WHILE-program p decides whether its argument d equals nil.
   d. WHILE-program p contains at least one assignment operation.
   e. WHILE-program p when given as input WHILE-program W terminates, where
     W is a WHILE-program with semantics:

$$[\![W]\!]^{\text{WHILE}} (d) = \begin{cases} \ulcorner n \div m \urcorner & \text{if } d = \ulcorner [n, m] \urcorner \text{ and } m \neq 0, \\ \text{undefined} & \text{otherwise} \end{cases}$$

     and binary operation $\div$ denotes integer division (i.e. returns an integer).
   f. WHILE-program p returns $\ulcorner 3 \urcorner$ for any input.

   The WHILE-program p mentioned here can be always regarded as an element of
   $\mathbb{D}$, since we can represent programs as abstract syntax trees (using our encoding
   from Sect. 6.3).

3. Let us consider the Tiling Problem and how sensitive it is to changes of the tile
   types.

   a. Consider the following three tile types:

   Show that with those tile types one can tile a 3×3 square.

b. Assume that we swap the south facing colour of the second and third tile type, i.e. red and green, and thus the following tile types (color figure online):

Show that one cannot even tile a 3×3 square correctly with those new tile types.

4. Try to write some short WHILE-programs that compute large numbers (in unary representation) disregarding their input.

5. (Ackermann-Péter function)[13] Consider the *Ackermann* function defined as follows

$$A(n, m) = \begin{cases} m + 1 & \text{if } n = 0 \\ A(m - 1, 1) & \text{if } m > 0 \text{ and } n = 0 \\ A(m - 1, A(m, n - 1)) & \text{if } m > 0 \text{ and } n > 0 \end{cases}$$

a. Show that $A(0, y) = y + 1$
b. Show that $A(1, y) = y + 2$
c. Show that $A(2, y) = 2y + 3$
d. Show that $A(3, y) = 2^{y+3} - 3$
e. Implement the Ackermann function in WHILE. Obviously, you need to replace the recursion. For instance, you could try to translate the recursive program into an iterative one using a stack to store intermediate values.
f. Do we know whether there exists an $m \in \mathbb{N}$ such that $BB(n) > A(n, n)$ for all $n > m$?

References

1. Bar-Hillel, Y., Perles, M., Shamir, E.: On formal properties of simple phrase structure grammars. Z. Phonetik, Sprachwiss. Kommunikationsforsch. **14**, 143–192 (1961)
2. Berger, R.: Undecidability of the domino problem. Mem. AMS **66**, 1–72 (1966)
3. Brady, A.H.: The determination of the value of Rado's noncomputable function $\Sigma(k)$ for four-state turing machines, Mathematics of Computation, **40**(162), 647–665 (1983)
4. Calcagno, C., Distefano, D., Dubreil, J., Gabi, D., Hooimeijer, P., Luca, M., O'Hearn, P., Papakonstantinou, I., Purbrick, J., Rodriguez, D.: Moving fast with software verification. In: Havelund, K., Holzmann, G., Joshi, R. (eds.) NASA Formal Methods. Lecture Notes in Computer Science, pp. 3–11. Springer International Publishing (2015)
5. Cantor, D.C.: On the ambiguity problem in Backus systems. J. ACM **9**(4), 477–479 (1962)

---

[13] This function is named after Wilhelm Friedrich Ackermann (29 March 1896–24 December 1962), a German mathematician, and Rózsa Péter, born Politzer, (17 February 1905–16 February 1977), a Hungarian mathematician.

6. Chomsky, N., Schützenberger, M.P.: The algebraic theory of context-free languages. In: Braffort, P., Hirschberg, D. (eds.) Computer Programming and Formal Systems, pp. 118–161. North Holland, Amestradam (1963)
7. Culik II, K.: An aperiodic set of 13 Wang tiles. Discrete Math. **160**(1–3), 245–251 (1996)
8. Floyd, R.W.: On ambiguity of phrase structure languages. Commun. ACM **5**(10), 526–532 (1962)
9. Hopcroft, J.E., Ullman, J.D.: Introduction to Automata Theory, Languages and Computation. Addison-Wesley, Reading (1979)
10. Lin, S., Radó, T.: Computer studies of Turing machine problems. J. ACM **12**(2), 196–212 (1965)
11. Markoff, A.: On the impossibility of certain algorithms in the theory of associative system. Academ. Sci. URSS. **55**, 83–586 (1947)
12. Matiyasevich, Y.V.: Hilbert's Tenth Problem. MIT Press, Cambridge (1993)
13. Marxen, H., Bundtrock, J.: Attacking the Busy Beaver 5. Bull. EATCS **40**, 247–251 (1990)
14. Michel, P.: The Busy Beaver Competition: a historical survey. Available via DIALOG. arXiv:0906.3749 (2012). Accessed 5 June 2015
15. Novikov, P.S.: On the algorithmic unsolvability of the word problem in group theory. Proc. Steklov Inst. Math. (in Russian). **44**, 1–143 (1955)
16. Pierce, B.: Types and Programming Languages. MIT Press, Cambridge (2002)
17. Post, E.: Recursive unsolvability of a problem of thue. J. Symb. Logic **12**(1), 1–11 (1947)
18. Presburger, M.: Über die Vollständigkeit eines gewissen Systems der Arithmetik ganzer Zahlen, in welchem die Addition als einzige Operation hervortritt.(English translation: Stansifer, R.: Presburger's Article on Integer Arithmetic: Remarks and Translation (Technical Report). TR84-639. Ithaca/NY. Dept. of Computer Science, Cornell University) Comptes Rendus du I congrès de Mathématiciens des Pays Slaves, 1929, pp. 92–101 (1930)
19. Radó, T.: On non-computable functions. Bell Syst. Tech. J. **41**(3), 877–884 (1962)
20. Rice, H.G.: Classes of recursively enumerable sets and their decision problems. Trans. Amer. Math. Soc. **74**, 358–366 (1953)
21. Silva, V.D., Kroening, D., Weissenbacher, G.: A survey of automated techniques for formal software verification. IEEE Trans. Comput.-Aided Des. Integr. Circuits Syst. **27**(7), 1165–1178 (2008)
22. Soare, R.I.: The history and concept of computability. In: Griffor, E.R. (ed.) Handbook of Computability Theory, pp. 3–36. North-Holland, Amestradam (1999)
23. Turing, A.M.: On computable numbers, with an application to the Entscheidungsproblem. Proc. Lond. Math. Soc. Sec. Ser. **42**, 230–265 (1936)
24. Wells, J.: Typability and type-checking in system F are equivalent and undecidable. Ann. Pure Appl. Logic **98**(1–3), 111–156 (1999)

Chapter 10
Self-referencing Programs

Can programs be reflective in the sense that they refer to themselves? If so, how can one define their semantics?

In WHILE there is no language feature that admits recursive macros or any direct reference to the program itself. This raises some questions. Can we write self-referencing programs in WHILE at all? More concretely, can we write a program, for instance, that returns itself, i.e. its own abstract syntax tree? And when is a program that refers to itself well-defined? Under which assumptions does it have a semantics?

Self-reference can lead to inconsistencies and paradoxes (as we have seen in the proofs of the undecidability of the Halting problem), so this question is more than justified. Such self-reference of programs is often called *reflection* and the corresponding program a reflexive program. Modern programming languages sometimes come with reflection libraries that allow for self-inspection and even manipulation. Those features are often used for flexibility, to make programs easily adaptable, such that a complex modification of an already existing larger software system can be achieved with a limited number of local changes. These changes depend then on the structure of the software system, for instance classes, in case the programming language is object-oriented. In those situations, using reflective methods can lead to elegant and localised solutions, at least more elegant than manually editing thousands of lines of code or interfaces globally.

Reflection is the ability of a running program to examine itself and its software environment, and to change what it does depending on what it finds [3, p. 5]. The representation of "self" is often referred to as *metadata*.

A fascinating fact is that the meaning of such reflective programs can be defined with the help of an old theorem, *Kleene's Recursion theorem*, that Stephen Kleene[1]

---

[1] Stephen Cole Kleene (5 January 1909– 25 January 1994) was a famous American mathematician. He might be known already from a course on finite automata. He invented regular expressions and was a student of Alonzo Church. He is the co-founder of mathematical logic and recursion theory.

© Springer International Publishing Switzerland 2016
B. Reus, *Limits of Computation*, Undergraduate Topics in Computer Science,
DOI 10.1007/978-3-319-27889-6_10

proved in 1938 and published later in [5]. One of many astonishing implications of Kleene's recursion theorem[2] is that

> …any acceptable programming system is "closed under recursion" [4, p. 222].

This then explains the name of the theorem. The above statement suggests that recursion can be done by reflection, it does not have to be built into the language. The recursion theorem has many applications. Before looking into it (Sect. 10.2) and its applications (Sect. 10.3) in detail, we first discuss another result of Kleene's that will be used in this context, and which has interesting applications in its own right.

10.1 The S-m-n Theorem

The so-called "S-*m*-*n*" theorem, sometimes also called *parameterization theorem*, can be regarded as justification for partial evaluation as an optimisation technique. The reasons for the strange name will be explained soon. In terms of metaprogramming (i.e. programming with programs) the theorem basically states the existence of *specialisers*.

Theorem 10.1 (S-1-1 Theorem) *For any programming language* L *with pairing and* programs as data *there exists a program* spec *such that*

$$\left[\!\left[\left[\!\left[\texttt{spec} \right]\!\right]^{\mathrm{L}} (p, s) \right]\!\right]^{\mathrm{L}} (d) = \left[\!\left[p \right]\!\right]^{\mathrm{L}} (s, d)$$

where (_, _) *denotes pairing for a general language* L.[3]

This program spec is a *specialiser* from L to L as first mentioned in Chap. 6. Recall from Chap. 6 that a specialiser takes a program and some of its input and produces a new program that needs fewer input parameters. The above equation specifies the resulting program's behaviour. The specialiser applies the one argument to the given program that requires two arguments, using it as its first argument. In doing so it constructs the result program with one remaining argument, otherwise preserving the program's semantics. One might say the specialiser is "freezing the first argument of the given program".

We can easily prove the S-1-1 theorem for L = WHILE by simply constructing the program spec[WHILE] as given in Fig. 10.1. The program in Fig. 10.1 performs the specialisation, but it does not do any *partial evaluation* as it does not look into the code (kept in variable code) of the given program. Such optimisations by *partial evaluation* are sometimes possible.

---

[2]This is often also called Kleene's Second Recursion Theorem because there is another recursion theorem that guarantees the existence of "smallest" fix points for recursive enumeration operators.

[3]In WHILE we know how to express pairing, namely as a list via ⌜[_, _]⌝ since we know the datatype is $\mathbb{D}$.

```
spec read PS   {             (* input list of form [P, S] *)
     P := hd PS;             (* P is program   [X,B,Y] *)
     S := hd tl PS;          (* S is input *)
     X := hd P;              (* X is input var of P *)
     B := hd tl P;           (* B is statement block of program P *)
     Y := hd tl tl P;        (* Y is output var of P *)
     expr:= [cons,[quote, S],[cons,[var, X],[quote,nil]]]
     newAsg := [:=, X, expr];
                      (* assemble asgmnt: "X := [S, X]" *)
     newB := cons newAsg B;
                      (* add assignment to old code *)
     newProg := [X, newB, Y]
                      (* assemble new program *)
}
write newProg
```

Fig. 10.1 A simple specialiser for WHILE

Imagine a program p to be specialised that has two arguments encoded by input variable XY where X has been assigned hd XY and Y been assigned hd tl XY using the standard list encoding of tuples. After p has been "specialised" with partial input 3 the resulting program will have input variable, say A, and an extra assignment XY:=[3,A] added to it. The latter implements the passing of the first argument provided. Therefore, if the program contained an assignment of the form Z:=<add>[X,4] (in case no other assignments have been made to X) we'd know that this was equal to Z:=<add>[3,4], and we could simplify/optimise the assignment to become Z:=7. This technique is called *partial evaluation* because just the parts that become amenable to evaluation by the additional provided partial input are evaluated statically, and not the entire program and its expressions as is the case at runtime.

In the above example, an assignment Z:=<add>[Y,4] could not be partially evaluated as this corresponds to Z:=<add>[A,4] and we don't know what value A has until we run the specialised program.

So partial evaluation can be regarded as "efficient program specialisation" [4, p. 96].

Now the S-m-n theorem is just a generalisation of the S-1-1 theorem (where $m = n = 1$). In the S-1-1 case we provided one extra argument to a program p that originally had $1 + 1 = 2$ arguments. Note that in WHILE programs the arguments must be encoded as one single list because WHILE programs formally only have one input parameter. Now we generalise the number of parameters. The number m refers to the number of arguments provided for specialisation, whereas number n refers to the number of arguments that remain after specialisation. The program p used for specialisation then must have used $m + n$ arguments, the resulting specialiser spec must have $m + 1$ arguments (one of which is p), and the specialised program must have, of course, n arguments.

There are some fascinating results that can be obtained with the help of specialisers and the power of self-application. We do not have the space to go into details but [4, Chap. 6] offers a detailed survey and explanations. Here are some highlights called "Futamura" projections [2], named after their discoverer:

- an interpreter and an appropriate specialiser can be combined to obtain the effect of a compiler (first Futamura projection)
- specialising an appropriate specialiser and an interpreter produces a compiler (second Futamura projection)
- composing three appropriate specialisers in the right way results in a compiler generator that can transform interpreters into compilers (weird but true) (third Futamura projection). Even weirder, a compiler generator obtained by the 3rd projection is self-generating, that means that there is a specialiser applied to which the generator produces itself.

These are also called the "cornerstones of partial evaluation". Anyone interested in compilation and optimisation is highly recommended to study those in detail. They are quite fascinating and highlight again the importance of the concept of self-application in programming. Neil Jones states, however, that it is *"far from clear what their pragmatic consequences are."* Since this is an introductory book on computability and complexity we leave the exploration of partial evaluation to the reader, and we swiftly move on.

10.2   Kleene's Recursion Theorem

The following theorem guarantees the existence of self-referencing programs, that we can also call "recursive" since with the help of self-reference one can implement recursion. This will be discussed in more detail later. Let us first state Kleene's theorem [5]. Recall the for a generic language we use notation $(\_,\_)$ for pairs.

Theorem 10.2 (Kleene's Recursion Theorem [5]) *Let* L *be an* acceptable programming language. *For any* L-*program* p, *there is a* L-*program* q *such that for all input* $d \in L$-*data we have:*

$$[\![q]\!]^{L}(d) = [\![p]\!]^{L}(q, d)$$

How does this deal with reflective programs and what is an acceptable programming language? The reflective program q is defined via the help of p which has an extra parameter to represent "self". So p may regard q as its own *metadata* representation.

Definition 10.1 An *acceptable* programming language is a programming language L (in the sense we have already provided earlier) for which the following holds:

1. L has a self-interpreter (or universal program in the sense of Definition 7.1) and therefore must have programs as data as well as pairing;
2. The S-*m*-*n* theorem (see Sect. 10.1) holds for L;
3. L is Turing-complete, i.e. it computes all programs that a Turing machine can compute.[4]

What can the Recursion Theorem be used for? A famous example is the following:

Example 10.1 Consider a *self-reproducing program*, also called a *quine*.[5] Such a program is supposed to produce—often print as most languages provide print statements—its own source code *without* using any input[6] or knowledge of its own file name. In our case, where L is WHILE, the program must return its own abstract syntax tree. Quines are notoriously difficult to write and this also holds for WHILE. Though it is a non-trivial task to write a *quine*,[7] it is easy to *prove* that such a self-reproducing program must exist with the help of the above theorem. Just choose p to be a program that satisfies:

$$[\![p]\!]^{\text{WHILE}}\,[x, y] = x$$

which is achieved by program p that writes as output hd XY if XY is its input variable. As usual, we do not write explicit encoding brackets for list $[x, y]$.

The proof of Kleene's (second) recursion theorem is relatively short but very clever indeed.

Proof We start with program p that takes two arguments the first of which is considered to be a program itself. We know there is a specialiser spec for the programming language (as it is acceptable). We can therefore apply any program, say r, to itself, i.e. to its data representation r, with the help of the specialiser:

$$\underbrace{[\![\text{spec}]\!]^{\text{L}}\,(r, r)}_{\text{self application of } r}$$

which returns an L program that "performs" this self-application. Then we can pass this resulting program as input to our p to obtain a function

$$f(r, d) = [\![p]\!]^{\text{L}}\,([\![\text{spec}]\!]^{\text{L}}\,(r, r)\,,\ d) \tag{10.1}$$

where f is now a function mapping L-data to L-data. Since L is acceptable, from Definition 10.1 we know that f must be computable by an L-program p^{self}. In short:

---

[4]This is a relatively mild assumption of a language that we are prepared to call a programming language.

[5]After the American philosopher and logician Willard Van Orman Quine (June 25, 1908–December 25, 2000).

[6]So one cannot simply pass the program's source as its own input.

[7]This is subject of Exercises 7 and 8, respectively.

$$\llbracket p^{\texttt{self}} \rrbracket^{\text{L}} = f \tag{10.2}$$

Now we simply define the program q we are looking for as another self-application, this time of $p^{\texttt{self}}$.

$$q = \underbrace{\llbracket \texttt{spec} \rrbracket^{\text{L}} (p^{\texttt{self}}, p^{\texttt{self}})}_{\text{self application of } p^{\texttt{self}}} \tag{10.3}$$

This program q returns an L-program, and this is actually what we will choose for the program required by the theorem. We have to show that it has the required property.

$$
\begin{aligned}
\llbracket q \rrbracket^{\text{L}}(d) &= \llbracket \llbracket \texttt{spec} \rrbracket^{\text{L}} (p^{\texttt{self}}, p^{\texttt{self}}) \rrbracket^{\text{L}}(d) & \textit{use Eq. 10.3, the definition of q} \\
&= \llbracket p^{\texttt{self}} \rrbracket^{\text{L}}(p^{\texttt{self}}, d) & \textit{use S-1-1 Theorem 10.1 with p and } s = p^{\texttt{self}} \\
&= f(p^{\texttt{self}}, d) & \textit{use Eq. 10.2, definition } p^{\texttt{self}} \\
&= \llbracket p \rrbracket^{\text{L}}(\llbracket \texttt{spec} \rrbracket^{\text{L}} (p^{\texttt{self}}, p^{\texttt{self}}), d) & \textit{use Eq. 10.1, definition of f} \\
&= \llbracket p \rrbracket^{\text{L}}(q, d) & \textit{use Eq. 10.3, definition q backwards}
\end{aligned}
$$

This is yet another neat proof achieved with self-application,[8] two helpings of self-application actually.

10.3   Recursion Elimination

There are many applications of this theorem. A classic one is recursion elimination. So let us return to the specific language WHILE (so L := WHILE). Figure 10.2 shows an attempt of implementing the factorial function by a program, called fac that makes use of a macro call to mult and also calls itself (fac) recursively.

Fig. 10.2 Recursive fac
not definable in WHILE

```
fac read n {
    if n {
        A   := <fac> tl n;
        Res := <mult> [n, A]
    }
    else
    {
      Res := 1
    }
    }
    write Res
```

---

[8]Readers familiar with *Alonzo Church*'s untyped λ-calculus will recognise the similarity with the construction of the recursion operator $Yf = (\lambda x \cdot f(x\ x))\ (\lambda x \cdot f(x\ x))$ which also uses two self-applications.

Fig. 10.3 `fac2` using
reflection for recursion

```
fac2 read qn {
   q := hd qn;
   n := hd tl qn;
   if n {
      A := <u> [q, tl n];
      Res := <mult> [n, A]
   }
   else {
     Res := 1
   }
}
write Res
```

We have seen already that in WHILE we cannot define such a recursive definition directly, even in extensions of WHILE because recursive macros are not supported and thus we cannot call `fac` as suggested. We first need to be able to compile the macro away. However, with the help of an extra parameter that represents the program `fac` and using the universal program (self-interpreter) for WHILE, u, we can write program `fac2` without a recursive macro as displayed in Fig. 10.3.

Program `fac2` takes two arguments the first of which is a program, namely the meta-representation of the factorial program. But `fac2` is not yet the factorial function we want. Applying Kleene's recursion theorem, however, we obtain a program `facRefl` such that:

$$[\![\texttt{facRefl}]\!]^{\text{WHILE}} (d) = [\![\texttt{fac2}]\!]^{\text{WHILE}} [\texttt{facRefl}, d]$$

and this program does implement the factorial function. We have used the self-interpreter here to make the recursion on the program source explicit.[9]

The proof of Kleene's recursion theorem is *constructive* which means that the proof tells us how the program q actually looks like. It turns out that, despite the elegance of the proof by self application, the resulting program is not very efficient. For instance, to compute the factorial of n, self application leads to a self-interpreter running a self-interpreter running a self-interpreter and so on for n levels, which leads to a significant slow-down in runtime. This is the reason why many interpreters provide explicit reflective means in terms of syntactic operators to refer to the program's own source (as abstract syntax tree) and to the self-interpreter.[10] As a consequence, only one level of self-interpreter needs to be running at any time (as it can be called itself repeatedly).

In compiled languages the situation is more difficult, as the compiled program usually dropped all reference to source code and consists of executable binaries only. Often this is also in the spirit of the language (as reflective mechanisms may lead to

---

[9]More on implementing recursion indirectly, on patterns, and also how to verify such programs, can be found e.g. in [1].

[10]In LISP we have for instance `quote` and `unquote`.

execution overhead even if one does not use reflection), and C or C++ are example cases.[11] In cases where compiled code is executed in a runtime environment like a virtual machine, reflective libraries provide access to metadata produced by this runtime system[12] intentionally for the purpose of reflective programming. This is possible because the intermediate (byte) code has more information attached to it than the binaries produced by direct compilation.

What Next?

Now that we have proved so many results about computability expressed in terms of WHILE, the question remains whether this choice was particularly lucky. Maybe with another programming language or machine model we could have actually programmed a decision procedure for HALT? Or maybe for another programming language we would not have obtained the recursion theorem? In the next chapter we will justify our choice and see that our results are independent of the particular programming language (as long as some basic criteria are met). This brings us to the Church-Turing thesis mentioned in the Introduction and covered in the next chapter.

Exercises

1. Prove the S-2-1 theorem by programming the corresponding specialiser that will be an adaptation of the one of Fig. 10.1.
2. Enrich the specialiser from Fig. 10.1 by some form of partial evaluation. For instance, substitute all occurrences of hd cons E F by E and all occurrences of tl cons E F by F. The specialiser has to run through the entire program body and whenever a corresponding pattern has been detected, replace it. This requires a stack.
3. Use recursion elimination (with the help of Recursion Theorem 10.2) to implement in WHILE the following function *fib*. *Fib* returns the nth *Fibonacci number* defined below:

$$\begin{aligned} fib(0) &= 1 \\ fib(1) &= 1 \\ fib(n+2) &= fib(n+1) + fib(n) \end{aligned}$$

   Use unary encoding of numbers for simplicity.
4. Use the Recursion Theorem 10.2 to show that there exists a WHILE-program that can recognise itself. More precisely, show that there is a program q such that

$$[\![q]\!]^{\text{WHILE}}(d) = \begin{cases} \text{true} & \text{if } d = q \\ \text{false} & \text{otherwise} \end{cases}$$

---

[11]"C++ is lean and mean. The underlying principle is that you don't pay for what you don't use." Bjarne Stroustrup in [6].

[12]E.g. meta representation of classes in Java.

Note that for this question, you do not have to write program q but just show its existence.

5. Find out what the acronym PHP stands for and explain in which sense it is self-referential.
6. (Quine's paradox) We have mentioned the philosopher and logician *Willard Quine* in this chapter. He is also famous for the following statement:

> "Yields falsehood when preceded by its quotation" yields falsehood when preceded by its quotation.

Explain why this is a paradox and where the self-reference occurs. Note that "this statement yields falsehood" is a posh way of saying "this statement is false".
7. Write a WHILE-program that returns its own encoding as object.
8. Write a Java, Haskell, or C program that prints itself.

References

1. Charlton, N.A., Reus, B.: Specification patterns for reasoning about recursion through the store. Inf. Comput. **231**, 167–203 (2013)
2. Futamura, Y.: Partial evaluation of computation process - an approach to a compiler-compiler. High. Order Symb. Comput. **12**, 381–391 (1999) (Reproduction of the 1971 paper)
3. Forman, I.R., Forman, N.: Java Reflection in Action. Manning Publications, Greenwich (2004)
4. Jones, N.D.: Computability and Complexity: From a Programming Perspective. MIT Press, Cambridge (Also available at http://www.diku.dk/neil/Comp2book.html) (1997)
5. Kleene, S.C.: Introduction to Metamathematics. North-Holland, Amsterdam (1952)
6. Toms, M.: An e-mail conversation with Bjarne Stroustrup. Overload J. #2. Available via DIALOG http://accu.org/index.php/journals/1356 (1993). Accessed 5 June 2015

Chapter 11
The Church-Turing Thesis

For computability considerations, does it matter which
programming languages we use?

So far, we have successfully identified quite a few undecidable problems and non-computable functions. We also have identified some recipes how to show that a problem is decidable or undecidable. We have looked at a weaker property than decidability, namely semi-decidability.

But we have done that using a particular notion of computation, in other words, a particular choice of what "effective procedures" are. We had chosen WHILE-programs. The WHILE-language had been picked because it is simple, but at the same time rich enough to allow one to write complicated algorithms with relative ease.

But how do we know that the problems that we have shown to be undecidable by WHILE-programs are not decidable if we used another computational device or language? That the results obtained in previous chapters do not depend on the choice of WHILE is something we still have to show, and we will do this now in this chapter. Naturally, we will have a look at a few more possible notions of computation: Turing machines (Sect. 11.3), the flowchart language GOTO (Sect. 11.4), counter machines (Sect. 11.6), register machines (Sect. 11.5), and cellular automata (Sect. 11.7). The former four models are instances of machine languages, for which we generally discuss a framework for semantics in Sect. 11.2.

In Sect. 11.8 we provide arguments that those languages are all equally powerful. The claim that *all reasonable formalisations* of the (more or less) intuitive notion of computation are equivalent is called *Church-Turing thesis*, which will be addresses first in Sect. 11.1.

© Springer International Publishing Switzerland 2016
B. Reus, *Limits of Computation*, Undergraduate Topics in Computer Science,
DOI 10.1007/978-3-319-27889-6_11

11.1   The Thesis

The famous *Church-Turing thesis* appears in all kinds of variations and disguises. It had been originally suggested by Turing and Church independently. They both had introduced a notion of effective procedure, namely the Turing machine (then called *a*-machine [26]) and the λ-calculus terms [5],[1] respectively. And both claimed that any form of computation can be described with their preferred choice.

In its original "Turing" form, the thesis thus states that:

> A function is 'effectively calculable' (that is algorithmically computable or effectively mechanisable) if, and only if, it is computable by a Turing machine.

To justify our choice of WHILE instead of the notoriously tedious to program Turing machines, we would therefore need to show that WHILE is Turing-complete (see Definition 2.1), i.e. every Turing machine can be compiled into a WHILE -program. A more liberal formulation of the thesis (also used by [15, pp. 8–9]) is the following:

Definition 11.1 All reasonable formalizations of the intuitive notion of effective computability are equivalent.

This formulation "factors out" the concrete Turing machine model and simply focuses on the consequence that all models must be equivalent (as they are all equivalent to Turing machines). Whatever version the reader might prefer, they are all describing a *thesis* and not a *theorem* (and thus cannot be proved). The reason is that any version refers to a vaguely defined notion of what a computational model is or what computable means: the first version talks about "algorithmically computable", the second one uses "reasonable formalisation of the intuitive notion of effective computability". The mentioned notions of computation are not formally defined like Turing machines are or the models we will discuss further below. The term "reasonable" implicitly refers to the fact that the corresponding notion of computability must be powerful enough to express Turing machines. So for instance, finite automata would not count as reasonable notion of computation.

Turing experts like *Andrew Hodges*[2] correctly emphasise that many of the theses that these days come with the label "Church-Turing Thesis" are very free interpretations of it. These variations can be interpreted in very different ways, but we do not want to go into that. The interested reader is kindly referred to Andrew Hodges' [13] and Jack Copeland's [7] relevant entries in *The Stanford Encyclopedia of Philosophy* or their respective blogs.

Although we cannot prove a thesis, we shall now provide *evidence for the Church-Turing thesis*, and there is plenty of it. In order to do that, we will introduce a few more notions of computation very precisely and formally, and then argue precisely (without formally proving anything) that they are equivalent:

---

[1]Church acknowledges in [5] Kleene's contributions.

[2]Andrew Hodges (born 1949) is a British mathematician and author of the bestselling biography "Alan Turing—The Enigma" [14] the original 1983 edition of which is the loose (unauthorised) basis for the stage play "Breaking the Code" in 1986 and the movie "The Imitation Game" in 2014.

- Turing machines TM (already briefly mentioned in Chap. 2)
- Register machines RAM
- Counter machines CM
- a flow chart language GOTO
- an imperative language WHILE (already discussed at length in Chaps. 3–5)
- Cellular Automata CA

Another language mentioned earlier is the λ-calculus that Alonzo Church had used for his definition of what computable means (namely computable by a λ-term) [5]. We won't discuss the λ-calculus in this book, explaining its absence from the above list.

We split the list above into two types of models: machine-based ones and others. The machine-based ones include TM, GOTO, CM, RAM as they all have in common that they use programs that consist of labelled instructions. The WHILE-language is of a different nature, and different from all of them are cellular automata.

11.2 Semantic Framework for Machine-Like Models

Before we define syntax and semantics of the languages not encountered so far, we first present a framework that can be instantiated to describe the semantics of all machine like models.

A machine program p consists of a list of labelled instructions:

$$1 : I_1; \quad 2 : I_2; \quad \ldots \quad m{-}1 : I_{m-1}; \quad m : I_m;$$

where the labels are natural numbers that in the above case range from 1 to m. The instruction of a program p with label ℓ is often referred to as $p(\ell)$. So in the program above $p(2) = I_2$.

The semantics of machine languages are easier to describe than those of high-level programming languages like WHILE since an instruction of a machine language has a small effect compared to a command of a high-level language, and there is no complex control flow operation like a while loop. A machine language program permits just jumps, which can be executed by simply changing the program counter. Consequently, we can describe the semantics by a one-step operational semantics for instructions.

Assuming we know the syntax of the machine language, i.e. the instruction set, all we need to define the interpretation of programs, that is the semantics of the machine language, are the following four ingredients:

- a definition of what a store is for the semantics of the language. The store, or memory, will typically contain the values of variables or registers or anything that is supposed to store data.
- a function *Readin* that takes the input and produces the initial store of the machine.
- a function *Readout* that takes the final store and reads out the result (output).

- the description of the semantics of the instructions for the machine (language) written as $p \vdash s \rightarrow s'$ which means that the machine program p transits from configuration s into configuration s' in one step. In other words, it provokes a state change from s to s'. The machine instructions are simple and thus such a transition can be described without defining a more complex relation as for the semantics of commands in WHILE.

Based on the third item, the one-step operational semantics $p \vdash s \rightarrow s'$, one defines the many-step-semantics $p \vdash s \rightarrow^* s'$ which shall denote that machine program p transits from state s into state s' in zero, one or many steps. A state is a tuple consisting of the current instruction label to be executed next and the store (or memory). So, a machine state is, e.g., $(3, \sigma)$ where σ is a concrete store for the type of machine in question.

If we have those three ingredients we can define the semantics of a machine program p as follows:

Definition 11.2 (*General framework for machine model semantics*) Let p be a machine program with m instructions.

$$\begin{aligned} [\![p]\!](d) = e \ \text{ iff } \ &\sigma_0 = Readin(d) &\text{and} \\ &p \vdash (1, \sigma_0) \rightarrow^* (m+1, \sigma) &\text{and} \\ &e = Readout(\sigma) \end{aligned}$$

In other words σ_0 describes the initial store and thus $(1, \sigma_0)$ the initial state of the machine as execution starts with machine instruction 1. We have reached the final state if the next instruction is beyond the last instruction of the program. As m is the last instruction of our program p, we know that if we are at instruction $m + 1$ the program has terminated. One then reads out the result from the final store σ using *Readout*.

In the following section we will instantiate this framework by describing the ingredients for the framework in each case: the syntax and then the four ingredients for the semantic framework.

11.3   Turing Machines TM

As mentioned in the introduction, in 1936 Alan Turing published his seminal paper [26] in which he defined so-called *a*-machines that were later named in his honour. These machines perform the simplest kind of computation Turing could imagine. He writes in [26]: The machine "...*is only capable of a finite number of conditions ...which will be called configurations. The machine is supplied with a "tape" (the analogue of paper) running through it, and divided into sections, (called "squares") each capable of bearing a "symbol". At any moment there is just one square ...which is in the machine. We may call this square the "scanned square". The symbol on the scanned square may be called the "scanned symbol". The "scanned*

Table 11.1 Syntax of TM

Instruction	Explanation
right$_j$	Move right
left$_j$	Move left
write$_j$ S	Write S
if$_j$ S goto ℓ_1 else ℓ_2	Conditional jump (read)

symbol" *is the only one of which the machine is, so to speak "directly aware"".*
The machine performs an action, changing its configuration, depending on the configuration it is currently in. We follow the ideas of Turing, splitting the several tape movements and printing instructions that Turing allows to happen in one rule into subsequent instructions. Each symbol dependent configuration change will still be represented by a single particular jump instruction. Our presentation follows [15, Chap. 3] as this will simplify the time measurement later in the complexity part.

We instantiate the framework from Definition 11.2 for Turing machines:

Syntax   Instructions are described in Table 11.1.

The j subscript indicates which tape the instruction is intended for.

Store   The stores for a Turing machine are its tapes. For Turing machines we also need to be aware of the head position on each tape which is indicated by an underscore. So for k-tape machines the store is a k-tuple of the form

$$(\mathrm{L}_1 \underline{\mathrm{S}_1} \mathrm{R}_1, \mathrm{L}_2 \underline{\mathrm{S}_2} \mathrm{R}_2, \dots, \mathrm{L}_k \underline{\mathrm{S}_k} \mathrm{R}_k)$$

where the L_i and R_i are strings of symbols of the tape alphabet and S_i are the currently scanned tape symbols belonging to the tape alphabet. The tape alphabet contains a special blank symbol B that the tape is initialised with and that is never considered part of the input or output and a finite number of other symbols. In Turing's original definition the only other symbols were 0 and 1. Allowing more (finitely many) symbols does not change much.[3]

Readin

$$Readin(\mathrm{x}) = (\underline{\mathrm{B}}\mathrm{x}, \underline{\mathrm{B}}, \underline{\mathrm{B}}, \dots, \underline{\mathrm{B}})$$

In other words, the input is stored on the first tape, the head is put left of the first input character and all other tapes are initially blank.

Readout

$$Readout(\mathrm{L}_1 \underline{\mathrm{S}_1} \mathrm{R}_1, \mathrm{L}_2 \underline{\mathrm{S}_2} \mathrm{R}_2, \dots, \mathrm{L}_k \underline{\mathrm{S}_k} \mathrm{R}_k) = Prefix(\mathrm{R}_1)$$

where $Prefix(\mathrm{R}_1\mathrm{B}\mathrm{R}_2) = \mathrm{R}_1$ provided that R_1 does not contain any blank symbols B. In other words, the output is the word just right of the current head position on tape 1 up to, and excluding, the first blank.

Semantics   In Table 11.2 below, the semantics for Turing machines with 1 tape is presented (it is pretty obvious how to generalise that to k tapes). Note that p is

---

[3]This will be discussed in Exercise 9.

Table 11.2 One-step operational semantics for TM

$$\begin{aligned}
p \vdash (\ell, L\underline{S}S'R) &\to (\ell+1, LS\underline{S}'R) &&\text{if } p(\ell) = \texttt{right}\\
p \vdash (\ell, L\underline{S}) &\to (\ell+1, LS\underline{B}) &&\text{if } p(\ell) = \texttt{right}\\
p \vdash (\ell, LS'\underline{S}R) &\to (\ell+1, L\underline{S}'SR) &&\text{if } p(\ell) = \texttt{left}\\
p \vdash (\ell, \underline{S}R) &\to (\ell+1, \underline{B}SR) &&\text{if } p(\ell) = \texttt{left}\\
p \vdash (\ell, L\underline{S}R) &\to (\ell+1, L\underline{S}'R) &&\text{if } p(\ell) = \texttt{write S'}\\
p \vdash (\ell, L\underline{S}R) &\to (\ell_1, L\underline{S}R) &&\text{if } p(\ell) = \texttt{if S goto } \ell_1 \texttt{ else } \ell_2\\
p \vdash (\ell, L\underline{S'}R) &\to (\ell_2, L\underline{S'}R) &&\text{if } p(\ell) = \texttt{if S goto } \ell_1 \texttt{ else } \ell_2 \text{ and } S \neq S'
\end{aligned}$$

the program $\ell_1 : I_1; \ldots \ell_m : I_{\ell_m};$. The second and fourth line of Table 11.2 deal with the cases where one moves right over the right end of the tape, and left over the left end of the tape, respectively. Since there should be no "end" of the tape (the tape is supposed to be potentially infinite) we accordingly can never "fall off the tape" in any direction. What happens is that we just write a blank symbol if we go to a tape cell we have never visited before.

Example 11.1 The one-tape Turing machine program in Fig. 11.1 computes the function that takes as argument a word over alphabet $\{0, 1\}$ and returns 0 if the input word contains an odd number of 1's. It returns 1 if the input word contains an even number of 1's. For instance, if the input is 0101110, the output is 1 as the input word contains an even number (namely 4) symbols 1. The instructions 1–5 of the program are executed while an even number of 1 symbols (including zero) has been detected, whereas instructions 6–9 are executed while an odd number of 1 symbols has been detected. The jumps at instruction 3 and 9, respectively, allow movement between these two program states. Finally, instructions 10–12 deal with writing the result 1 and instructions 13–14 with writing the result 0. The input has been already deleted at the point of writing the result. These conceptual states of the program can

```
 1: right;
 2: if B goto 10 else 3;  // B detected -> result = 1
 3: if 0 goto 4 else 6;
 4: write B;              // 4: 0 detected
 5: if 0 goto 1 else 1;   // jump to 1
 6: write B;              // 6: 1 detected
 7: right;
 8: if B goto 13 else 9;  // B detected -> result = 0
 9: if 0 goto 6 else 4;
10: write 1;
11: left;
12: if 0 goto 15 else 15;// jump to end
13: write 0;
14: left;
15:                       // end of program
```

Fig. 11.1 TM program checking input for even number of 1s

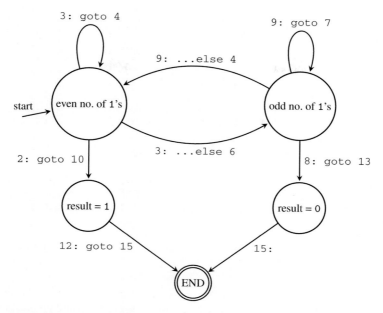

Fig. 11.2 Internal state automaton for TM program of Fig. 11.1

be visualised as a finite automaton in Fig. 11.2 which is much more readable than the program itself. Of course, the representation is owed to the fact that the task can actually be carried out by a finite state automaton. The reader is cordially reminded that this is exactly the reason why we use WHILE as our programming language of choice and not Turing machines as already pointed out in the *Preface* on p. xxx. More complicated Turing machine programs are, naturally, even less readable.

11.4 GOTO-Language

Next, we discuss a so-called *flow chart language* originally introduced in [12] to investigate partial evaluation. This language uses the same datatype as WHILE, namely binary trees. The main difference is that it does not provide a complex abstract control flow mechanism like a while loop but instead a jump, also known as *goto* command, which lends the language its name. Such jumps are known from machine languages but also the good old language BASIC (beginner's all purpose instruction code) [17] that was the only language shipped with the first basic desk

Table 11.3 Syntax of GOTO

Instruction	Explanation
X := nil	Assign nil
X := Y,	Assign variable
X := hd Y	Assign hd Y
X := tl Y,	Assign tl Y
X := cons Y Z	Assign cons Y Z
if X goto ℓ_1 else ℓ_2	Conditional jump

top computers in the early eighties.[4] In GOTO we cannot build complex expressions as we can do in WHILE. So all arguments to operators hd, tl, and cons, resp., are variables.

Syntax   In the following X, Y, and Z denote program variables. Instructions are as described in Table 11.3.

Store   The store is exactly the same as for WHILE-programs: the store is a set of bindings, i.e. pairs of GOTO variable names and elements of $\mathbb{D}$. As before, we write stores as list of pairs, e.g. $\{X : d, \ldots Z : e\}$ using the same conventions as for WHILE-stores.

Readin   The input and output variable are both fixed as X.

$$Readin(d) = \{X : d\}$$

In other words, the initial store maps the input variable to X, all other variables of the program are initialised with nil. The latter is a consequence of how the variable lookup, $\sigma(X)$, is implemented.

Readout

$$Readout(\sigma) = \sigma(X)$$

In other words, the result of the program is the value of variable X.

Semantics   We give the one step semantics in Table 11.4. Note that $\sigma[X := d]$ abbreviates the store that is the same as σ with the one exception that the value for variable X is now set to be d. As before with other machines, we assume that p is the program $\ell_1 : I_1; \ldots \ell_m : I_{\ell_m};$.

Example 11.2 The program that reverses a list known from Fig. 3.4 can also be written as a GOTO program as outlined in Fig. 11.3. It is apparent that the WHILE-program is more readable than the GOTO-program.

---

[4]The language was designed by John Kemeny and Thomas Kurtz in 1964 [8] and was one of the first intended for interactive use.

Table 11.4 One-step operational semantics for GOTO

$$p \vdash (\ell, \sigma) \rightarrow (\ell+1, \sigma[\mathtt{X} := \mathrm{nil}]) \quad \text{if } p(\ell) = \mathtt{X} \mathtt{:=nil}$$
$$p \vdash (\ell, \sigma) \rightarrow (\ell+1, \sigma[\mathtt{X} := \sigma(\mathtt{Y})]) \quad \text{if } p(\ell) = \mathtt{X} \mathtt{:=Y}$$
$$p \vdash (\ell, \sigma) \rightarrow (\ell+1, \sigma[\mathtt{X} := d]) \quad \text{if } p(\ell) = \mathtt{X} \mathtt{:=hd\ Y} \text{ and } \sigma(\mathtt{Y}) = \langle d.e \rangle$$
$$p \vdash (\ell, \sigma) \rightarrow (\ell+1, \sigma[\mathtt{X} := \mathrm{nil}]) \quad \text{if } p(\ell) = \mathtt{X} \mathtt{:=hd\ Y} \text{ and } \sigma(\mathtt{Y}) = \mathrm{nil}$$
$$p \vdash (\ell, \sigma) \rightarrow (\ell+1, \sigma[\mathtt{X} := e]) \quad \text{if } p(\ell) = \mathtt{X} \mathtt{:=tl\ Y} \text{ and } \sigma(\mathtt{Y}) = \langle d.e \rangle$$
$$p \vdash (\ell, \sigma) \rightarrow (\ell+1, \sigma[\mathtt{X} := \mathrm{nil}]) \quad \text{if } p(\ell) = \mathtt{X} \mathtt{:=tl\ Y} \text{ and } \sigma(\mathtt{Y}) = \mathrm{nil}$$
$$p \vdash (\ell, \sigma) \rightarrow (\ell+1, \sigma[\mathtt{X} := \langle d.e \rangle]) \quad \text{if } p(\ell) = \mathtt{X} \mathtt{:=cons\ Y\ Z} \text{ and } \sigma(\mathtt{Y}) = d \text{ and } \sigma(\mathtt{Z}) = e$$
$$p \vdash (\ell, \sigma) \rightarrow (\ell_1, \sigma) \quad \text{if } p(\ell) = \mathtt{if\ X\ goto\ } \ell_1 \text{ else } \ell_2 \text{ and } \sigma(\mathtt{X}) \neq \mathrm{nil}$$
$$p \vdash (\ell, \sigma) \rightarrow (\ell_2, \sigma) \quad \text{if } p(\ell) = \mathtt{if\ X\ goto\ } \ell_1 \text{ else } \ell_2 \text{ and } \sigma(\mathtt{X}) = \mathrm{nil}$$

Fig. 11.3 GOTO-program to reverse a list

```
1: if X goto 2 else 6;
2: Y := hd X;
3: R := cons Y R;
4: X := tl X;
5: if R goto 1 else 1;
6: X := R;
```

11.5 Register Machines **RAM** and **SRAM**

RAM actually stands traditionally for *Random Access Memory*, i.e. memory that can be directly addressed (and one does not have to run through the storage sequentially to find the data like e.g. on a Turing machine tape). However, we read it as *Random Access Machine*. It is the machine type that comes closest to the actual CPUs in modern computing devices. It was first introduced under the name *URM* (Unlimited Register Machine) in [24] with a slightly different syntax (that rather resembles assembly code) and two jump instructions, one unconditional and one conditional that has no "else" part. There are numerous different variations of the RAM suggested in the literature.

Since we deal with computability and do not want to restrict the memory used,[5] we allow an unbounded number of registers which are indexed by natural numbers. Moreover, each register contains a natural number of arbitrary size.[6] So we have an *idealised register* machine in the same way as we had idealised WHILE and GOTO where variables can store arbitrarily large trees. Although our real-world gadgets always have a finite number of registers that store numbers of a certain maximum size, this is not a problem for our considerations of computability (without resource bounds). We will consider computations with resource bounds later in the complexity part of the module.

It is interesting, however, that the machines we use daily have limited number of registers with limited word size, and are thus actually finite state automata, but of course enormously large and complex ones. About those we know they decide

---

[5] After all, we have not restricted the length of the tapes for Turing machines.

[6] We have not restricted the size of trees that can be stored in variables of WHILE-programs either.

exactly the regular languages and are therefore not Turing-complete. This seems to be a contradiction at first. One must consider, however, that we use idealised computing models that allow us to abstract away from details. Note that for negative results this idealisation is not a problem, because if something cannot be computed even with an idealised machine it won't be computable with a real, but limited, machine either.

As for most machine languages there are instructions to move data from one register to another, there are conditional jumps depending on whether the value of a register is 0 or not, and there are operations like increase or decrease of a register, addition or multiplication of registers (storing the result in a third register). The concept of *indirect addressing* is supported in assignments. Indirect addressing allows a program to access an unbounded number of registers beyond the fixed number of registers mentioned in the program explicitly. Moreover, those indirectly addressed registers can depend on the input of the program. Indirect addressing provides much flexibility at runtime. In particular, it allows the efficient implementation of arrays.

The successor RAM model, called SRAM, is like the RAM machine but does not allow binary operations on registers.

Syntax   In the following Xi denotes the ith register. Instructions are listed in Table 11.5.

The extra angle brackets in assignments indicate indirect addressing. The meaning will be clear from the semantics below. Note that Xi := Xi$\dot{-}$1 denotes the *decrement operation on natural numbers* which stipulates that $0\dot{-}1 = 0$. In order to highlight this, one writes the dot on top of the minus symbol. Extra operations of RAM not available in SRAM are the last two binary operations in Table 11.5.

Store   Since registers store natural numbers and are indexed by natural numbers, a store maps register (indices) to their values. Thus a store actually maps natural numbers to natural numbers and therefore is a function of type $\mathbb{N} \to \mathbb{N}$ where $\mathbb{N}$ denotes the type of natural numbers.

Table 11.5 Syntax of SRAM and RAM

Instruction	Explanation
Xi := 0	Reset register value
Xi := Xi+1	Increment register value
Xi := Xi$\dot{-}$1	Decrement register value
Xi := Xj	Move register value
Xi := <Xj>	Move content of register addressed by Xj
<Xi> := Xj	Move into register addressed by Xi
if Xi=0 goto ℓ_1 else ℓ_2	Conditional jump
Available only in RAM	
Xi := Xj + Xk	Addition of register values
Xi := Xj * Xk	Multiplication of register values

Table 11.6 One-step operational semantics for SRAM and RAM

$$p \vdash (\ell, \sigma) \rightarrow (\ell+1, \sigma[i := \sigma(i)+1]) \quad \text{if } p(\ell) = \texttt{Xi:=Xi+1}$$
$$p \vdash (\ell, \sigma) \rightarrow (\ell+1, \sigma[i := \sigma(i)-1]) \quad \text{if } p(\ell) = \texttt{Xi:=Xi}\dot{-}\texttt{1} \text{ and } \sigma(i) > 0$$
$$p \vdash (\ell, \sigma) \rightarrow (\ell+1, \sigma[i := 0]) \qquad\quad \text{if } p(\ell) = \texttt{Xi := Xi}\dot{-}\texttt{1} \text{ and } \sigma(i) = 0$$
$$p \vdash (\ell, \sigma) \rightarrow (\ell+1, \sigma[i := \sigma(j)]) \quad\; \text{if } p(\ell) = \texttt{Xi:=Xj}$$
$$p \vdash (\ell, \sigma) \rightarrow (\ell+1, \sigma[i := 0]) \qquad\quad \text{if } p(\ell) = \texttt{Xi:=0}$$
$$p \vdash (\ell, \sigma) \rightarrow (\ell_1, \sigma) \qquad\qquad\qquad \text{if } p(\ell) = \texttt{if Xi = 0 goto } \ell_1 \texttt{ else } \ell_2 \text{ and } \sigma(\texttt{Xi}) = 0$$
$$p \vdash (\ell, \sigma) \rightarrow (\ell_2, \sigma) \qquad\qquad\qquad \text{if } p(\ell) = \texttt{if Xi = 0 goto } \ell_1 \texttt{ else } \ell_2 \text{ and } \sigma(\texttt{Xi}) \neq 0$$
$$p \vdash (\ell, \sigma) \rightarrow (\ell+1, \sigma[i := \sigma(\sigma(j))]) \quad \text{if } p(\ell) = \texttt{Xi:=<Xj>}$$
$$p \vdash (\ell, \sigma) \rightarrow (\ell+1, \sigma[\sigma(i) := \sigma(j)]) \quad \text{if } p(\ell) = \texttt{<Xi>:=Xj}$$

$$p \vdash (\ell, \sigma) \rightarrow (\ell+1, \sigma[i := \sigma(j)+\sigma(k)]) \quad \text{if } p(\ell) = \texttt{Xi:=Xj+Xk}$$
$$p \vdash (\ell, \sigma) \rightarrow (\ell+1, \sigma[i := \sigma(j) \times \sigma(k)]) \quad \text{if } p(\ell) = \texttt{Xi:=Xj*Xk}$$

Readin   The input and output variable are both fixed as register 0.

$Readin(n)$ is the function that maps 0 to n and all other numbers to 0. In other words, the initial store moves n into the register 0, all other registers contain initially 0.

Readout

$$Readout(\sigma) = \sigma(0)$$

The result of the program is the value of register 0.

Semantics   We give the one-step semantics in Table 11.6. Note that $\sigma[i := n]$ abbreviates the store that is the same as σ with the one exception that the value for register i is now set to be number n. As before with other machines, we assume that p is the program $\ell_1 : \texttt{I}_1; \dots \ell_m : \texttt{I}_{\ell_m};$. The indirect addressing for $\texttt{Xj}$ in $\texttt{Xi:=<Xj>}$ is semantically expressed by the "double feature" of looking up a register value: $\sigma(\sigma(j))$: the value found in register j, $\sigma(j)$, is used itself as register name/number and then the value of this register, $\sigma(\sigma(j))$, is used. The indirection refers to this fact that the value is not directly taken from register j, but from the register *addressed by the value in* register j. An analogous indirection happens in the semantics of $\texttt{<Xi>:=Xj}$ when the indirect address is used for the register into which the value from $\texttt{Xj}$ is moved. This feature is most important to implement pointer arrays.

Example 11.3 The RAM program in Fig. 11.4 computes the factorial function.

Fig. 11.4 RAM program computing the factorial function

```
1: X1 := 1;
2: if X0=0 goto 6 else 3;
3: X1 := X1 * X0;
4: X0 := X0 - 1;
5: if X2=0 goto 2 else 2;
6: X0 := X1;
```

11.6   Counter Machines CM

Counter machines are much simpler than register machines. Since the registers for such machines only permit increment and decrement operations and a test for 0, and not any arbitrary register transfer or even more complicated binary operations, the registers for such machines are called *counters*.

Counter machines that only use *two* counters are called "two counter machines", and the class of such machines is abbreviated 2CM. On closer inspection, the counter machine model is actually a subset of the SRAM model where operations on registers are extremely limited. Therefore the remainder of this definition should read familiar:

Syntax   In the following Xi denotes the *i*th counter. Instructions are listed in Table 11.7.

Store   Like for register machines, a store is a function of type $\mathbb{N} \to \mathbb{N}$ where $\mathbb{N}$ denotes the type of natural numbers.

Readin   The input (and output) counter is fixed as counter 0.

Readin(*n*) is the function that maps 0 to *n* and all other numbers to 0. In other words, the initial store moves *n* into counter 0, all other counters contain initially 0.

Readout $Readout(\sigma) = \sigma(0)$.

The result of the program is the value of counter 0.

Semantics

We give the one step semantics in Table 11.8. As before, $\sigma[i := n]$ abbreviates the store that is the same as σ with the one exception that the value for counter i is now set to be number *n* and we assume that p is the program $\ell_1 : I_1; \ldots \ell_m : I_{\ell_m};$.

Example 11.4   The program outline in Fig. 11.5 computed the remainder of integer division by 2, also known as the "modulo 2" function. More precisely, it returns 1 if the input is an even number and 0 if the input is an odd number.

Table 11.7   Syntax of CM

Instruction	Explanation
Xi := Xi+1	Increment register value
Xi := Xi$\dot{-}$1	Decrement register value
if Xi=0 goto ℓ else ℓ'	Conditional jump

Table 11.8   One-step operational semantics of CM

$$p \vdash (\ell, \sigma) \to (\ell+1, \sigma[i := \sigma(i)+1]) \quad \text{if } p(\ell) = Xi := Xi + 1$$
$$p \vdash (\ell, \sigma) \to (\ell+1, \sigma[i := \sigma(i)-1]) \quad \text{if } p(\ell) = Xi := Xi \dot{-} 1 \text{ and } \sigma(i) > 0$$
$$p \vdash (\ell, \sigma) \to (\ell+1, \sigma[i := 0]) \quad \text{if } p(\ell) = Xi := Xi \dot{-} 1 \text{ and } \sigma(i) = 0$$
$$p \vdash (\ell, \sigma) \to (\ell', \sigma) \quad \text{if } p(\ell) = \text{if } Xi = 0 \text{ goto } \ell' \text{ else } \ell'' \text{ and } \sigma(Xi) = 0$$
$$p \vdash (\ell, \sigma) \to (\ell'', \sigma) \quad \text{if } p(\ell) = \text{if } Xi = 0 \text{ goto } \ell' \text{ else } \ell'' \text{ and } \sigma(Xi) \neq 0$$

Fig. 11.5 CM program
computing "mod 2"

```
1:  if X0=0 goto 6 else 2;
2:  X0 := X0 - 1;
3:  if X0=0 goto 7 else 4;
4:  X0 := X0 - 1;
5:  if X0=0 goto 6 else 2;
6:  X0 := X0 + 1;
```

11.7 Cellular Automata

The fascination of (two-dimensional) *cellular automata* has gripped students since John Horton Conway[7] devised *Game of Life* [18] (often simply called *Life*) in 1970. It shot to immediate fame through publication and promotion in the "Mathematical Games" section of *Scientific American* [9, 10].

The original concept of cellular automata is usually associated with the mathematician and computing pioneer John von Neumann[8] [27].

The classic version of *Life* consists of an infinite two-dimensional orthogonal grid of square cells, each of which can be in one of two possible states: alive or dead. Each cell interacts with its eight neighbour cells. At each tick of the clock, the following transitions happen:

1. Any live cell with fewer than two live neighbours dies (under-population).
2. Any live cell with two or three live neighbours lives on to the next generation (survival).
3. Any live cell with more than three live neighbours dies (overcrowding).
4. Any dead cell with exactly three live neighbours becomes a live cell (reproduction).

The game starts with a *seed* of live cells at time tick zero, describing the 0th generation, i.e. which cells in the two-dimensional grid are initially alive and which are dead. It is important that the rules above are applied to all cells *simultaneously* at each time tick.

Apparently, the syntax (and semantics) of cellular automata is very different from the notions of computation we have seen so far. This makes it also quite difficult to program a specific desired behaviour and restricts applicability. Nevertheless, cellular automata are fascinating the reason being twofold. First of all, it is astounding that complex behaviour emerges from very simple rules. Secondly, computation rules of cellular automata are *local* and *decentralised*. There is no central control like in any of our other notions of computation where the semantics always allowed the program

---

[7]Born 26 December 1937, John Conway is a British mathematician working in the area of finite groups, number theory, combinatorial game theory and others. He is a Fellow of the Royal Society and professor in Applied and Computational Mathematics at Princeton University.

[8]John von Neumann (December 28, 1903–February 8, 1957) was a Hungarian-born American scientist famous for contributions to various fields, e.g. mathematics, game theory, statistics, and of course computing that all had significant impact. The single-processor, stored-program computer architecture is, for instance, now known as von Neumann machine (or architecture).

to access the entire memory of a store or tape.[9] In cellular automata, rules act locally on neighbourhoods of cells.

In a short interview on Youtube,[10] Conway explains the rules of the "game" and points out universality. He also points out that *Game of Life* owed its name to the desire to simulate some basic features of "real life of basic organisms". More aspects of *Game of Life* are covered in [1], while [16, 20] can provide an overview of the subject more generally, and [21] provides some insights into two-dimensional cellular automata.

Applications of cellular automata can be found in simulations of dynamical systems in physics (e.g. flow systems, gas behaviour, growth of crystals or snow flakes), natural systems (like expansion of forest fires) as well as image processing (for instance to express blurring). Cellular automata can also be used to produce natural (self-similar) patterns for computer generated art (images and music). In [19], for instance, it is discussed how a cellular automaton can generate natural looking cave systems that can be used as levels of a computer game.

Game of Life is just one particular instance of a (two-dimensional) cellular automaton. One can, of course, imagine a different rule for cells to develop and one could also imagine one-dimensional grids and more than just two states for each cell. In order to be able to talk about those generalisations, we provide now a generic semantics for cellular automata.

Definition 11.3 A cell lattice is a (possibly infinite) set of cells C together with a binary operation $+$ that encodes "walks across the cell lattice" or linear transformations of a cell.

Definition 11.4 The *neighbourhood template* for a cell lattice C is a sequence of cells in C that describe the relative walks from any cell to its neighbouring cells, i.e. the cells that are considered adjacent. If $T = \langle c_1, c_2, \ldots, c_n \rangle$ is the neighbourhood template, the concrete neighbouring cells for $c \in C$ can then be computed by $c + c_1, c + c_2, \ldots, c + c_n$ where $+$ is the binary operator provided by the cell lattice. The *size of a neighbourhood* is determined by the length of the template T.

Example 11.5 For a two-dimensional discrete coordinate system $C = \mathbb{N} \times \mathbb{N}$, the so-called *nearest neighbours* neighbourhood (also called Moore[11] neighbourhood) has the following template

$$T_{\text{Moore}} = \left\langle \begin{array}{ccc} (-1, 1), & (0, 1), & (1, 1), \\ (-1, 0), & (0, 0), & (0, 1), \\ (-1, -1), & (0, -1), & (-1, -1) \end{array} \right\rangle$$

and the concrete neighbours of cell (m, n), i.e. $(m, n) + T_{\text{Moore}}$, are displayed accordingly in Fig. 11.6.

---

[9]Thus, cellular automata are often referred to as "non-standard computation".

[10]http://www.youtube.com/watch?v=E8kUJL04ELA.

[11]Edward Forrest Moore (November 23, 1925–June 14, 2003) was an American professor of mathematics and computer science, who among other things defined the Moore finite state machine.

Fig. 11.6 Moore
neighbourhood of (n, m)

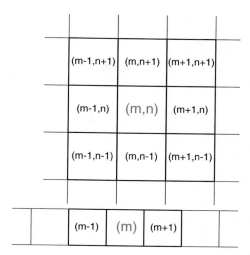

Fig. 11.7 Nearest
neighbourhood of (m)

For a one-dimensional coordinate system $C = \mathbb{N}$, *"nearest neighbours"* has the
following template

$$T_{\text{nearest}} = \langle -1, 0, 1 \rangle$$

and the concrete neighbours of cell m, i.e. $(m) + T_{\text{nearest}}$, are displayed accordingly
in Fig. 11.7.

We use only one neighbourhood template for each automaton, in which case
neighbourhoods are called *uniform*. This could be further relaxed to get even more
general definitions of cellular automata. For uniform neighbourhoods it makes sense
to talk about their size:

Example 11.6 Let us consider the size of neighbourhoods described in
Example 11.5. For the Moore neighbourhood we easily see that $|T_{\text{Moore}}| = 9$. For
the one-dimensional nearest neighbourhood we immediately see that $|T_{\text{nearest}}| = 3$.

Definition 11.5 A *cellular automaton* is a tuple (C, T, Σ, u) where:

- $(C, +)$ is the (unbounded)[12] lattice of cells,
- T is the neighbourhood template of size n,
- Σ is the finite set of states a cell can assume with a distinguished state, called the
  quiescent state 0 (denoting a 'blank cell'), and
- $u : \Sigma^n \rightarrow \Sigma$ is the *local rule*, i.e. the update function that describes how to
  compute the new state of a cell c from values for all cells of the neighbourhood of
  c. There is one condition on u, namely that it maps n quiescent states to a quiescent
  state.[13]

---

[12]This is in analogy with unbounded size of Turing machine tapes.

[13]So nothing can be created out of thin air, i.e. in an entirely "dead" neighbourhood.

Definition 11.6 Given a cellular automaton $\Delta = (C, T, \Sigma, u)$ with $T = \langle c_1, c_2, \ldots c_n \rangle$, its *global map* is the transition map from one configuration to the next which has type $(C \to \Sigma) \to (C \to \Sigma)$ and is defined as follows:

$$(F_\Delta\, v)(c) = u \langle v(c + c_1), v(c + c_2), \ldots, v(c + c_n) \rangle$$

It describes the new state of each cell depending on the states of its neighbours (and only its neighbours!). An *initial configuration* v_0 for Δ is a configuration $C \to \Sigma$ for which only finitely many cells $c \in C$ are initially not quiescent, i.e. $v_0(c) \neq 0$ for finitely many c only.[14]

The semantics of Δ is a map $[\![\Delta]\!]^{CA} : (C \to \Sigma) \to \mathbb{N} \to (C \to \Sigma)$ that takes an initial configuration v_0 of type $C \to \Sigma$, called the *seed* of Δ, and a number of time ticks $i \in \mathbb{N}$, and returns the result configuration of the cellular automaton if run in the initial configuration for i time ticks. It is formally defined as follows:

$$([\![\Delta]\!]^{CA}\, v_0)(i) = \begin{cases} v_0 & \text{if } i = 0 \\ F_\Delta(([\![\Delta]\!]^{CA}\, v_0)(i-1)) & \text{otherwise} \end{cases}$$

A cellular automaton has no natural notion of termination. One can apply the global map as many times as one wishes. In case of oscillators (see Fig. 11.8d) one will then just oscillate between several certain configurations. But it can also be that $([\![\Delta]\!]^{CA}\, v_0)(k) = ([\![\Delta]\!]^{CA}\, v_0)(k+1)$ for a certain $k \in \mathbb{N}$ in which case the configuration became a still that does not change any further and one might say that the computation "stabilizes".[15] In order to compare cellular automata with other notions of computations, one stipulates that one of the states Σ be a distinguished *accepting state*. An input is then supposed to be accepted, i.e. it is supposed to be in the set that is to be (semi-)decided once the accepting state has fired in a predefined cell c_{result}. The accepting state must be persistent in the sense that u will leave an accepting state alone.

11.7.1   2D: Game of Life

The cellular automaton *Game of Life* discussed above can now be instantiated from the general definition as follows.

---

[14]This corresponds to the definition of the initial state for a Turing machine where only a finite part is not blank (the input word). This condition could be dropped but then we would have to consider also Turing machine computation with infinite non-empty initial tape or Turing machines with oracles which is not covered in this book.

[15]The concept of "stabilizing computation" is also important for Chemical Reaction Networks in Sect. 22.4.

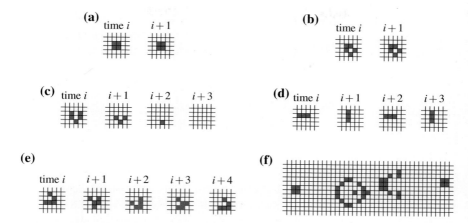

Fig. 11.8 Game of life patterns. **a** Static pattern 'block'. **b** Static pattern 'boat'. **c** Stages of collapsing 'V'. **d** Oscillating stages of the 'blinker'. **e** Diagonal movement of the 'glider' in 4 stages. **f** 'Gosling's glider gun'

Definition 11.7 (*Game of Life*) *Game of Life* is a specific instance of a two-dimensional cellular automaton (C, T, Σ, u) with $(C, +) = (\mathbb{N} \times \mathbb{N}, (m, n) + (l, k) = (m+n, l+k))$, cell states $\Sigma = \{0, 1\}$, Moore neighbourhood T as described in Example 11.5, Fig. 11.6, and local rule u defined as follows:

$$
u \begin{pmatrix} v_{(-1,1)}, & v_{(0,1)}, & v_{(1,1)} \\ v_{(-1,0)}, & v_{(0,0)}, & v_{(0,1)} \\ v_{(-1,-1)}, & v_{(0,-1)}, & v_{(-1,-1)} \end{pmatrix} = \begin{cases} 1 & \text{if } \sum_{i,j \in \{-1,0,1\}} v_{(i,j)} = 3 \\ & \text{or} \\ & \sum_{i,j \in \{-1,0,1\}:(i,j) \neq (0,0)} v_{(i,j)} = 3 \\ & \text{and } v_{(0,0)} = 1 \\ 0 & \text{otherwise} \end{cases}
$$

Despite the simplicity of the above rules for *Game of Life*, some complex behaviour may evolve (in the shape of patterns). A few simple patterns are depicted in Fig. 11.8: There are *oscillators* like "blinker" (see Fig. 11.8d), which changes in intervals of even and odd time ticks,[16] or *stills* like "block" (see Fig. 11.8a) and "boat" (see Fig. 11.8b) which do not change at all over time, unless their local neighbourhood changes. Some configurations disappear over time like the "collapsing V" (see Fig. 11.8c) which disappears after three ticks (again if there is no stimulus coming from the outside).

Probably the most fascinating features of *Life* are the moving patterns, also called "spaceships". An example is the "*glider*" in Fig. 11.8e, that moves one cell diagonally after four time ticks. Even more amazingly and famously, there exist patterns that create such gliders in regular intervals: the "*Gosper Glider Gun*" (see Fig. 11.8f) that "shoots" gliders (a form of "spaceships", see Fig. 11.8e) in regular intervals every 30

---

[16]Other patterns may need a longer interval to repeat their original state.

time ticks.[17] It was discovered by *Bill Gosper*[18] as an answer to Conway's question whether the black (alive) cells set of a *Life* configuration can grow without limit.

There is an ongoing discussion whether cellular automata are more suitable to explain computational phenomena in nature than more traditional notions of computations, but this discussion is beyond the scope of this book.

11.7.2   1D: Rule 110

At first glance, a one-dimensional cellular automaton may seems a pretty weak model of computation. But this is wrong. On the contrary, one-dimensional automata are very flexible and how expressive they are will be discussed in Sect. 11.8. Stephen Wolfram[19] is renowned for his systematic analysis of one-dimensional automata and developed a classification for one-dimensional cellular automata with $\Sigma = \{0, 1\}$ and $T = T_{\text{nearest}}$, called *elementary cellular automata* [29].

Definition 11.8 (*Rule 110*) The cellular automaton *Rule 110* is a specific instance of a one-dimensional cellular automaton (C, T, Σ, u) with $(C, +) = (\mathbb{N}, (m) + (n) = (m + n))$, cell states $\Sigma = \{0, 1\}$, nearest neighbourhood template T as described in Fig. 11.7, and local rule u defined as follows:

$$u \langle v_{-1}, v_0, v_1 \rangle = \left\{ \begin{array}{l} 1 \text{ if } \langle v_{-1}, v_0, v_1 \rangle \in \{\langle 1, 1, 0 \rangle, \langle 1, 0, 1 \rangle, \langle 0, 1, 1 \rangle, \langle 0, 1, 0 \rangle, \langle 0, 0, 1 \rangle\} \\ 0 \text{ otherwise} \end{array} \right\}$$

For the purpose of investigating their computational powers, we distinguish different classes of cellular automata:

Definition 11.9 The class of all cellular automaton as defined in Definition 11.5 is denoted CA. The class of all cellular automata with $C = \mathbb{N} \times \mathbb{N}$, of which *Game of Life* is an example setting $T = T_{\text{Moore}}$ and $\Sigma = \{0, 1\}$, is denoted CA2. The class of all cellular automata with $C = \mathbb{N}$, of which *Rule 110* is an example setting $T = T_{\text{nearest}}$ and $\Sigma = \{0, 1\}$, is denoted CA1.

---

[17]An excellent and efficient simulator for *Life* is "Golly" available for all (including mobile) platforms, see [11].

[18]Ralph William Gosper (born 1943) is an American mathematician and programmer famous for his puzzles and among other things his discoveries of moving oscillators in *Life*.

[19]Stephen Wolfram (born 29 August 1959) is a British physicist, computer scientist and entrepreneur, most famous for being the chief designer of *Mathematica* published by Wolfram Research whose CEO he is as well.

11.8 Robustness of Computability

We show that the formally defined models of computation introduced above are all equivalent w.r.t *what* can be computed in them.[20] In other words, we give evidence for the Church-Turing thesis. How do we do that? Well, we must show that whenever we can decide a problem with a program (machine) of one kind we can also do it with a program (machine) of the second kind. In other words, we need to be able to translate programs between the languages, preserving the semantics of the program in question. But, hang on a second, this is exactly what we said a compiler does (in Chap. 6 about Programs as data).

11.8.1 The Crucial Role of Compilers

A compiler is, in a way, a constructive proof, that the target language of the compiler simulates the source language.[21] For every program in the source language there is a program in the target language with the same semantics (the same input/output behaviour).

The set of compilers from (language) S to (language) T— where S-data = T-data, that is source and target language use the same datatype[22]— written in (language) L is the set of L-programs that for all S-programs p as input produce program $[\![comp]\!]^L$ (p) which is a T-program that when run on any input d behaves like the original program p. More formally, we can define the set of such compilers as follows:

$$\left\{ comp \in L\text{-program} \,\middle|\, \forall p \in S\text{-program}, \forall d \in S\text{-data}. \; [\![p]\!]^S (d) = [\![\,[\![comp]\!]^L (p)\,]\!]^T (d) \right\}$$

Observe that we feed d to both, the original S-program p, and the compiled T program $[\![comp]\!]^L$ (p). This is why we require that S-data = T-data so that d can be in both. It simply has to be if we don't compile the data-types as well.

Compilers can be composed. If we have a compiler comp1 from language L1 to language L2 written in S and a compiler comp2 from L2 to language L3 written in L2 we get a compiler comp3 from L1 to language L3 written in S (see Exercise 4). The fact that we can compose compilers is very important for what follows below.

---

[20]Issues of how long it takes are covered in the next part of the book about complexity. Other issues, for instance, how easy or convenient it is to program in those languages are not systematically studied here.

[21]Or, seen from the other direction, that the source language can be simulated by the target language.

[22]Without any problem one can generalise this by also encoding the data type of the source language in the data type of the target language. However, this significantly complicates the presentation, so we leave this out. More can be found in [15].

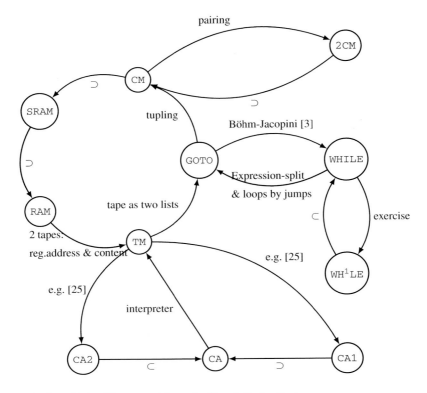

Fig. 11.9 Compilation Diagram: evidence for Church-Turing thesis

11.8.2   *Equivalence of Models*

To give some evidence for the Church-Turing thesis, we need to show that all the introduced notions of computation are equivalent. As we just have seen, it is enough to argue that we can compile each language into each other. Since we do not want to need to write 10×10 compilers (since we have 10 computational models on show) and since we know that we can compose compilers, we rather sometimes use circular compilation. For instance, we can show that we can compile CM to SRAM to RAM to TM to GOTO to CM. Since we can compose compilers, we thus can compile CM to GOTO and also GOTO to CM.

The diagram in Fig. 11.9 gives an overview regarding the compilers needed. Where we use L ⊂ L′ this indicates that the language L is already a subset of L′ and thus actually no compilation is needed.[23]

Below we give short explanations of the compilation processes involved. For more information consult [15, Chap. 8] or the exercises of this chapter.

---

[23]In other words, the compilation is the identity function.

- WHILE $\rightarrow$ WH^{1}LE (one variable as list of variables)
  The idea of this compiler has been mentioned earlier. We store all values for the variables of the source program in a list that is stored in the one allowed variable of the target program. One has to produce extra code that accesses the right element in the list but this is clearly possible.
- WHILE $\rightarrow$ GOTO
  The main issues is to compile away complex tree expressions as they can appear in WHILE and the while loop itself. The while loop can clearly be compiled into a conditional jump (this is a classic compiler exercise). Complex expressions need to be split into several individual commands that manipulate only variables (as provided by GOTO). For instance, X:=cons hd nil tl X will be compiled into Z:=nil; Y:=hd Z; U:=tl X; X:=cons Y U. Any (potentially nested) while loop can be compiled away using several jumps.
- GOTO $\rightarrow$ WHILE (translation of flowcharts according to Böhm-Jacopini [3])
  The idea is that we explicitly model the program counter (as seen in the semantics of GOTO) which is stored in a variable. The WHILE-program then only needs one loop that checks whether the current program (instruction) counter is still within the program. If it is, execute the instruction with the corresponding label. Instructions change the program counter and this is how jumps can be executed. All other GOTO instructions are actually also WHILE-instructions.
- TM $\rightarrow$ GOTO (tape as two lists)
  As Turing machines use words over a tape alphabet this requires a change of data representation. But a tape symbols can be easily encoded in $\mathbb{D}$ and words are simply lists of such encoded symbols. The tape of the machine is represented as two lists. The first element of the second list is the currently scanned symbol on the tape. The first list is considered to contain the symbols left of the head in reverse order of direction. The first element of that list is the content of the cell left to the head, the second element the content of the cell left of the cell left to the head and so on. Clearly, each instruction of a Turing machine can then be easily simulated by one or several GOTO instructions.
- GOTO $\rightarrow$ CM (tupling functions)
  The main idea here is to encode each GOTO variable by a CM counter. This requires a change of data representation, as GOTO uses datatype $\mathbb{D}$ whereas CM works with natural numbers $\mathbb{N}$. To this end, one defines a coding function $c : \mathbb{D} \rightarrow \mathbb{N}$ as follows:

$$c(\mathtt{nil}) = 0 \qquad c\langle \mathtt{d}_1.\mathtt{d}_2 \rangle = 1 + 2^{c(\mathtt{d}_1)} \times 3^{c(\mathtt{d}_2)}$$

This encoding is appropriate since for each code there is clearly at most one tree with this encoding (thanks to the use of prime numbers 2 and 3).
One can simulate GOTO instructions by several CM instructions that use the above encoding. One also needs to show that operations on numbers like

$$m(x) = \max\{ y \mid \text{there is a } z \text{ such that } x = c^y \times z \}$$

are CM computable. This is necessary to encode the tree operations (using $c = 2$ and $c = 3$).

- RAM $\rightarrow$ TM (registers as (address, content) on two tapes)
  The idea here is relatively simple: each register is encoded as a pair of two numbers: its address and its content. These pairs of numbers are then stored on the Turing machine tape. Numbers are written on the tape in binary representation. It is best to use 4 tapes for the simulation. One tape for the addresses, one tape for the contents, one tape as "accumulator", and one tape for intermediate results (scratch). But one can in principle do the same with one tape.

- CM $\rightarrow$ 2CM (pairing)
  The idea is to use one of the two counters to store the values of all the counters of the original CM program. We use prime numbers to encode their different values in one (see also GOTO $\rightarrow$ CM). So if the original program has counters X0, X1,..., Xk and at any given point in the execution they have values $x0, x1, \ldots, xk$, respectively, then the first of the two counters of the 2CM program, let's call it Y, will contain the following:

$$Y = 2^{x0} \times 3^{x_1} \times \cdots \times h^{xk}$$

where h is the $(k+1)$th prime number. Now any instruction of the original program can be compiled into one or several of the new 2CM machine. Of course initially the input d must be encoded as

$$Readin(d) = \{Y : 2^d \times 3^0 \times 5^0 \cdots \times h^0, X : 0\} = \{Y : 2^d, X : 0\}$$

and before the program returns, the result must be computed from the two counters as follows:
$$Readout(\sigma) = x \quad \text{if } \sigma(Y) = 2^x \times c$$

- TM $\rightarrow$ CA2
  This has been done already by van Neumann and others [27] but a more efficient translation has been suggested by [25]. One can, however, go further and may wonder whether the *single particular two-dimensional* CA2 *Game Of Life* can compute the same as any Turing Machine can. The latter requires the TM to be encoded in the *seed* of this particular CA2, proving the *universality* of the cellular automaton which is much harder than compiling a given TM into any CA2. Already in 1974, Wainwright observed that there are "Life patterns that can imitate computers." [28] He showed how to control the spatial location of gliders such that glider streams can be used as signals, how to duplicate them, and how to implement logical gates[24] for those signals. In 1982, Elwyn R Berlekamp, John

---

[24]OR, AND, and NOT gates.

H Conway, and Richard K Guy proved that *Game of Life* is Turing-complete, i.e. one can simulate Turing machines [2].

Much later a concrete universal register machine (slightly different variant of the RAM we used) was implemented in *Life* by Paul Chapman in 2002 [4]. Moreover, a fully universal Turing machine was created[25] by Paul Rendell [22, 23]. Since the rules of the game are fixed, in order to create register machines or Turing machines, one has to find the right seed pattern which is extremely complex. To obtain and manage such complex behaviour, one uses several layers of abstractions, by first implementing logic gates, registers etc. Needless to say, those constructions are very involved creations, their execution is extremely slow. Chapman, for instance, writes about his register machine: *"Version 0 of the URM is large (268,096 live cells in an area of 4,558 × 21,469)"* [4]. Rendell reports that running the universal Turing machine for a TM with 6 instructions and two symbols that doubles string "101010" (with result "101010101010") needed 116 million life generations to complete [22]. Therefore, the universal machines in *Life* discussed above are not usable to compute any reasonable register machine or Turing machine program, respectively.

- CA → TM

This direction is relatively straightforward, if tedious, as one "simply" has to implement the semantics for the cellular automaton. One can use the tape of the TM to simulate the cell space of the cellular automaton. Of course, all local updates have to be simulated sequentially on the TM, which is not at all efficient. It is also important that the initial configuration (seed) is finite.

- TM → CA1

It turns out that even the one-dimensional version of a cellular automaton introduced is sufficient to simulate TMs as shown in [25], which actually presents several constructions using different sizes of neighbourhoods and sate set Σ. Like in the two-dimensional case, one might wonder whether even a single *elementary* (and thus extremely simple) cellular automaton is expressive enough to simulate any TM. It turns out that the cellular automaton defined in Definition 11.8 can simulate any Turing machine as shown in [6] (and already suspected in [29]). It is worth pointing out though that there is some "magic" involved in the sense that the initial configuration in this case is *not* finite. The simulated TM and its input tape need to be encoded as three repeated patterns interwoven with a repeated (unbounded) background pattern.

What Next?

In this chapter we have seen various different notions of computation other than WHILE, and we have seen that they are all equivalent in terms of what one can compute with them. Earlier, we have covered some fundamental topics in computability to answer questions of the kind: *"is a certain problem computable/decidable?"* and have found various problems that are undecidable. In the next part of the book we only

---

[25]See also: http://www.youtube.com/watch?v=My8AsV7bA94.

deal with decidable problems but ask new questions of the sort: "*is there enough time to solve this problem?*" This will lead to quite deep questions about what is feasibly computable by any means of computation.

Exercises

1. For machine-like languages (computation models) L (where L is either RAM, SRAM, CM, GOTO, TM) we defined a "framework" to describe the computation model in question and its semantics. Each L-program p is a labelled sequence of instructions. To define the semantics formally, we need a definition of L-store (that contains values during computation), of input and output functions *Readin* and *Readout*, and of the one-step transition relation of the operational semantics $p \vdash s \to s'$ where p is an L-program and s, s' are L-states i.e. pairs of the form (ℓ, σ) where ℓ is a label and $\sigma \in$ L-store.

   a. What is the exact type of the functions *Readin* and *Readout*, respectively?
   b. Given a L-program p and input d, when is $[\![p]\!]^L (d) = e$?
   c. Why is the WHILE-language not an instance of this framework?

2. Show that predecessor can already be computed in RAM even if there were no command Xi := Xi⁻1.

3. Write the program even in GOTO that takes as input a (unary encoding of a) natural number and returns the (encoding of) true if the number is even and false if the number is odd.

4. Assume we have a compiler comp1 from language L1 to language L2 given as WHILE-program and a compiler comp2 from L2 to language L3 given as WHILE-program. Write a WHILE-program comp3 (use extended language) that implements a compiler from L1 to L3.

5. Explain in more detail than sketched in Sect. 11.8.2 how to compile from WHILE to GOTO.

6. Explain in more detail than sketched in Sect. 11.8.2 how to compile from GOTO to WHILE.

7. In the diagram of equivalences in Fig. 11.9 there is a compiling function f specified as follows:
   f: GOTO-programs → CM-programs
   such that for all $p \in$ GOTO-programs and $d \in$ GOTO-data

   $$[\![p]\!]^{GOTO} (d) = [\![f(p)]\!]^{CM} (c(d))$$

   where c is the necessary compilation for data values.

   a. Explain what the type of the function c must be, i.e. what does it map into what?
   b. Explain (informally at least) how you would define such a map.

c. Sketch how the compilation function f works. The best way to do this is to explain how the various GOTO instructions can be compiled into (sequences of) CM-instructions such that the semantics of the original program is preserved.

8. Show that a Turing machine with m tapes cannot compute more functions than a Turing machine with one tape.
9. Show that a (one tape) Turing machine with an alphabet of size $n + 1$ (including the special blank symbol) can be compiled into an equivalent (one tape) Turing machine with 3 symbols (including the special blank symbol).
10. Convince yourself that the *Gosling Glider Gun* actually produces gliders by using a *Life* simulator like e.g. [11].
11. Compute the behaviour of the following pattern by using the local rule u of *Game of Life*:

References

1. Adamatzky, A. (ed.): Game of Life Cellular Automata. Springer, Berlin (2010)
2. Berlekamp, E.R., Conway, J.H., Guy, R.K.: Winning Ways for Your Mathematical Plays: Volume 2. Academic Press, New York (1982)
3. Böhm, C., Jacopini, G.: Flow diagrams, turing machines and languages with only two formation rules. Commun. ACM **9**(5), 366–371 (1966)
4. Chapman, P.: Life universal computer. Available via DIALOG: http://www.igblan.free-online.co.uk/igblan/ca/index.html. 23 June 2015
5. Church, A.: An unsolvable problem of elementary number theory. Am. J. Math. **58**(2), 345–363 (1936)
6. Cook, M.: Universality in elementary cellular automata. Complex Syst. **15**, 1–40 (2004)
7. Copeland, J.: The Church-Turing Thesis. In: References on Alan Turing. Available via DIALOG. http://www.alanturing.net/turing_archive/pages/Reference%20Articles/The%20Turing-Church%20Thesis.html. 2 June 2015 (2000)
8. Dartmouth College Computation Center: A manual for BASIC, the elementary algebraic language design for use with the dartmouth time sharing system. 1 Oct 1964. Available via DIALOG: http://www.mirrorservice.org/sites/www.bitsavers.org/pdf/dartmouth/BASIC_Oct64.pdf. 23 June 2015
9. Gardner, M.: The fantastic combinations of John Conway's new solitaire game life. Sci. Am. **223**, 120–123 (1970)
10. Gardner, M.: On cellular automata, self-reproduction, the garden of eden and the game of life. Sci. Am. **224**(2), 112–117 (1971)
11. Golly. An open-source, cross-platform application for exploring Conway's Game of Life. Available via DIALOG. http://golly.sourceforge.net. 30 Nov 2015
12. Gomard, C.K., Jones, N.D.: Compiler generation by partial evaluation: a case study. In: Ritter, G. (ed.) Proceedings of the IFIP 11th World Computer Congress, pp. 1139–1144, North-Holland (1989)

13. Hodges, A.: Alan turing in the stanford encyclopedia of philosophy: introduction. Available via DIALOG. http://www.turing.org.uk/publications/stanford.html. 23 June 2015
14. Hodges, A.: Alan Turing: The Enigma, Vintage (1992)
15. Jones, N.D.: Computability and complexity: From a Programming Perspective (Also available online at http://www.diku.dk/neil/Comp2book.html.). MIT Press, Cambridge (1997)
16. Kari, J.: Cellular automata: a survey. Theor. Comput. Sci. **334**, 3–33 (2005)
17. Kemeny, J.G., Kurtz, '.T.E.: Back to BASIC: The History, Corruption and Future of the Language. Addison-Wesley Publishing, USA (1985)
18. Koenig, H., Goucher, A., Greene, D.: Game of Life News. Available via DIALOG: http://pentadecathlon.com/lifeNews/index.php. 19 October 2015
19. Kun, J.: The cellular automaton method for cave generation. Blog post. Available via DIALOG. http://jeremykun.com/tag/cellular-automata/. 21 July 2015
20. Mitchell, M.: Computation in cellular automata: A selected review. In Gramss, T., Bornholdt, S., Gross, M., Mitchell, M., Pellizzari, T. (eds.) Nonstandard Computation, pp. 95–140. VCH Verlagsgesellschaft (1998)
21. Packard, N.H., Wolfram, S.: Two-dimensional cellular automata. J. Stat. Phys. **38**(5–6), 901–946 (1985)
22. Rendell, P.: A universal turing machine implemented in conway's game of life. Available via DIALOG. http://rendell-attic.org/gol/utm/index.htm. 21 July 2015
23. Rendell, P.: turing universality of the game of life. In: Adamatzky, A. (ed.) Collision-Based Computing, pp. 513–539, Springer, Berlin (2002)
24. Shepherdson, J.C., Sturgis, H.E.: Computability of recursive functions. J. ACM **10**(2), 217–255 (1963)
25. Smith III, A.R.: Simple computation-universal cellular spaces. J. ACM **18**(3), 339–353 (1971)
26. Turing, A.M.: On computable numbers, with an application to the Entscheidungsproblem. Proc. London Math. Soc. series 2 **42** (Parts 3 and 4), pp. 230–265 (1936)
27. Von Neumann, J., Burks, A.W.: Theory of self-reproducing automata. IEEE Trans. Neural Netw. **5**(1), 3–14 (1966)
28. Wainwright, R.T.: Life is universal! Proceedings of the 7th conference on Winter simulation WSC'74, Vol 2, pp. 449–459, ACM (1974)
29. Wolfram, S.: Computation theory of cellular automata. Commun. Math. Phys. **96**(1), 15–57 (1984)

Part II
Complexity

"I argue that computational complexity theory—the field that studies the resources (such as time, space, and randomness) needed to solve computational problems—leads to new perspectives on the nature of mathematical knowledge, . . . the foundations of quantum mechanics, . . . and several other topics of philosophical interest."

Scott Aaronson,

"Why Philosophers Should Care About Computational Complexity" in Copeland, B.J., Posy, C.J. and Shagrir, O.: Computability: Gödel, Turing, Church, and beyond, MIT Press, 2012

Chapter 12
Measuring Time Usage

How do we measure runtime of programs? How do we measure runtime of WHILE-programs independently of any assumptions on the hardware they run on?

In the second part of this book we investigate the limits of computing regarding runtime. What problems take too long to solve? From a practitioner's point of view this very much depends on the hardware in use, the compiler and its ability to optimise code, and most importantly maybe the programmer's skill to code efficiently. Of course, this is *not* what we want to capture. We are interested in the *inherent limitations* of problems regarding their feasibility. A problem is feasible if it can be computed in a "reasonable" (practically acceptable) amount of time. This is vague, and we will have to make this more precise in the following chapters. Instead of measuring real time we will introduce some abstract notion of time.

The field of *Computational Complexity* probably started with the seminal paper by Hartmanis[1] and Stearns[2] [2] who also introduced this term.[3]

There is another parameter that program runtime depends on that we have not mentioned already above, namely the size of the input. Even complex tasks can be solved quickly by a program if the data to be considered is small. For instance, a small array can be quickly sorted or searched, whatever inefficient algorithm one uses to do so. We will therefore be looking at *asymptotic time complexity* which measures how the runtime of a program increases when the input increases. Such a relationship can be neatly expressed by a function that maps the size of input to the abstract cost of runtime. Size and runtime will be measured in terms of natural numbers. For instance for WHILE, the input is always a binary tree, and its size can be measured by the function *size* we introduced in Definition 3.2.

---

[1] Juris Hartmanis was born July 5, 1928 in Latvia and was a professor at Cornell University.

[2] Richard Edwin Stearns was born July 5, 1936 and was a professor at University at Albany (State University of New York).

[3] They both received the 1993 ACM Turing Award "in recognition of their seminal paper which established the foundations for the field of computational complexity theory".

© Springer International Publishing Switzerland 2016
B. Reus, *Limits of Computation*, Undergraduate Topics in Computer Science,
DOI 10.1007/978-3-319-27889-6_12

But we want to measure time usage not only for for WHILE-programs but also for the other models of computation we have discussed in Chap. 11. Regarding the size of input, for GOTO that uses trees we can use function *size* again, whereas for RAM SRAM, CM and 2CM the data types used are the natural numbers. If we take the unit representation of a number as measure then the size of a number n is clearly just n. However, we can also take the binary representation of a number (as modern computers do), but then the size of n is just $1 + \log_2 n$, where $\log_2$ denotes the (integer) binary logarithm of n. Consider the binary representation of $15 = 2^3 + 2^2 + 2^1 + 2^0$ which is 1111. Then $4 = 1 + \log_2 15$. There is an exponential gap in size between a number and its logarithm, and so it is much harder to provide good asymptotic runtime when taking binary representation as input which is, however, standard. We will therefore have to consider the data representation of the input when determining runtime complexity of algorithms.

Finally for TM, we note that the input of a Turing machine is a word w written on the input tape. One can easily measure the size of w as its length. When w encodes a number, again it is important whether the number is represented unary or binary.

In Sect. 12.1 the various notions of time measure for our machine models of computation are defined, whereas Sect. 12.2 defines a measure for WHILE. One can then compare such *timed* programming languages (Sect. 12.3) to see how compilation affects runtime.

12.1   Unit-Cost Time Measure

For imperative languages (and machine models) with labelled instructions (which is all our models besides WHILE) we can use the following measure, called *unit-cost time measure*, which counts each instruction as a "unit" of time, thus not distinguishing between any differences in time the various instructions may need to execute. The cost of the computation is then the sum of the costs of the instructions executed.

Definition 12.1 (*Unit-cost measure*) For an imperative language L, the function $time^{\mathrm{L}} : \mathrm{L}\text{-program} \to \mathrm{L}\text{-data} \to \mathbb{N}_\perp$ is defined a follows: For any $p \in \mathrm{L}\text{-program}$, and $d \in \mathrm{L}\text{-data}$ we define:

$$time^{\mathrm{L}}_p(d) = \begin{cases} t+1 & \text{if } p \vdash s_1 \to s_2 \cdots \to s_t \text{ where } s_1 = Readin(d) \text{ and } s_t \text{ terminal} \\ \perp & \text{otherwise} \end{cases}$$

With any completed program execution we thus associate as runtime the number of transition steps the program with the given input takes according to the program's semantics. It should be pointed out that reading the input takes one step implicitly, as the first state is indexed by 1 and so after one step we're in state 2 and so on. Therefore, the cost for a program executing one instruction is actually 3.

It is important that the time a program takes for execution obviously depends on the input. For different inputs it can (and very often will) take different time to execute the program. Since the time depends on the input, the running time measure is a function from L-data to natural numbers. It is a *partial function* since the program may not terminate on some input.

For cellular automata CA, the time-unit measure is already built into the semantics, so to speak. Remember that for a CA Δ, its semantics $[\![\Delta]\!]^{CA}$ requires an initial seed configuration as well as a natural number that describes the time ticks to iterate the global map on the initial seed. This number argument naturally acts as a time measure. In the case where cellular automata are used to decide membership or compute a function (this happens in particular when comparing cellular automata with other notions of computation, see Chap. 14) one may identify certain configurations as "accepting" ones and thus measure how many times the global map has to be applied to get to such an accepting configuration.

Fairness of Time Measures

Is the unit-cost measure always appropriate? It is worth pointing out again that the unit-cost measure treats all instructions the same way, discounting what those instructions actually achieve. If those instructions "did a huge amount of work" in terms of the size of the input then the measure would be questionable.

If we consider Turing machines TM, we know (as Turing explained himself in his seminal paper) that each instruction can only do very little: move left, right, write one symbol on the tape or jump depending on the symbol read. So for TM the unit-cost measure is fine. For GOTO we have to consider that we can only move data values around in variables as we cannot have proper expressions in assignments. Instead we need to use assignments of the form X:=cons Y Z or X:=hd Y. Those cannot contribute much to the size of a tree. The cons case, for instance, uses two trees already constructed (so the time for their construction should be already accounted for) and adds just one node to it. The hd operation just reduces the size of a tree by one. Thus, it is justified to count each assignment for one "unit" of time only.[4]

The situation is a bit more complicated for RAM despite the fact that—as for GOTO—most instructions only can move data around (or jump): it is the binary operations though that could create large numbers in one single assignment. Since this computation model deals with numbers as input and output, we could construct polynomially large numbers in constant time with the help of those binary operators + and $*$. In order to get a "reasonably realistic" time measure we would have to employ logarithmic measure of the size of numbers (how many digits needed to represent them). This is entirely possible and not too difficult but left to the interested reader for further study [3, Sect. 16.5.2], [1] (see also Exercise 7).

This issue is also the motivation to look at SRAM (Successor RAM) that does not allow us to use any binary operators on registers. On top of moving the register content, one can only increment (X:=Y+1) and decrement (X := Y $\dot{-}$ 1). Here the unit cost measure is realistic and adequate.

For counter machines CM we somewhat have the opposite problem. A counter can only be incremented or decremented by one (X:=X+1 and X := X $\dot{-}$ 1), respectively

---

[4]Whether it is one or two does not make a difference anyway which we will see later.

in an instruction, and counters cannot be copied in one step (at cost "1") from one counter to the other (like in SRAM). Any copying action will need to deconstruct a counter (using decrement) and build another one (or several) of the same value (using increment). Therefore, counter machines are too weak to compute interesting results in "reasonable" time and are thus disregarded in time complexity.[5]

Finally, for WHILE the unit-cost time measure is definitely not adequate as it would disregard the time it costs to build a tree used in an assignment. Consider for instance the assignment X:=[nil,nil,nil,...,nil] which stores a tree of arbitrary size (depending on how many nils we put in the list) in X in just one assignment. It would incur just the cost "1". This surely is not "fair" or adequate. We thus need to carefully consider the size of tree expressions in the commands of a WHILE-program.

12.2 Time Measure for WHILE

In this section we define a time measure for core WHILE. Any programs using extensions have to be translated into the core language first before we can analyse their runtime.

We first define a measure for expressions. The slogan is *"count operations and constants in expressions of commands as well"*. Note that for a fair measure of time complexity we exclude the equality "=" from expressions (equality is not in core WHILE). Any such comparison can be programmed already in WHILE using just comparisons with nil.[6] Therefore, we only have to define the time measure for programs in the original language, not the extensions of Chap. 5.

Definition 12.2 (*Time measure for* WHILE-*expressions*) The function $\mathcal{T}$ maps WHILE-expressions into the time it takes to evaluate them and thus has type:

$$\mathcal{T} : \text{WHILE-Expressions} \rightarrow \mathbb{N}$$

It is defined inductively according to the shape of expressions as follows:

$$
\begin{array}{lll}
\mathcal{T}\,\text{nil} & = 1 & \text{special atom} \\
\mathcal{T}\,\text{X} & = 1 & \text{where X is variable} \\
\mathcal{T}\,\text{hd}\,\text{E} & = 1 + \mathcal{T}\,\text{E} & \\
\mathcal{T}\,\text{tl}\,\text{E} & = 1 + \mathcal{T}\,\text{E} & \\
\mathcal{T}\,\text{cons}\,\text{E}\,\text{F} & = 1 + \mathcal{T}\,\text{E} + \mathcal{T}\,\text{F} &
\end{array}
$$

In other words, for atoms like nil and variables we require just 1 unit of time as the atom is a basic constant and lookup of a variable should not cost much either.

---

[5]They are however usable for space complexity considerations.

[6]See Exercise 7 in Chap. 5.

For the head hd and tail tl expressions, respectively, we measure the time for evaluating their argument(s) and the time for the operation itself which is 1. For the binary tree constructor, we count 1 for the constructor and add the time needed to evaluate the subexpressions E and F. So $\mathscr{T}$ E describes the time it takes to evaluate a WHILE-expression E in terms of units of time. Those units of time are given as natural numbers. Expressions do not have side effects so, although their value depends on the store, their time measure does not, thus their evaluation always terminates. Therefore, $\mathscr{T}$ is a *total* function.

Example 12.1 Consider the expression cons hd X tl nil. According to the equations of Definition 12.2 we compute:

$$
\begin{aligned}
\mathscr{T}\,\text{cons hd X tl nil} &= 1 + \mathscr{T}\,\text{hd X} + \mathscr{T}\,\text{tl nil} \\
&= 1 + (1 + \mathscr{T}\text{X}) + (1 + \mathscr{T}\text{nil}) \\
&= 1 + (1 + 1) + (1 + 1) \\
&= 5
\end{aligned}
$$

Next, we define a measure for WHILE-commands. We will write $\text{S}\vdash^{time}\sigma \Rightarrow t$ to indicate that it takes t units of time to run statement list S in store σ. In analogy with the semantics of WHILE (due to the possibility of non-termination), we describe the above relation as the smallest relation fulfilling some rules.

Definition 12.3 (*Time measure for WHILE-commands statement lists*) For a store σ the relation $\text{S}\vdash^{time}\sigma \Rightarrow t$ states that executing WHILE-statement list S in store σ takes t units of time. Deviating minimally from the grammar in Fig. 3.2, we allow statement lists to be empty here to avoid unnecessary distinction between empty and non-empty blocks. The relation $\bullet\vdash^{time}\_ \Rightarrow \_ \subseteq \text{StatementList} \times Store \times \mathbb{N}$ is defined as the smallest relation satisfying the rules in Fig. 12.1.

Let us look at the rules in Fig. 12.1 and explain them in more detail. First, we consider statement lists that just contain one command: for an assignment, we measure the time it takes to evaluate E and then add one to account for the time it takes to perform the assignment. In case of the conditional, we have to first measure the time it takes to evaluate E, add one for the time it takes to check whether it is nil or not and then add the execution time for the branch that is executed according to the result of E. As for the semantics of WHILE, we have two possible cases for the while loop: either the loop terminates or the body is executed (at least once). If for while E {S} it is the case that E evaluates to nil, we measure the time it takes to evaluate E and add one for the test of being nil. If, however, the guard evaluates to true (not nil) we need to unfold the loop (according to the semantics discussed earlier), so we measure again the time it takes to evaluate E, adding one for the equality test, but we now also have to add the time for executing the body S followed, again, by the entire loop.

For statement lists, we distinguish two cases. The empty statement list, ε, requires no computation whatsoever and thus costs zero units. For sequential composition of

$$\texttt{X := E} \qquad\qquad \vdash^{time} \sigma \Rightarrow t+1 \quad\ \text{if}\quad \mathscr{T}\texttt{E} = t$$

$$\texttt{if E \{S}_\texttt{T}\texttt{\} else \{S}_\texttt{E}\texttt{\}} \vdash^{time} \sigma \Rightarrow t+1+t' \ \text{if}\ \begin{array}{l}\mathscr{E}[\![\texttt{E}]\!]\sigma = \text{nil},\quad \mathscr{T}\texttt{E}=t \ \text{and}\\ \texttt{S}_\texttt{T} \vdash^{time} \sigma \Rightarrow t'\end{array}$$

$$\texttt{if E \{S}_\texttt{T}\texttt{\} else \{S}_\texttt{E}\texttt{\}} \vdash^{time} \sigma \Rightarrow t+1+t' \ \text{if}\ \begin{array}{l}\mathscr{E}[\![\texttt{E}]\!]\sigma \neq \text{nil},\quad \mathscr{T}\texttt{E}=t \ \text{and}\\ \texttt{S}_\texttt{E} \vdash^{time} \sigma \Rightarrow t'\end{array}$$

$$\texttt{while E \{S\}} \qquad \vdash^{time} \sigma \Rightarrow t+1 \quad \text{if}\ \ \mathscr{E}[\![\texttt{E}]\!]\sigma = \text{nil},\quad \mathscr{T}\texttt{E}=t$$

$$\texttt{while E \{S\}} \qquad \vdash^{time} \sigma \Rightarrow t+1+t' \ \text{if}\ \begin{array}{l}\mathscr{E}[\![\texttt{E}]\!]\sigma \neq \text{nil},\quad \mathscr{T}\texttt{E}=t \ \text{and}\\ \texttt{S;while E \{S\}} \vdash^{time} \sigma \Rightarrow t'\end{array}$$

$$\vdash^{time} \sigma \Rightarrow 0$$

$$\texttt{C;S} \qquad\qquad \vdash^{time} \sigma \Rightarrow t+t' \quad \text{if}\quad \begin{array}{l}\texttt{C} \vdash^{time} \sigma \Rightarrow t, \ \texttt{C} \vdash \sigma \to \sigma'\\ \text{and } \texttt{S} \vdash^{time} \sigma' \Rightarrow t'\end{array}$$

Fig. 12.1 Time measure for WHILE-commands

a command C and a statement list S, we define the time measure to be the added measure for executing C and S. In order to be able to measure the time for S we need to know the (intermediate) store in which to begin the execution, σ'. We get this from actually computing the result state of C run in σ.

Definition 12.4 (*Time measure for* WHILE-*programs*) For a WHILE-program $p = $ read X {S} write Y we define the time measure

$$time_\bullet^{\text{WHILE}} : \text{WHILE-program} \to \mathbb{D} \to \mathbb{N}_\perp$$

i.e. the time it takes p to run with input d as follows:

$$time_p^{\text{WHILE}}(d) = \begin{cases} t+2 \ \text{iff}\quad \texttt{S} \vdash^{time} \sigma_0^p(d) \Rightarrow t \\ \perp \quad \text{otherwise} \end{cases}$$

Note that the program argument is written as a subscript here to enhance readability.

So the time it takes for program p to run on input d is the time it takes its body S to execute in the initial state for input d plus one for reading the input plus one for writing the output. If the program p does not terminate on input d then the time measure $time_p^{\text{WHILE}}(d)$ is undefined.

Example 12.2 Consider the WHILE-program simple:

```
simple read X { Y := cons X X } write Y
```

We can compute the time measure $time^{\text{WHILE}}_{\text{simple}}(d)$ for any input $d \in \mathbb{D}$ as follows:

$$time^{\text{WHILE}}_{\text{simple}}(d) = t + 2 \quad \text{iff} \quad \text{Y} := \text{cons X X} \vdash^{\text{time}} \sigma^{\text{simple}}_0(d) \Rightarrow t$$

$$\text{Y} := \text{cons X X} \vdash^{\text{time}} \sigma^{\text{simple}}_0(d) \Rightarrow t + 1 \quad \text{iff} \quad \mathscr{T} \text{cons X X} = t$$

$$\mathscr{T} \text{cons X X} \;=\; 1 + \mathscr{T} \text{X} + \mathscr{T} \text{X} = 1 + 1 + 1 = 3$$

It follows that $time^{\text{WHILE}}_{\text{simple}}(d) = 2 + (3 + 1) = 6$ independently of the concrete shape of the initial state and thus independently of input d.

12.3   Comparing Programming Languages Considering Time

We can now equip a programming language L (which consists of syntax and semantics, see Definition 6.1) with a time measure and call the result a *timed* programming language.

Definition 12.5 (*Timed programming language*) A timed programming language consists of:

1. Two sets, namely L-programs and L-data
2. A function $[\![\_]\!]^{\text{L}} :$ L-programs $\rightarrow$ (L-data $\rightarrow$ L-data$_\perp$) that describes the semantics of L and
3. A function $time^{\text{L}}_\bullet :$ L-programs $\rightarrow$ (L-data $\rightarrow$ $\mathbb{N}_\perp$), the time measure for L, such that for every $p \in$ L-programs and $d \in$ L-data we have that $[\![p]\!]^{\text{L}}(d)\uparrow$ if, and only if, $time^{\text{L}}_p(d)\uparrow$.

Note that the last condition in (3) just prescribes the timing function (time measure) to be undefined on those inputs for which the semantics prescribes that the result is undefined (because the program does not terminate). So the time function returns undefined exactly for those inputs for which the program does not terminate. This makes sense since we cannot measure run time for non-terminating programs in a sensible way.

In Chap. 11 we discussed compilers and how programs could be compiled into equivalent programs of another language (computational model). If we can compile language L into language M (preserving program semantics), we can say that M simulates L. For every L-program we can get an M-program, by compilation, that "does the same job". Now we want to add timing information. To keep things simple, we assume that both languages have the same datatype.[7] We thus define the following simulation relations for languages:

---

[7] In cases where the data types are different one can still apply the same techniques but the translations get more complicated as one needs to encode data values and this is somewhat distracting. For details see e.g. [3].

Definition 12.6 (*Simulation relation*) Suppose we are given two *timed* programming languages L and M with L-data = M-data. We define:

1. L$\preceq^{ptime}$M if for every L-program p there is an M-program q such that $[\![p]\!]^{\mathrm{L}} = [\![q]\!]^{\mathrm{M}}$ and a polynomial $f(n)$ such that for all $d \in$ L-data

$$time_q^{\mathrm{M}}(d) \leq f(time_p^{\mathrm{L}}(d))$$

   In words: M can simulate L up to polynomial difference in time (Or alternatively: L can be simulated by M up to polynomial difference in time.)
2. L$\preceq^{lintime}$M if for every L-program p there is a constant $a_p \geq 0$ and an M-program q such that $[\![p]\!]^{\mathrm{L}} = [\![q]\!]^{\mathrm{M}}$ such that for all $d \in$ L-data

$$time_q^{\mathrm{M}}(d) \leq a_p \times time_p^{\mathrm{L}}(d)$$

   In words: M can simulate L up to linear difference in time. Accordingly, a_p is called the overhead factor. It can be less than 1 (speedup) or (more likely) greater than one (slowdown). The factor can depend on the program in question, hence the subscript p in a_p.
3. L$\preceq^{lintime-pg-ind}$M if there is a constant $a \geq 0$ and for every L-program p an M-program q such that $[\![p]\!]^{\mathrm{L}} = [\![q]\!]^{\mathrm{M}}$ such that for all $d \in$ L-data

$$time_q^{\mathrm{M}}(d) \leq a \times time_p^{\mathrm{L}}(d)$$

   In words: M can simulate L up to a program-independent linear time difference (or overhead factor).

The equations above can be explained in words as follows:

1. The time it takes for simulating the execution of program q with any input (worst-case analysis!) is less than or equal to a polynomial function of the time it takes the original program p to run with the same input.
2. The time it takes for simulating the execution of program q with any input (worst-case analysis!) is less than or equal to a constant program dependent factor a_p times the time it takes the original program p to run with the same input.
3. The time it takes for simulating the execution program q with any input (worst-case analysis!) is less than or equal to a constant program independent factor a times the time it takes the original program p to run with the same input. Observe how this establishes a *stronger*, i.e. more restrictive, definition of "simulation up to linear time", where the linear overhead factor does not actually depend on the program p in question, but only on the programming language itself.

We can now define a *simulation equivalence* between timed programming languages as simply the symmetric closure of simulation (just like equality is the symmetric closure of $\leq$):

Definition 12.7 (*Simulation equivalence*) Suppose we are given two *timed* programming languages L and M with L-data = M-data. We define:

1. $L \equiv^{lintime-pg-ind} M$ if, and only if, $L \preceq^{lintime-pg-ind} M$ and $M \preceq^{lintime-pg-ind} L$. In words: L and M are *strongly linearly equivalent*.
2. $L \equiv^{lintime} M$ if, and only if, $L \preceq^{lintime} M$ and $M \preceq^{lintime} L$. In words: L and M are *linearly equivalent*.
3. $L \equiv^{ptime} M$ if, and only if, $L \preceq^{ptime} M$ and $M \preceq^{ptime} L$. In words: L and M are *polynomially equivalent*.

What Next?

We have seen how to measure time usage of programs and how to compare languages w.r.t. their expressivity considering also the run time of programs. Having done that, we can try to define classes of programs with similar runtime "needs" as well as classes of problems that can be decided in "similar time". In the next chapter we will do exactly that and use the above definitions in order to describe complexity *classes*.

Exercises

1. Consider the program `myloop` in Fig. 12.2. What is its time measure $time^{WHILE}_{myloop}(d)$ for any input $d \in \mathbb{D}$?
2. Assume L and M are timed programming languages. Assume further that $p \in L^{time(f)}$ and that $M \succeq^{lintime} L$. It follows that $p \in M^{time(g)}$ for some g. Give the definition of a suitable g and explain why it satisfies $p \in M^{time(g)}$.
3. Consider the program `reverse` from Fig. 3.4 in Sect. 3.4.2: What is its time measure, $time^{WHILE}_{reverse}(d)$, for any input $d \in \mathbb{D}$?
4. Derive cost measures for extended WHILE-features by means of translation. Measure the cost of the translated features and use this information to define the measure for the extension directly.

   a. Macro calls: i.e. define $X := <p>E \vdash^{time} \sigma \Rightarrow t$ where p is a program of which one knows the time measure $time^{WHILE}_p$ and E is an expression.
   b. List expressions, i.e. define $\mathscr{T}[\, E_1, \dots E_n\,]$ where all E_i are expressions.

5. Consider program `size` in Fig. 12.3 that computes the size of the input tree: What is the time measure $time^{WHILE}_{size}(d)$ for all $d \in \mathbb{D}$?
   Hint: use the fact that a binary tree with n leaves must have $n - 1$ inner nodes $\langle l.r \rangle$. Moreover, variable `Stack` *is used to keep those subtrees of the input that still need visiting, and each such subtree is popped from the stack only once inside the while loop. Consider the worst case scenario as we are using worst-case complexity.*

```
myloop read X {
  while X { X:= tl X }
}
write Y
```

Fig. 12.2 Sample program for Exercise 1

```
size read X {
   Stack := cons X nil;
   s := nil;
   while Stack {
      c := hd Stack;
      Stack := tl Stack;
      if  c   { Stack := cons hd c  cons tl c Stack }
      else    { s := cons nil s  }
   }
}
write s
```

Fig. 12.3 Program size for Exercise 5

6. Show that all three timed simulation relations $\preceq^{ptime}$, $\preceq^{lintime}$, and $\preceq^{lintime-pg-ind}$ are transitive. Recall that a binary relation $\_ \circ \_$ is transitive if, and only if, $x \circ y$ and $y \circ z$ implies that $x \circ z$ for all appropriate x, y, z.
7. Develop the logarithmic time measure for RAM that accounts for the size of numbers in registers. The size is measured in the number of digits, i.e. logarithmically. Consider the computation of addresses for the instructions with indirect addressing. More about this can be found in [1].

References

1. Cook, S.A., Reckhow, R.A.: Time-bounded random access machines. J. Comput. Syst. Sci. **7**, 354–375 (1973)
2. Hartmanis, J., Stearns, R.E.: On the computational complexity of algorithms. Trans. Am. Math. Soc. **117**, 285–306 (1965)
3. Jones, N.D.: Computability and Complexity: From a Programming Perspective. MIT Press, Cambridge (1997) (Also available online at http://www.diku.dk/neil/Comp2book.html.)

Chapter 13
Complexity Classes

How do we classify problems according to the time it takes to solve them? What are the relationships between those classes and how do we prove them?

In the previous chapter we have seen how to measure the runtime of programs. We wish to extend this notion to problems. We classify problems according to the time it takes to decide them. We will focus on decision problems as time keeping in those cases does not have to account for the time to construct a solution. Any limits for decision problems will therefore automatically also be limits for function problems (in the sense of Definition 2.13).

One can define a class of problems decidable by a program of a certain language (a certain type of "effective procedure") in a certain amount of time. For that, one needs time bounds, i.e. upper bounds of runtime w.r.t. input size (Sect. 13.1). Then one can abstract away from the concrete time bound and use time bounds of a certain type: linear, polynomial, exponential. The problem classes defined in Sect. 13.2 will allow us to differentiate between problems in a precise, but not too fine-grained way. In Sect. 13.2 relationships between complexity classes that use different kinds of "effective procedures" are established by straightforwardly "lifting" simulation relations (see Definition 12.6) between languages to relations between complexity classes.

In Sect. 13.4 we will consider a specific notation for runtimes that abstracts away from little details and can replace the notion of "up to a constant factor" by something more precise. Thanks to this abstraction one can focus on runtime of *algorithms* rather than on specific programs that encode algorithms. The runtime of programs may depend on certain choices of language, style, data structures which can have some important effect on the runtime, but *asymptotically* will always be dwarfed by the slow-down caused by enormously large input.

© Springer International Publishing Switzerland 2016
B. Reus, *Limits of Computation*, Undergraduate Topics in Computer Science,
DOI 10.1007/978-3-319-27889-6_13

13.1   Runtime Bounds

Runtime bounds are total functions from natural numbers to natural numbers. They describe the time (i.e. number of time units) a program may take at most for a specific input in terms of the size of the input. Note that $\bot \not\le n$ for any n, so programs with a time bound f are necessarily always terminating.[1] As we are interested in *worst-case complexity,* we consider upper bounds.

Since we only look at *asymptotic time complexity* we observe how the time bound (and thus the time needed at most by a program) grows when the input grows. In Fig. 13.1 several functions are plotted that describe such time bounds. According to the above, we are interested in the behaviour when we move (far) right on the x-axis. The time needed is supposed to increase, the question is by how much?

Figure 13.1 highlights that logarithmic and linear time bounds are good and exponential ones rather bad. In the following, recall that we abbreviate by $|d|$ the size of a data value $d \in$ L-data. Recall also that we have defined this function for $\mathbb{D}$ explicitly in Definition 3.2.

Definition 13.1 (*programs with bounds*) Given a timed programming language L and a total function $f : \mathbb{N} \to \mathbb{N}$, we define four classes (sets) of *time bounded programs*:

1. $\text{L}^{time(f)} = \{p \in \text{L-programs} \mid time_p^{\text{L}}(d) \le f(|d|) \text{ for all } d \in \text{L-data}\}$
   This is the set of programs that have a runtime that is bounded by function f.
2. $\text{L}^{ptime} = \bigcup_{p \text{ is a polynomial}} \text{L}^{time(p(n))}$
   This set is the union of all set of L programs that have a polynomial function p as time bound. Recall that a polynomial function is of the form

$$c_k \times n^k + c_{k-1} \times n^{k-1} + \cdots + c_2 \times k^2 + c_1 \times k + c_0.$$

3. $\text{L}^{lintime} = \bigcup_{k \ge 0} \text{L}^{time(k \times n)}$
   This set is the union of all set of L programs that have a linear function f as time bound. *Linear functions* are of the form $f(n) = k \times n$.
4. $\text{L}^{exptime} = \bigcup_{k \ge 0} \text{L}^{time(2^{p(n)})}$
   Where $p(n)$ is a polynomial. This set is the union of all set of L programs that have an exponential function f as time bound. Exponential functions are of the form $f(n) = 2^{p(n)}$ so we allow not only n in the exponent but any polynomial of n as well.

In this chapter we focus on polynomial time. Why do we do this? Why don't we like exponential time bounds for algorithms? And what about linear time?

Let us have a closer look at the growth of linear, polynomial, and exponential functions. Assume that you have various algorithms for a problem and that a unit in the unit cost measure is in real time 10^{-9} s. Consider how much time it takes (what the result of the time bound functions is) if we consider functions $f(n) = n$, $f(n) = n^2$, $f(n) = n^3$, $f(n) = 2^n$ and $f(n) = 3^n$. Table 13.1 from [2] shows the

---

[1]This is of course sufficient since we are interested in runtime of decision procedures.

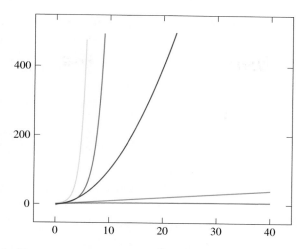

Fig. 13.1 Graph of $\log_2 x$ (*magenta*), x (*green*), x^2 (*blue*), 2^x (*red*), 3^x (*yellow*)

Table 13.1 Growth of various time bounds

size = n	n (μs)	n^2 (μs)	n^3	2^n	3^n
10	0.01	0.1	0.01 ms	1 μs	59 μs
30	0.03	0.9	24.3 ms	1 s	2.4 days
50	0.05	2.5	0.31 s	13 days	2.3×10^5 centuries

time bound expressed in the corresponding "real time" (instead of natural numbers) to convey a better feeling for the large numbers involved.[2]

We have only used polynomials with small exponents. But this is somewhat justifiable as for most known (useful) algorithms with polynomial runtime, the polynomials have a small degree. The degree of the time bound corresponds often to the depth of a nested loop, and there is only a certain depth of nesting needed (and comprehensible). It is obvious that exponential time bounds are too generous and do not guarantee good runtime behaviour. But there is another reason why polynomial time is considered in complexity theory which will be discussed in the next chapter.

13.2   Time Complexity Classes

Now that we have defined classes of programs with time bounds, we are interested in defining classes of problems that can be decided within certain time bounds.

One problem we run into here is that different languages, machine models, i.e. models of computation, use different data-types. But we want to compare those classes of problems so we need to unify the domain of discourse of those problems. Let us simply take the words over alphabet {0, 1} as this domain, i.e. {0, 1}*. In a

---

[2]It seems humans can much better grasp large amounts of time than large natural numbers. Maybe this is due to evolution?

way they correspond to words of bits which is how information is stored on modern
devices anyway. It is also clear that Turing machines take such words as inputs
(restricting the alphabet accordingly). In languages with binary tree type $\mathbb{D}$ we need
to encode those words as lists of $\ulcorner 0 \urcorner$ and $\ulcorner 1 \urcorner$ which is not a problem. The word 010
corresponds then to the list $[\,0, 1, 0\,]$ or $[\,\text{nil}, [\text{nil}, \text{nil}], \text{nil}\,]$. For register machines we
can interpret lists of 0 and 1 as binary numbers and represent them in the decimal
system as natural numbers.

Below we define time complexity classes for problems. We use boldface short
abbreviations to denote complexity classes of problems as it is standard in the literature.[3]

Definition 13.2 (*complexity classes*) Given a timed programming language L and a
total function $f : \mathbb{N} \to \mathbb{N}$, we define four (complexity) classes of problems:

1. The class of *problems L-decidable in time f* is:
   $\mathbf{TIME}^{L}(f) = \{A \subseteq \{0, 1\}^* \mid A \text{ is decided by some } p \in L^{time(f)}\}$
   In other words, this is the class of all problems (or sets) A about words over
   alphabet $\{0, 1\}$ that are decided by an L-program with a runtime that is bounded
   by time bound (function) f.
2. The class $\mathbf{P}^{L}$ of *problems L-decidable in polynomial time* is:
   $\mathbf{P}^{L} = \{A \subseteq \{0, 1\}^* \mid A \text{ is decided by some } p \in L^{ptime}\}$
   In other words, this is the class of all problems (or sets) A about words over
   alphabet $\{0, 1\}$ that are decided by an L-program with a runtime that is bounded
   by a polynomial.
3. The class $\mathbf{LIN}^{L}$ of *problems L-decidable in linear time* is:
   $\mathbf{LIN}^{L} = \{A \subseteq \{0, 1\}^* \mid A \text{ is decided by some } p \in L^{lintime}\}$
   In other words, this is the class of all problems (or sets) A about words over
   alphabet $\{0, 1\}$ that are decided by an L-program with a runtime that is bounded
   by a linear function.
4. The class $\mathbf{EXP}^{L}$ of *problems L-decidable in exponential time* is:
   $\mathbf{EXP}^{L} = \{A \subseteq \{0, 1\}^* \mid A \text{ is decided by some } p \in L^{exptime}\}$
   In other words, this is the class of all problems (or sets) A about words over
   alphabet $\{0, 1\}$ that are decided by an L-program with a runtime that is bounded
   by an exponential function.

We often drop the superscript indicating the language in question and simply use
LIN, **P**, or **EXP** if we implicitly understand which programming language we are
referring to.

Another justification to drop the superscript would be if we were allowed not
to care which programming language we use. In the next section we will justify
exactly that. We will look at the "robustness" of the class **P**, i.e. how $\mathbf{P}^{L}$ relate for
different languages L, and we will meet yet another thesis that is similar to the
Church–Turing Thesis.

From Definition 13.2 it follows immediately for a fixed language L:

---

[3]In [1] the classes are called PTIME, LINTIME, NPTIME (there is no EXPTIME), respectively.

Proposition 13.1 *For any notion of "effective procedure"* L, *it holds that*

$$LIN^{\text{L}} \subseteq P^{\text{L}} \subseteq EXP^{\text{L}}$$

13.3   Lifting Simulation Properties to Complexity Classes

Proposition 13.1 tells us the simple relationship between complexity classes of problems decided using the same language or notion of computation. Next, we would like to establish relationships between complexity classes that use different kinds of effective procedures. It turns out we can use the simulation relations from Definition 12.6 and lift them straightforwardly to relations between complexity classes.

Lemma 13.1 L $\preceq^{lintime}$ M *implies* $LIN^{\text{L}} \subseteq LIN^{\text{M}}$, *and as a consequence,* L $\equiv^{lintime}$ M *implies* $LIN^{\text{L}} = LIN^{\text{M}}$.

   This means that if L *can be simulated by* M *up to linear time difference then every problem in* LIN^{L} *is already in* LIN^{M}. *And then, as a consequence, that if* L *and* M *are linearly equivalent then* LIN^{L} *and* LIN^{M} *are the same.*

Proof Let us prove this (as it is not difficult): Assume that we have a problem $A \in \textbf{LIN}^{\text{L}}$. This means there is some L-program that decides A in linear time, the latter means more precisely that $time_p^{\text{L}}(d) \leq c \times |d|$ for some constant $c \geq 0$ and all $d \in \mathbb{D}$. By definition L $\preceq^{lintime}$ M implies that there exists an M-program q such that $[\![q]\!]^{\text{M}} = [\![p]\!]^{\text{L}}$, and program q is at most a linear factor slower than p, in other (more precise) words: $time_q^{\text{M}}(d) \leq a_p \times time_p^{\text{L}}(d)$ for some (program dependent) $a_p \geq 0$ and all data $d \in \mathbb{D}$.
Now combining those two results about the runtime of q we get for all d that:

$$time_q^{\text{M}} \leq a_p \times time_p^{\text{L}}(d) \leq a_p \times c \times |d|$$

From this follows (as $\leq$ is an order, in particular transitive)

$$time_q^{\text{M}} \leq (a_p \times c) \times |d|$$

which demonstrates that program q has a linear time bound, with linear factor $a_p \times c$. Secondly, q decides A as $[\![q]\!]^{\text{M}} = [\![p]\!]^{\text{L}}$ and p decides A. Thirdly, q is an M-program, so putting all this information about q together we have shown that $A \in \textbf{LIN}^{\text{M}}$ according to the definition of $\textbf{LIN}^{\text{M}}$. And therefore we have shown that $\textbf{LIN}^{\text{L}} \subseteq \textbf{LIN}^{\text{M}}$.
For the second part assume L $\equiv^{lintime}$ M which by definition means L $\preceq^{lintime}$ M and M $\preceq^{lintime}$ L. Applying the first result to both these simulation statements, we obtain $\textbf{LIN}^{\text{L}} \subseteq \textbf{LIN}^{\text{M}}$ as well as $\textbf{LIN}^{\text{M}} \subseteq \textbf{LIN}^{\text{L}}$ which implies $\textbf{LIN}^{\text{M}} = \textbf{LIN}^{\text{L}}$.

   We get a similar result for polynomial time:

Lemma 13.2 L $\preceq^{ptime}$ M *implies* $P^{\text{L}} \subseteq P^{\text{M}}$, *and as a consequence,* L $\equiv^{ptime}$ M *implies* $P^{\text{L}} = P^{\text{M}}$.

The proof is left as Exercise 10.

The next little result describes how the different simulation relations are related to each other:

Lemma 13.3 $L \preceq^{lintime} M$ *or* $L \preceq^{lintime-pg-ind} M$ *implies* $L \preceq^{ptime} M$.

Proof We can prove the lemma as follows: we distinguish two cases according to the "or" in the assumption. First $L \preceq^{lintime} M$. We need to show $L \preceq^{ptime} M$. Unfolding the definition we need to show that for any L-program p there is an M-program q and a polynomial f such that that $[\![q]\!]^M = [\![p]\!]^L$, and $time_q^M(d) \leq f(time_p^L(d))$ for all data $d \in \mathbb{D}$. Unfolding the assumption we have such a required q with $[\![q]\!]^M = [\![p]\!]^L$ but with $time_q^M(d) \leq a \times time_p^L(d)$ for a constant $a \geq 0$. If we choose $f(n) = a \times n$ this yields the requested time constraint $time_q^M(d) \leq f(time_p^L(d))$. But $f(n) = a \times n$ is clearly a polynomial (of degree 1). Actually, every linear function is by default also a polynomial.

The case for $L \preceq^{lintime} M$ is done analogously but with the exception that the constant a is program-independent.

Lemma 13.4 $L \equiv^{lintime} M$ *implies* $\boldsymbol{P}^L = \boldsymbol{P}^M$.

Proof This is a simple consequence of the previous two lemmas as $L \preceq^{lintime} M$ implies $L \preceq^{ptime} M$ which, in turn, implies $\boldsymbol{P}^L \subseteq \boldsymbol{P}^M$. If we consider both directions, we get the desired result.

13.4 Big-O and Little-o

Computer Science students usually come across the so-called "*Big-O*" notation,[4] short $\mathcal{O}$, in their favourite algorithms and data structures course. Here "$\mathcal{O}$" stands for the "order of growth rate" of a function, ignoring (discounting) constant factors and only looking at what the function does in the limit (ignoring the initial values on the "left" of the x-axis so to speak). When measuring runtime it makes sense to disregard the concrete time measure of a program as this depends on many little factors like style of coding or even optimisations. Abstracting away from such constants allows one to focus on how the runtime increases with the size of the input independent of minor implementation decisions or from the speed of a specific processor or machine that is used to run the program. This will become most helpful in the following chapters.

In our quest for infeasible problems, constant factors are only relevant for small values anyway. For instance, it makes a big difference to us whether a program runs for 1 or 50 seconds. However, if we consider large input and thus long running

---

[4]This notation as well as "little-o" discussed later, is by *Edmund Georg Hermann Landau* (14 February 1877–19 February 1938), a German mathematician who worked mainly in the field of analytic number theory. "$\mathcal{O}$" stands for order.

times, it won't make any difference to us whether a program runs for 1 century or 50 centuries, both are equally useless to us.

The "Big-O" notation will be useful to indicate the quality of an algorithm in terms of asymptotic worst-case runtime.

We say $g(n) \in \mathcal{O}(f(n))$, or equivalently $g \in \mathcal{O}(f)$, to mean that $g(n)$ approximately (up to a constant factor) grows at most as quickly as $f(n)$. So g can grow even more slowly than f.

Definition 13.3 *(Big-O)* Let $f : \mathbb{N} \to \mathbb{N}$ be a function. The *order of* f, short $\mathcal{O}(f)$, is the set of all functions defined below:

$$\{g : \mathbb{N} \to \mathbb{N} \mid \forall n > n_0.\ g(n) \leq c \times f(n) \text{ for some } c \in \mathbb{N}\backslash\{0\} \text{ and } n_0 \in \mathbb{N}\}$$

In other words, $\mathcal{O}(f)$ are those functions that up to constant factors grow at most as fast as f (or, in other words, not faster) and are thus at least "not worse" a runtime bound than f (maybe even "better"). For $g \in \mathcal{O}(f)$ we also say g is $\mathcal{O}(f)$.

In Definition 13.3 the n_0 specifies the upper bound for small numbers ignored for the purpose of asymptotic complexity considerations, and c allows one to ignore constant differences that in the limit are negligible as well. Recall that $\lim_{n \to \infty} \frac{c}{n} = 0$ whatever constant c is.

Example 13.1 Consider the following examples of $\mathcal{O}(\_)$:

1. n is $\mathcal{O}(n^2)$ as it grows at most like n^2. In this case the constant factor can be one as obviously $n \leq n^2$ for all n.
2. $3n^2$ is $\mathcal{O}(n^2)$, as it grows like n^2 up to a constant factor. This can be seen by setting n_0 from Definition 13.3 to 0 and c to 3.
3. $34n^2 + 23n + 12$ is $\mathcal{O}(n^2)$ as it grows like n^2 up to a constant factor. This can be seen by setting n_0 from Definition 13.3 to 0 and c to 69. This suffices as we have $34n^2 + 23n + 12 \leq 69n^2$ for all $n > 0$ because $35n^2 \geq 23n + 12$ for $n \geq 1$.

The $\mathcal{O}(\_)$ notation is widespread to explain the complexity of algorithms. Also there, constant factors are ignored since they might depend on details of the coding of the algorithm or the language it is implemented in.

Example 13.2 A very famous example is sorting where the input is an array of data values and its size is usually measured as the length of the array. It is well known that for worst case complexity, merge sort is in $\mathcal{O}(n \log_2 n)$ whereas quick sort is in $\mathcal{O}(n^2)$. For average case complexity both are in $\mathcal{O}(n \log_2 n)$. We will only be considering worst case complexity in the following.

For a concrete program p one can give a concrete time bound like $time_p^{\text{WHILE}}(2n^2)$ but for *algorithms* this is difficult as they can be implemented by various programs that differ slightly. The "$\mathcal{O}(\_)$" notation will take care of this. On the one hand, when talking about algorithms, we might therefore say something like

(The runtime of) quick sort is $\mathcal{O}(n^2)$.

where n is understood to be the size of the array to be sorted (i.e. the input). But on the other hand, when talking about concrete programs, we are able to say

The WHILE-program p that performs quick sort is in WHILE$^{time(9n^2+37)}$.

although we might only be interested in the fact that

The WHILE-program p that performs quick sort is in WHILE$^{time(f)}$ were $f(n) \in \mathscr{O}(n^2)$.

From these examples we see the advantage of allowing the use of "Big-O" and would like to make it available to the definition of complexity class **TIME**:

Definition 13.4 (**TIME**(f)) We define another complexity class using "Big-O" as follows:

$$\textbf{TIME}(\mathscr{O}(f)) = \bigcup_{g \in \mathscr{O}(f)} \textbf{TIME}(g)$$

Lemma 13.5 *If $g \in \mathscr{O}(f)$ then* **TIME**$(g) \subseteq$ **TIME**$(\mathscr{O}(f))$.

Proof Exercise 14.

To express the dual concept of $\mathscr{O}(\_)$, namely that a function g grows asymptotically *much faster* than a function f, one uses the "Little-o" notation defined as follows:

Definition 13.5 (*Little-o*) Let f and g be functions of type $\mathbb{N} \to \mathbb{N}$. Then $o(g)$ are those functions that eventually grow much slower than f. Formally we can define this as follows:

$f \in o(g)$ iff for all $0 < \varepsilon \in \mathbb{R}$ there exists $N \in \mathbb{N}$ s.t. $\varepsilon \times g(n) \geq f(n)$ for all $n \geq N$

The above definition is equivalent to

$$f \in o(g) \iff \lim_{n \to \infty} \frac{f(n)}{g(n)} = 0$$

which shows that f grows ultimately much more slowly than g, or in other words, g grows faster than f, as the limit of the quotient $\frac{f(n)}{g(n)}$ approaches zero when n grows very large.

Example 13.3 Consider the following examples of $o(\_)$:

1. n is in $o(n^2)$ as it grows much more slowly than n^2. This follows from the fact that $\lim_{n \to \infty} \frac{n}{n^2} = \lim_{n \to \infty} \frac{1}{n}$ which is 0.
2. $3n^2$ is *not* in $o(n^2)$, as it does not grow more slowly. This can be seen by looking at the limit $\lim_{n \to \infty} \frac{3n^2}{n^2}$ which is 3 and thus not 0.
3. n^2 is *not* in $o(34n^2 + 23n + 12)$ as $\lim_{n \to \infty} \frac{n^2}{34n^2+23n+12} = \frac{1}{34}$ and thus not 0.

Sometimes we know more than just an upper bound for the runtime of algorithms. For instance, we might know that upper and lower asymptotic bounds up to some multiplicative constants coincide. In this case we know that there is, asymptotically speaking, no better algorithm than the one we have. In this situation, one uses the Θ (pronounce "Big-Theta") notation:

Definition 13.6 (*Big-Θ*) Let $f : \mathbb{N} \to \mathbb{N}$ be a function. The *asymptotically tight bound of* f, short $\Theta(f)$, is

$$\left\{ g : \mathbb{N} \to \mathbb{N} \;\middle|\; \begin{array}{l} \forall n > n_0.\, c_1 \times f(n) \le g(n) \le c_2 \times f(n) \\ \text{for some } c_1, c_2 \in \mathbb{R}\backslash\{0\} \text{ and } n_0 \in \mathbb{N} \end{array} \right\}$$

In other words $\Theta(f)$ are those functions that grow approximately *at the same rate* as f.

Example 13.4 For instance, $34n^2+23n+12$ is in $\Theta(n^2)$ as $n^2 \le 34n^2 + 23n + 12 \le 69n^2$ for all $n \in \mathbb{N}$. In this case the n_0 of the definition Definition 13.6 is 0, c_1 is 1 and c_2 is 69.
Similarly $2n + \sqrt{n} \in \Theta(n)$ as $n \le 2n + \sqrt{n} \le 3n$. In this case n_0 of the definition Definition 13.6 is 0, c_1 is 1 and c_2 is 3.

The Θ notation is also widespread explaining the complexity of algorithms. Constant factors are again ignored since they might depend on details of the coding of the algorithm or the language it is implemented in.

Note that $\mathcal{O}(\_)$, $o(\_)$ and $\Theta(\_)$ can be applied to average complexity as well as to worst-case complexity.[5]

What Next?

We have defined time complexity classes and shown some results regarding how to transfer simulation results of computational models to the complexity classes for those different models. This allows us now to analyse the robustness of computability within certain time bounds by comparing the same complexity class in different models of computation. In the next chapter we will do this and meet an "extended version of the Church–Turing thesis" that involves time complexity.

Exercises

1. Consider the program `simple` from Example 12.2 in Chap. 12. Give a time bound for `simple`, i.e. a function f such that for any input $d \in \mathbb{D}$:

$$\text{time}^{\text{WHILE}}_{\text{simple}}(d) \le f(|d|)$$

2. Give a time bound for the program `myloop` from Exercise 1 in Chap. 12 where its runtime measure has been given.

---

[5]As the reader may already have guessed there is also a notation for lower bounds Ω (called "Big-Omega"), dual to $\mathcal{O}(\_)$, but we don't need it for our considerations.

3. Give a time bound for the program `size` from Exercise 5 in Chap. 12 where a time measure has been given. Explain why this time bound is in $\mathcal{O}(n)$.
4. Consider the WHILE-program `simple` from Exercise 1 in Chap. 12, where a time bound has been given already. In which of the following classes is program `simple`?

    a. $\text{WHILE}^{\text{time}(2n)}$
    b. $\text{WHILE}^{\text{time}(4n)}$
    c. $\text{WHILE}^{\text{lintime}}$
    d. $\text{WHILE}^{\text{ptime}}$

5. Give a time bound for the program `reverse` from Fig. 3.4 in Sect. 3.4.2. Its time measure has been computed in Exercise 3 Chap. 12.
6. Consider the program `reverse` from Fig. 3.4 in Sect. 3.4.2, for which a time bound has been given in Exercise 5. In which of the following classes is program `reverse`?

    a. $\text{WHILE}^{\text{time}(2n)}$
    b. $\text{WHILE}^{\text{time}(10n)}$
    c. $\text{WHILE}^{\text{lintime}}$
    d. $\text{WHILE}^{\text{ptime}}$

7. Show that $\mathbf{LIN}^{\text{L}} \subseteq \mathbf{P}^{\text{L}}$ for any L.
8. What is the difference between WHILE^{ptime} and $\mathbf{P}^{\text{WHILE}}$?
9. Show that $\mathbf{P}$ is closed under union and intersection.
10. Explain the following statements by unfolding their definition:

    a. Explain the meaning of $\mathbf{LIN}^{\text{L}}$ and $\mathbf{P}^{\text{L}}$.
    b. Explain the meaning of $\text{L}_1 \preceq^{\text{ptime}} \text{L}_2$.
    c. Explain the meaning of $\mathbf{P}^{\text{L}_1} \subseteq \mathbf{P}^{\text{L}_2}$.

11. Prove Lemma 13.2, i.e. prove that $\text{L} \preceq^{ptime} \text{M}$ implies $\mathbf{P}^{\text{L}} \subseteq \mathbf{P}^{\text{M}}$, and as a consequence, $\text{L} \equiv^{ptime} \text{M}$ implies $\mathbf{P}^{\text{L}} = \mathbf{P}^{\text{M}}$.
12. Consider the following functions of type $\mathbb{N} \to \mathbb{N}$.

    a. $f(n) = 3n^2 \log n$
    b. $f(n) = n^3 + n^2 + 1$
    c. $f(n) = 12n^2 + 146$
    d. $f(n) = 12n^5$
    e. $f(n) = 2^n$
    f. $f(n) = 34n + 34$

    a. Which of the functions above is in $\mathcal{O}(n^2)$?
    b. Which of the functions above is in $o(n^3)$?
    c. Which of the functions above is in $\Theta(n^2)$?

13. Analyse the size of the different encodings of natural numbers in $\mathbb{D}$ described in Sect. 3.3.3:

   a. Let $\ulcorner n \urcorner$ be the unary encoding of n. Compute $|\ulcorner n \urcorner|$ in terms of n. Provide a function f such that $|\ulcorner n \urcorner| \in \mathcal{O}(f(n))$.

   b. Let $\ulcorner n \urcorner^{10}$ be the decimal encoding of n. Compute $|\ulcorner n \urcorner^{10}|$ in terms of n. Provide a function f such that $|\ulcorner n \urcorner^{10}| \in \mathcal{O}(f(n))$.

   c. Let $\ulcorner n \urcorner^2$ be the binary encoding of n. Compute $|\ulcorner n \urcorner^2|$ in terms of n. Provide a function f such that $|\ulcorner n \urcorner^2| \in \mathcal{O}(f(n))$.

14. Explain why multiplication on unary numbers (see Exercise 8 in Chap. 5) has the following time complexity:

$$\text{time}_{\text{mult}}^{\text{WHILE}}(\ulcorner [m, n] \urcorner) \in \mathcal{O}(|\ulcorner m \urcorner| \times |\ulcorner n \urcorner|)$$

Is mult in $\text{WHILE}^{\text{lintime}}$?

15. Prove Lemma 13.5, i.e. show that if $g \in \mathcal{O}(f)$ then $\textbf{TIME}(g) \subseteq \textbf{TIME}(\mathcal{O}(f))$.

References

1. Jones, N.D.: Computability and Complexity: From a Programming Perspective. MIT Press, Cambridge (1997) (Also available online at http://www.diku.dk/neil/Comp2book.html.)
2. Ong, C.-H.L.: Computational Complexity. Oxford University Programming Laboratory (1999) Available via DIALOG http://citeseerx.ist.psu.edu/viewdoc/download?doi=10.1.1.27.3433&rep=rep1&type=pdf. Accessed 10 June 2015

Chapter 14
Robustness of P

> *How robust is computability with polynomial time bounds? Is polynomially computable the same as feasibly computable?*

In the previous chapter we have seen how to measure the runtime of programs, and how to define complexity classes of problems accordingly. It is natural to ask what runtime is feasible or tractable. Which classes contain only problems that are feasible? It is also natural to ask "For complexity issues, does it matter which model of computation is used?" These questions will be addressed in this chapter with a specific focus on polynomial time computability. It will turn out that there is some evidence to consider polynomial time "good", i.e. programs that run in polynomial time are feasible. The picture is, however, not that black and white.

Robustness is an invariance property we have already discussed for computability in Chap. 11. What is computable (or decidable) is invariant, it does not depend on the model of computation chosen (see Church–Turing thesis). Now we consider whether time-bounded computation is (in a sense to be specified) invariant too.

We need a refined definition of robustness. Neil Jones suggests criteria for resource-bounded problem solving. He writes [8, Chap. 18, p. 271]

Ideally, resource-bounded problem solvability should be:

1. *invariant with respect to choice of machine model; (i.e. model of computation)*
2. *invariant with respect to size and kind of resource bound; (e.g. quadratic time etc.); and*
3. *invariant with respect to the problem representation; (e.g. the choice for the data structure to represent the problem domain, for instance directed graph by an incidence matrix or by adjacency lists should not make a complexity difference)*

The first item states that the defined complexity classes should be robust under compilation between models of computations (machine models/languages). And we will show this in this section for the models introduced previously (see Chap. 11).

© Springer International Publishing Switzerland 2016
B. Reus, *Limits of Computation*, Undergraduate Topics in Computer Science,
DOI 10.1007/978-3-319-27889-6_14

The second item simply says that complexity classes should be invariant under instances of the chosen kind of resource bound. It should not matter, for instance, *which* specific version of a quadratic time function is chosen if one defines a class for quadratic time.

The following three sections present three well-known (under various differing names) theses about computability in polynomial time. First of all, note that they are *theses* and not theorems. They cannot be proven. The reason is that they all refer to a formally undefined class of feasible or tractable problems and/or undefined notions of computation. This is, of course, intentional.

14.1 Extended Church–Turing Thesis

The first thesis we will consider is the so-called Extended Church–Turing Thesis:

Definition 14.1 (*Extended Church–Turing Thesis*) All notions of computation can simulate each other up to a polynomial factor.

In [1] the extended thesis reads as follows

> The Extended Church–Turing Thesis says that the time it takes to compute something on any one machine is polynomial in the time it takes on any other machine.

whereas in [2, p. 31] it is phrased

> Any function naturally to be regarded as *efficiently* computable is *efficiently* computable by a Turing machine.

The last version is more of a corollary of the first two, when identifying "efficiently computable" with "polynomial time computable" and identifying Turing machine computable as reference point.

There is significant doubt about this (i.e. all versions of this) extended thesis because highly parallel computer architectures may efficiently compute what a (sequential) Turing machine can't. We get back to this in Chap. 23.

14.2 Invariance or Cook's Thesis

As the *Extended Church–Turing Thesis* discussed above may not hold, there is a weaker version of the *Extended Church–Turing Thesis* for which we can find plenty of good evidence.

Definition 14.2 (*Invariance Thesis*) All "reasonable" *sequential* notions of computation can simulate each other up to a polynomial factor.

This is known as Cook's[1] *Thesis* in [8, p. 242] which phrases it as follows:

$\mathbf{P}^{\mathrm{L}}$ is the same class of problems for all reasonable sequential (that is, nonparallel) computational models L.

This means that whatever (reasonable) sequential computational model we use to decide problems, the difference in running time will be a polynomial factor. Similarly to the Church–Turing thesis, we can provide some evidence for this thesis, which is, again, widely believed in computer science. We can give evidence by showing that the various instances of **P** for those models we defined formally are equal. In other words that

$$\mathbf{P}^{\mathrm{SRAM}} = \mathbf{P}^{\mathrm{TM}} = \mathbf{P}^{\mathrm{GOTO}} = \mathbf{P}^{\mathrm{WHILE}} = \mathbf{P}^{\mathrm{WH^1LE}} = \mathbf{P}^{\mathrm{CA}}$$

It is important here that the notions of computations (models of computation) are *restricted to sequential* ones. This excludes *true parallel computation* that needs special treatment in complexity which we will address briefly below in Sect. 14.2.1. The cellular automata are not excluded since we stipulate that their seed is always finite.

14.2.1 Non-sequential Models

Until recently, most conventional computing devices used one CPU only, implementing a sequential model of computation. The effect of "concurrent tasks or threads" was a mere illusion produced by the operating system, more precisely its *scheduler*, that gives away time slices to various processes. If the CPU is fast enough and the time slices are fairly distributed, it seems to the human observer as if those processes were actually running in parallel.[2] Modern computers now use multi-core technology so there are a number of CPUs involved in computations (but usually a very small number like four, for instance, which obviously can ever only produce a constant speed-up and is thus not interesting for asymptotic complexity considerations.) There is also a growing trend in using *GPU*s (Graphical Processing Unit) and so-called *GPGPU*s (General Purpose Graphical Processing Unit) which are available in

---

[1]Stephen A Cook (born December 14, 1939) is a world renowned American–Canadian mathematician and theoretical computer scientist who proved some fundamental results in computational complexity. He won the Turing Award for his seminal work on Computational Complexity in 1982. We will encounter more of his work in following chapters.

[2]Compare this with the effect of moving still images very quickly—i.e. at a rate of about 16 images per second (modern cinema uses 24 per second)—in front of human observers, so that they think they see a moving picture (hence the term "motion picture" or "movie").

each modern day computer and can also be clustered, for parallel execution of easily parallelisable tasks like vector or matrix computations:

> If you were ploughing a field, which would you rather use: two strong oxen or 1024 chickens?
>
> Seymour Cray[3]

Also the emerging field of *distributed systems* where several computers (or agents) communicate via message passing appears to be the system architecture of the future. The study of *distributed algorithms* and programs is still in its infancy, at least when compared to sequential algorithms. Unfortunately, it is not even very obvious what the notion of "solving a problem by a distributed algorithm" should mean, let alone what the right notion of their runtime complexity should be. Distributed algorithms are algorithms that try to achieve one goal by distributing a task across an entire network of host computers. The main challenge is the coordination of work among all the nodes in the network. Such algorithms are common in networking and networking applications. The network is usually modelled by a graph. For an overview of the field consult, e.g., the textbooks [6, 10] and for the theory behaviour of distributed systems [7]. By contrast, *parallel algorithms* communicate directly via shared memory and not message passing. There is, for instance, a parallel version of RAM, called PRAM for "**P**arallel Random Access Machine", that can serve as a computational model for studying complexity of parallel systems with shared memory. There are many versions restricting the concurrent use of reads and writes. For an overview see e.g. [5].

14.2.2   Evidence for Cook's Thesis

We will show some auxiliary lemmas that relate the simulation property between languages to the inclusion relation between complexity classes. Once we have those lemmas, we can show inclusion (and equality) between complexity classes by timed simulation (or equivalence) between languages and thus by providing compilers that make certain guarantees about the runtime behaviour of their target programs w.r.t. source programs.

To provide evidence for Cook's thesis, we look at the various models of computation and use the compilers discussed in Chap. 11. We analyse the runtime of the compiled program and compare it to the runtime of the original program, obtaining statements of the form $L \preceq^{lintime} M$ or $L \preceq^{ptime} M$. With the auxiliary lemmas we can then derive results like

$$\mathbf{P}^{L} \subseteq \mathbf{P}^{M} \qquad \mathbf{LIN}^{L} \subseteq \mathbf{LIN}^{M}$$

---

[3]Seymour Roger Cray (September 28, 1925–October 5, 1996) "the father of supercomputing" was an American electrical engineer who founded Cray Computer Corporation and designed many of the fastest supercomputers between 1976 and 1995.

Listed below are the main results without detailed proof. The time differences between source and target program can be derived by inspecting the code that the compilers generate. More details can be found in [8] again.

Lemma 14.1

$$TM \preceq^{lintime-pg-ind} GOTO \preceq^{lintime-pg-ind} SRAM \preceq^{ptime} TM$$

Proof The result of compiling a Turing machine program into a GOTO program results in a GOTO program that is at most a constant (program-independent) linear factor slower than the original Turing machine program. Similarly, any GOTO program can be compiled into an SRAM program with only a (program-independent) constant slowdown. Compiling an SRAM program, however, into a Turing machine program will produce a Turing machine program that has a time bound that differs from the one of the original program by a polynomial. This means that the resulting Turing machine program may be significantly slower. This is not surprising as the Turing machine head must move across the tape to find the right encoded registers and register values (represented in a binary encoding on the Turing machine tape). By carefully monitoring how far the head moves in the worst case in terms of the size of the input,[4] one observes polynomial slow-down.

Lemma 14.2

$$TM \preceq^{lintime} CA \preceq^{ptime} TM$$

Proof It was shown by [11] (where details of the construction can be found) that a Turing machine can be simulated in linear time by a (one-dimensional) cellular automaton. A general cellular automaton with k-dimensional cell space can be simulated by a TM as the seed pattern is finite. Assume the CA in question uses $T(n)$ steps. The question is how large the non-blank cell space becomes that initially is, say, n^d large where d is the dimension of the automaton. In each step the non-blank cell space can only grow by one cell in each direction of each dimension. Thus after $T(n)$ steps the grid that is not blank covers at most the following number of cells:

$$\left(n + \sum_{k=1}^{T(n)} 2 \right)^d = (n + 2T(n))^d$$

If we assume each update of one cell takes time c, altogether for the simulating TM we obtain a runtime of $(n + 2T(n))^d \times c \times T(n)$ which describes a polynomial slowdown.

Theorem 14.1 *It holds that*

$$P^{CA} = P^{TM} = P^{GOTO} = P^{SRAM}$$

---

[4]See Exercise 6.

Proof From Lemma 14.1 with the auxiliary result Lemma 13.3 about simulation relations we can conclude that

$$\text{TM} \preceq^{ptime} \text{GOTO} \preceq^{ptime} \text{SRAM} \preceq^{ptime} \text{TM}$$

and thus

$$\text{TM} \equiv^{ptime} \text{GOTO} \equiv^{ptime} \text{SRAM} \equiv^{ptime} \text{TM} .$$

From Lemma 14.2 with the auxiliary result Lemma 13.3 about simulation relations we can conclude that TM $\preceq^{ptime}$ CA and thus TM $\equiv^{ptime}$ CA. By Lemma 13.2 we therefore get

$$\mathbf{P}^{\text{CA}} = \mathbf{P}^{\text{TM}} = \mathbf{P}^{\text{GOTO}} = \mathbf{P}^{\text{SRAM}}.$$

14.2.3 Linear Time

In Chap. 11 we have seen that we can compile between GOTO, WHILE, and WH^{1}LE. Let us now investigate the effect of those compilations on run time.

Lemma 14.3
$$GOTO \ \preceq^{lintime-pg-ind} WHILE \preceq^{lintime} WH^1 LE$$
$$WH^1 LE \preceq^{lintime-pg-ind} WHILE \preceq^{lintime} GOTO$$

Proof WH^{1}LE $\preceq^{lintime-pg-ind}$ WHILE holds trivially (with linear factor 1) as each WH^{1}LE program is already a WHILE program. WHILE $\preceq^{lintime}$ WH^{1}LE is discussed in Exercise 3 and GOTO $\preceq^{lintime-pg-ind}$ WHILE in Exercise 4. To see that WHILE $\preceq^{lintime}$ GOTO we need to recall from Sect. 11.8.2 how the compilation works. Since GOTO is quite similar to WHILE there are only two issues to address:

- Compile away the while-loops: while E {S} is replaced by a conditional jump plus an extra unconditional jump at the end of loop body S back to the instruction with the conditional jump. Obviously the extra jump adds only constant time per while-loop.
- Compile away complex expressions, "splitting" them into GOTO assignments that can add one operator on each expression only. So a complex expression E with $\mathscr{T} E = n$ uses at most $n - 1$ operators hd, tl or cons (see Exercise 5). To compile expression E into a GOTO program it needs to be built up using additional assignments X:=hd X, X:=tl X, or X:=cons X Y, which each cost 3 or 4 time units to execute. So instead of evaluating an expression E with $\mathscr{T} E = n$, we execute at most $n - 1$ assignments which never cost more than 4 time units. Therefore, instead of n units we spend at most $4n - 4$ units thus the slowdown remains linear.

With the help of previous observations we can also derive the following which describes to what extent **LIN** is "robust" under some compilation:

Theorem 14.2

$$LIN^{GOTO} = LIN^{WHILE} = LIN^{WH^1LE}$$

This proves that **LIN** is not as robust as **P** or one might argue it is not robust at all, since when compiling into Turing machines, for instance, one might lose linear time bounds. This also is evidence for the special role that **P** has in complexity theory.

14.3 Cobham–Edmonds Thesis

The Cobham–Edmons thesis is also known as *Cook-Karp thesis* in various textbooks, see e.g. [2, Chap. 1.2]:

Definition 14.3 (*Cobham–Edmons Thesis*) The tractable (feasible) problems are exactly those in **P**.

Again, this is only a thesis, because "tractable problems" is not a formally defined class. The statement is quite a strong one for which there is much less evidence than for Cook's thesis. It is true that polynomial functions have a much more benign rate of growth than exponential functions and as such are more welcome as time bounds. But there arise two questions:

1. Is every polynomial time bound really a good time bound indicating feasibility?
2. Is every time bound beyond polynomial really a bad time bound indicating intractability?

Well, one might answer both questions with "no". Regarding (1), consider polynomials with a very high degree, e.g. $f(n) = n^{20}$. Is a problem with this time bound really tractable? If we consider Table 13.1 from Sect. 13.1 and add $f(n) = n^{20}$ then for $n = 10$ we obtain a value of 31.68 *centuries*! Surely this is not feasible, and taking a larger exponent like n^{100} even less so. So in practice, "polynomial" is not automatically a synonym for "good". On the other hand, one could argue that the argument (1) is not relevant because one does not know any algorithms that have a polynomial time bound where the degree is higher than 5 or 6.

Regarding (2), there are algorithms with exponential worst case complexity, like the *simplex algorithm*. It was discovered in 1947 by George Dantzig[5] [4] for *linear programming* and on average it performs much better than polynomial algorithms. The simplex algorithm solves a problem that we discuss in more detail in Sect. 16.6. On the other hand, one could argue that argument (2) is not relevant because we

---

[5]George Bernard Dantzig (November 8, 1914–May 13, 2005), "the father of linear programming", was an American mathematician who made important contributions in particular to operations research and statistics. In statistics, Dantzig solved two open problems, which he had mistaken for homework after arriving late to a lecture. He told this story himself in [3, p. 301]. Apparently, this was also the basis for the Hollywood movie "Good Will Hunting".

are interested in worst-case complexity, not average-case, and because an alternative polynomial algorithm has been found [9] for linear programming.

In any case, the Cobham–Edmonds Thesis, also known as Cook–Karp Thesis, is slightly controversial.

What Next?

We now know that polynomial time is often identified with feasibility so we understand that certain problem classes are better than others. But how does this look when we consider individual functions as time bounds? Can we solve more problems with a larger time bound? Even if both time bounds are polynomial, or linear? This is the topic of the next chapter.

Exercises

1. Explain the difference between the various theses discussed in this chapter.
2. Show that from Theorems 14.1 and 14.2 it follows that $\mathbf{P}^{\text{WHILE}} = \mathbf{P}^{\text{GOTO}}$.
3. In Chap. 7 Exercise 7, a WHILE to WH^{1}LE compiler was developed. Explain why the following statement from Lemma 14.3 holds:

$$\text{WHILE} \preceq^{lintime} \text{WH}^1\text{LE}$$

   In other words, give a proof sketch for the part of Theorem 14.2 that says that $\mathbf{LIN}^{\text{WHILE}} = \mathbf{LIN}^{\text{WH}^1\text{LE}}$.
4. In Chap. 11 Exercise 6 a compiler from GOTO to WHILE was discussed. Explain why the following statement from Lemma 14.3 holds:

$$\text{GOTO} \preceq^{lintime-pg-ind} \text{WHILE}$$

5. Show that any "core" WHILE-expression E (without equality or literals) with $\mathscr{T}\text{E} = n$ contains at most $n - 1$ operators hd or tl, and at most $\frac{n-1}{2}$ operators cons.
   Hint: Prove this by induction on n.
6. In Sect. 11.8.2 a compiler from SRAM to TM has been very briefly sketched. Without carrying out a formal proof, provide (sound and convincing) arguments that the slowdown by this compilation is polynomial at most. In other terms give evidence for

$$\text{SRAM} \preceq^{\text{ptime}} \text{TM}$$

7. From results shown in this chapter derive that $\mathbf{EXP}^{\text{WHILE}} = \mathbf{EXP}^{\text{TM}}$.

References

1. Aaronson, S.: Quantum Complexity Theory Lecture 1. Available via DIALOG http://ocw.mit.edu/courses/electrical-engineering-and-computer-science/6-845-quantum-complexity-theory-fall-2010/lecture-notes/MIT6_845F10_lec01.pdf. Cited on 12 June 2015
2. Aaronson, S.: Quantum Computing since Democritus. Cambridge University Press, Cambridge (2013)
3. Albers, D.J., Reid, C.: An interview with George B. Dantzig: the father of linear programming. Coll. Math. J. **17**(4), 293–314 (1986)
4. Dantzig, G.B.: Linear Programming. Problems for the Numerical Analysis of the Future, Proceedings of Symposium on Modern Calculating Machinery and Numerical Methods, UCLA, 29–31 July 1948. Applied Mathematics Series, vol. 15, pp. 18–21. National Bureau of Standards (1951)
5. Eppstein, D., Galil, Z.: Parallel algorithmic techniques for combinatorial computation. Annu. Rev. Comput. Sci. **3**, 233–283 (1988)
6. Fokkink, W.: Distributed Algorithms–An Intuitive Approach. MIT Press, Cambridge (2013)
7. Hennessy, M.: A Distributed Pi-Calculus. Cambridge University Press, Cambridge (2007)
8. Jones, N.D.: Computability and complexity: From a Programming Perspective[6]. MIT Press, Cambridge (Also available online at http://www.diku.dk/~neil/Comp2book.html) (1997)
9. Karmarkar, N.: A new polynomial-time algorithm for linear programming. Combinatorica **4**(4), 373–395 (1984)
10. Lynch, N.A.: Distributed Algorithms. Morgan Kaufmann, San Francisco (1996)
11. Smith III, A.R.: Simple computation-universal cellular spaces. J. ACM **18**(3), 339–353 (1971)

Chapter 15
Hierarchy Theorems

> *Can you solve more problems with larger time bounds? How much larger do they need to be?*

Previously, we have defined classes of problems that can be decided (solved) by programs that run with certain time bounds. For instance, class $\mathbf{TIME}^L(f)$ is the class of problems that can be decided by an L-program with time bound f. The latter means that for a problem A in this class the question whether some $d \in$ L-data is in A can be decided by running an L-program p such that the running time is bounded by $f(|d|)$, i.e. $time_p^L(d) \leq f(|d|)$. Note that we decided that for all complexity classes of problems we fixed L-data to be (suitably encoded) words over $\{0, 1\}$.

An obvious question that now arises is: what is the hierarchy of those complexity classes. Is the class $\mathbf{TIME}^L(f)$ bigger than $\mathbf{TIME}^L(g)$ (i.e. $\mathbf{TIME}^L(f) \supsetneq \mathbf{TIME}^L(g)$) if f is, in some appropriate sense, "bigger" than g?[1] In other, words,

> Can we decide more problems if we have more time available to do so?

This is a very natural question to ask and the so-called *Hierarchy Theorems* give us some answers. We will look at linear, time hierarchies for WH^1LE in Sect. 15.1 and asymptotic hierarchies for all languages in Sect. 15.2. There appear to be certain gaps in the hierarchy as briefly discussed in Sect. 15.3.

One does encounter a significant challenge if one wants to show that a certain problem is *not* in a specific complexity class, as this this requires a proof that there is *no* algorithm that decides the problem with a given time bound or a time bound of a given kind. How can one confidently prove such a statement that involves reasoning about *all* algorithms? In other words, how can one prove lower complexity bounds? A lower bound states what asymptotic run time is at least needed to solve a problem. This turns out to be really tricky. According to Cook [2] "the record for proving lower bounds on problems of smaller complexity is appalling." This statement refers to natural problems.

---

[1]Note that $A \subsetneq B$ means that A is a *proper* subset of B in the sense that it is not equal to B but properly contained in A. In other words, B has more elements than A.

© Springer International Publishing Switzerland 2016
B. Reus, *Limits of Computation*, Undergraduate Topics in Computer Science,
DOI 10.1007/978-3-319-27889-6_15

Luckily, for *Hierarchy Theorems* the situation is slightly better, as one only needs
to construct one problem for which one can prove a lower bound and this problem does
not have to be *natural*. Moreover, a proof technique encountered already in previous
chapters, *diagonalisation*, will save the day once again. It is possible to construct
a program that involves runtime information with the help of a self-interpreter that
counts runtime. Using diagonalisation, i.e. a form of self-application, one can then
define the problem that shows the *Hierarchy Theorem* in question to be the one
decided by this particular program.

15.1   Linear Time Hierarchy Theorems

Let us first consider more in detail what "larger" means for time bounds. If we talk
about linear time bounds, i.e. linear functions, then it should be obvious that function
$f(n) = a \times n$ is "larger" than $g(n) = b \times n$ if $a > b$. Therefore, for linear bounds,
we can reformulate the above question.

> Do constants matter for linear time complexity (classes)?

We can thus rephrase the original question for the special cases of linear runtime as
follows

> Can L-programs decide more problems with a larger constant in their linear running time
> allowance?

or, more formally

$$\text{Does } a < b \text{ imply } \mathbf{TIME}^{\text{L}}(a \times n) \subsetneq \mathbf{TIME}^{\text{L}}(b \times n) \text{ ?}$$

or equivalently

$$\text{Does } a < b \text{ imply } \mathbf{TIME}^{\text{L}}(b \times n) \setminus \mathbf{TIME}^{\text{L}}(a \times n) \neq \emptyset \text{ ?}$$

We can answer this question for the language WH^1LE (recall that this is WHILE where
programs can only have one variable). In order to answer it we need to consider
universal programs with timeout parameter, called *timed universal programs*:

Definition 15.1 (*Timed universal program*) A WH^1LE program tu is a *timed universal program* if for all $p \in \text{WH}^1\text{LE}$-program, $d \in \mathbb{D}$ and $n \geq 1$:

1. If $time_p^{\text{WH}^1\text{LE}}(d) \leq n$ then $[\![\text{tu}]\!]^{\text{WH}^1\text{LE}}[p, d, n] = [\,[\![p]\!]^{\text{WH}^1\text{LE}}(d)\,]$
2. If $time_p^{\text{WH}^1\text{LE}}(d) > n$ then $[\![\text{tu}]\!]^{\text{WH}^1\text{LE}}[p, d, n] = \text{nil}$

This definition states that the effect of $[\![\text{tu}]\!]^{\text{WH}^1\text{LE}}[p, d, n]$ is to simulate p for either
$time_p^{\text{WH}^1\text{LE}}(d)$ or n steps, whatever is the smaller. If $time_p^{\text{WH}^1\text{LE}}(d) \leq n$, i.e. p terminates
within n steps, then tu produces a result that is not nil, namely a single element list

that contains as its only element p's result. If not, then `tu` produces nil indicating *"time limit exceeded"*. So a timed universal program has a built-in stopwatch. It is given an additional timeout parameter, and when the number of steps simulating the argument program has exceeded the timeout parameter, the interpreter stops and returns nil. As usual, we drop the encoding brackets around lists for the sake of readability.

Definition 15.2 (*Efficient timed universal program*) Assume L is a timed programming language with programs as data and pairing. A timed universal L-program `tu` is *efficient* if there is a constant k such that for all $p, d \in$ L-data and $n \geq 1$:

$$time_{\mathtt{tu}}^{\mathtt{L}}[p, d, n] \leq k \times \min\,(n,\,time_p^{\mathtt{L}}(d))$$

In other words, interpreting p using an efficient timed universal program is only a constant factor k slower than the running time of p where k is a *program-independent* constant.

The timed universal program still aborts program execution after the timeout parameter has been reached. That also means that the runtime of a timed universal program is bounded by the third input parameter (interpreted as a number) times some program-independent constant k.

Figure 15.1 shows how a timed universal program for $\mathtt{WH^1LE}$ can be programmed as a `WHILE`-program `tw`. The `STEP` macro is the one used already for the universal program in Sect. 7.1.2. We make good use of the `switch` statement here to check whether an actual time consuming step has been executed (which could also be translated away to get a pure `WHILE`-program).

Lemma 15.1 *Let p be a WH^1LE-program and $d \in \mathbb{D}$ such that $time_p^{WH^1LE}(d) = n$. During the execution of `tu` with input $[\,p, d\,]$ the STEP macro will be called at most $k \times n$ times where k is a constant.*

Proof Details can be found in [1] (or Exercise 1).

Theorem 15.1 *There is an efficient timed universal WH^1LE program.*

Proof To prove this we take the `WHILE`-program `tw` from Fig. 15.1 and translate it into a $\mathtt{WH^1LE}$ program which is always possible. The resulting universal program shall be called `tu`. It remains to show that `tu` is efficient. So we need to find two constants k_1 and k_2 such that $time_{\mathtt{tu}}^{\mathtt{WH^1LE}}([p, d, n]) \leq k_1 \times n$ in case the interpreter times out, and $time_{\mathtt{tu}}^{\mathtt{WH^1LE}}([p, d, n]) \leq k_2 \times time_p^{\mathtt{WH^1LE}}(d)$ in case the interpreter computes the result within the allowed time interval. Then one chooses the required linear factor k to be the maximum of k_1 and k_2. The first inequality simply follows from the construction as after having executed one of the special atoms { `quote`, `var`, `doHd`, `doTl`, `doCons`, `doAsgn`, `doIf`, `doWhile` } n times, the counter is zero and `tu` terminates. The second equation can be shown by close inspection of how the universal program interprets the original argument program via the `STEP` macro. By revisiting Sect. 7.1.2 one can convince oneself that each rewriting rule implemented

```
tw read X {                          (* X = [p, d, niln] *)
        B := hd tl hd X;             (* code block to be executed *)
        val := hd tl X;              (* initial value for variable *)
        Cntr := hd tl tl X;          (* Time bound *)
        DSt := nil;                  (* initial data stack *)
        CSt := B;                    (* initial code stack *)
        state := [CSt,DSt,val];(* initial state *)
        while CSt {
          if Cntr {                  (* continue interpretation *)
            switch hd hd CSt {
            case quote, var,
                 doHd, doTl,
                 doCons, doAsgn,
                 doIf, doWhile : { Cntr := tl Cntr }
                 }
            state := <STEP> state;
            CSt := hd state;
            val := hd tl tl state;
            X := [ val ]             (* wrap up potential result *)
            }                        (* end if *)
          else {                     (* out of time so stop *)
                CSt := nil;
                X := nil             (* and return nil *)
                }
          }                          (* end of while loop *)
}
write X
```

Fig. 15.1 Timed universal program tw for WH^{1}LE in WHILE

in STEP can be executed in constant time independent of size and shape of its input parameters. Furthermore, Lemma 15.1 gives the time bound $k_2 \times time_p^{\text{WH}^1\text{LE}}(d)$ for some k_2.

Now we are able to prove a concrete version of the question:

$$\text{Does } a > b \text{ imply } \mathbf{TIME}^{\text{L}}(a \times n) \subsetneq \mathbf{TIME}^{\text{L}}(b \times n)?$$

instantiating L to be WH^{1}LE.

Theorem 15.2 (Linear Time Hierarchy Theorem [5, Theorem 19.3.1]) *There is a constant b such that for all $a \geq 1$, there is a problem A in* $\mathbf{TIME}^{WH^1LE}(a \times b \times n)$ *that is not in* $\mathbf{TIME}^{WH^1LE}(a \times n)$.

This states the existence of a set (or problem) A that can be decided in linear time with bound $f(n) = (a \times b) \times n$, but not with time bound $f(n) = a \times n$. Therefore, in order to decide A, the bigger constant $a \times b$ is sufficient but a is insufficient.

Proof To show the existence of such a problem, we construct one by *diagonalisation*, a concept we have already used before (for the proof of the undecidability of the Halting problem).

Let us first sketch the idea of this proof. We define A via diagonalisation such that the following hold:

1. A is decidable by a $\mathrm{WH}^1\mathrm{LE}$ program $\texttt{decA}$
2. $\texttt{decA} \in \mathrm{WH}^1\mathrm{LE}^{\mathrm{time}(a \times b \times n)}$ for some b, in other words, $A \in \mathbf{TIME}^{\mathrm{WH}^1\mathrm{LE}}(a \times b \times n)$
3. if $A \in \mathbf{TIME}^{\mathrm{WH}^1\mathrm{LE}}(a \times n)$ with a decision procedure $p \in \mathrm{WH}^1\mathrm{LE}^{\mathrm{time}(a \times n)}$, then we get a contradiction if we ask $p \in A$ using p itself as decision procedure (diagonalisation), so we ask what $[\![p]\!]^{\mathrm{WH}^1\mathrm{LE}}(p)$ is.

In order to immediately ensure the first condition, we use a program, called $\texttt{diag}$ depicted in Fig. 15.2 to directly define A as the set accepted by $\texttt{diag}$. The use of the timed universal program (or self-interpreter) is essential here. As discussed earlier, we can compile $\texttt{diag}$ into a $\mathrm{WH}^1\mathrm{LE}$ program $\texttt{decA}$ and thus define

$$A = \{ d \mid [\![\texttt{decA}]\!]^{\mathrm{WH}^1\mathrm{LE}}(d) = \mathrm{true} \}$$

Some comments about $\texttt{diag}$ are in order: first of all, the computation of $a \times |\mathtt{X}|$ needs to be performed at runtime do as only at runtime do we know the value of variable $\mathtt{X}$. The computation makes use of macro calls. Program $\texttt{size}$ is supposed to compute the size of the argument and $\texttt{mult}$ computes the multiplication of two (encoded) natural numbers. The timed universal program for $\mathrm{WH}^1\mathrm{LE}$, $\texttt{tu}$, is also called per macro. It is worth recalling that while an interpreter might not terminate, a timed interpreter always does since it has a timeout parameter. Consequently, program $\texttt{diag}$, and thus $\texttt{decA}$, always terminate.

Next, we show the third condition, that $A \notin \mathbf{TIME}^{\mathrm{WH}^1\mathrm{LE}}(a \times n)$ by a diagonal argument. As before, we carry out a proof by contradiction and assume $A \in \mathbf{TIME}^{\mathrm{WH}^1\mathrm{LE}}(a \times n)$. By definition of class $\mathbf{TIME}^{\mathrm{WH}^1\mathrm{LE}}(a \times n)$ this means that there is a $\mathrm{WH}^1\mathrm{LE}$ program p that decides membership in A and also satisfies $\mathrm{time}_p^{\mathrm{WH}^1\mathrm{LE}}(d) \leq a \times |d|$ for all $d \in \mathbb{D}$. We now ask the question:

```
diag  read X {
  N := <size> X;
  Timebound := <mult> [N, "a"];
  Arg := [X, X, Timebound];
  X := <tu> Arg;          (* Run X on X for up to a×|X| steps *)
  if hd X {               (* if there is enough time *)
      X := false }        (* and result is true: return false *)
  else {                  (* if there is not enough time *)
      X := true }         (* or result is false: return true *)
}
write X
```

Fig. 15.2 Program $\texttt{diag}$ for diagonalisation

What happens if we run p on p itself (as we have done before and which is the reason why we call this a diagonal argument)?

In other words, we ask what is $[\![p]\!]^{\mathrm{WH}^1\mathrm{LE}}(p)$?

We use the fact that by assumption $time_p^{\mathrm{WH}^1\mathrm{LE}}(p) \leq a \times |p|$ which means that $\mathtt{tu}$ has enough time to simulate p on input p. By definition of the efficient universal program this means that:

$$[\![\mathtt{tu}]\!]^{\mathrm{WH}^1\mathrm{LE}}([\,p,p,a\times|p|\,]) = [\,[\![p]\!]^{\mathrm{WH}^1\mathrm{LE}}(p)\,]$$

Now assume that $[\![p]\!]^{\mathrm{WH}^1\mathrm{LE}}(p) = \mathsf{false}$. Then, by definition of $\mathtt{decA}$ we know that $[\![\mathtt{decA}]\!]^{\mathrm{WH}^1\mathrm{LE}}(p) = \mathsf{true}$. But then $\mathtt{decA}$ and p have different *semantics* and thus they can't both decide membership of A. In this case, the assumption must be wrong that p decides the membership of A and thus the assumption $A \in \mathbf{TIME}^{\mathrm{WH}^1\mathrm{LE}}(a \times n)$ must be wrong. Consequently, it must be true that $[\![p]\!]^{\mathrm{WH}^1\mathrm{LE}}(p) = \mathsf{true}$. But in this case, by definition of $\mathtt{decA}$ again, we know that $[\![\mathtt{decA}]\!]^{\mathrm{WH}^1\mathrm{LE}}(p) = \mathsf{false}$. But then, by the same argument as above, it must be wrong that p decides the membership of A and thus the assumption $A \in \mathbf{TIME}^{\mathrm{WH}^1\mathrm{LE}}(a \times n)$ must be wrong.

So in either case $A \in \mathbf{TIME}^{\mathrm{WH}^1\mathrm{LE}}(a \times n)$ must be wrong and therefore $A \notin \mathbf{TIME}^{\mathrm{WH}^1\mathrm{LE}}(a \times n)$.

It remains to show the second condition, namely that $A \in \mathbf{TIME}^{\mathrm{WH}^1\mathrm{LE}}(a \times b \times n)$ for some b. This is shown by first analysing the runtime of A's decision procedure $\mathtt{diag}$ which is a WHILE-program. We do this step by step:

- Assignment $\mathtt{N:=<size>\ X}$ requires the execution of $\mathtt{size}$ which is in $\mathcal{O}(|p|)$ (see Exercise 3 in Chap. 13) as $\mathtt{X}$ is $|p|$. Therefore, the assignment takes time $1 + k_s \times |p|$ for some constant k_s.
- Assignment $\mathtt{Timebound:=<mult>[N,"a"]}$: here $\mathtt{N}$ is $|p|$, a is a constant, and we know that macro $\mathtt{mult}$ runs in time $\mathcal{O}(|p| \times a)$ (see Exercise 13 in Chap. 13). Therefore, the time this assignment takes is $1 + k_m \times a \times |p|$ for some constant k_m.
- Assignment $\mathtt{X:=<tu>\ Arg}$: we recall that $\mathtt{tu}$ is efficient, so we can estimate the time it takes to execute this assignment is $1 + k_t \times \min(a \times |p|, time_p^{\mathrm{WH}^1\mathrm{LE}}(p)) \leq k_t \times a \times |p|$ for some constant k_t.
- The conditional takes either time $1 + 2 + 2$ (false) or $1 + 2 + 3$ (true) which is at most 6.

Putting now all the commands in $\mathtt{diag}$ together we can estimate:

$$time_{\mathtt{diag}}^{\mathrm{WHILE}}(p) \leq 2 + (1 + k_s \times |p|) + (1 + k_m \times a \times |p|) + (k_t \times a \times |p|) + 6$$

or, adding up the constants,

$$time_{\mathtt{diag}}^{\mathrm{WHILE}}(p) \leq (k_s + (k_m + k_t) \times a) \times |p| + 10$$

Since $|p| \geq 1$ (remember that `nil` has size 1) we obtain:

$$time_{\mathtt{diag}}^{\mathtt{WHILE}}(p) \leq a \times (k_s + k_m + k_t + 10) \times |p|$$

Let $c = k_s + k_m + k_t + 10$. From Lemma 14.3 we know that $\mathtt{WHILE} \preceq^{\text{lintime}} \mathtt{WH^1LE}$ so we can conclude for `decA` that:

$$time_{\mathtt{decA}}^{\mathtt{WHILE}}(p) \leq k' \times (a \times c \times |p|) \quad \leq \quad a \times (k' \times c) \times |p|$$

So the required b is $k' \times c$.

More details, as well as results for other variants of language $\mathtt{WH^1LE}$, can be found in [1]. Dahl and Hesselund [4] have computed b precisely for a specific version of WHILE and a specific time measure (which is is similar to ours), but b must necessarily differ for different variants of WHILE.

A rather intriguing open problem that remains [5] is whether a similar Linear Hierarchy Theorem can be proven for languages WHILE and GOTO? Note that for the latter languages we cannot apply the proof technique used above, since we do *not* get an *efficient* timed self-interpreter for those languages in the same way. The reason is that the number of variables in use is not program-independent but has an impact on the running time of the universal interpreter due to variable lookup (see Exercise 3).

Note that, however, for SRAM we could prove an analogous theorem (with either the unit-cost model or a logarithmic time model). For more information about that see [5, Chap. 19].

15.2   Beyond Linear Time

Let us now generalise the Hierarchy Theorem from linear time bounds to arbitrary ones that go beyond linear time. The original question we tackle in this chapter is now rephrased like so:

> Let f and g be time bound functions that are not both linear (as previously). Can L-programs decide more problems with time bound f than with time bound g if f grows (asymptotically) faster than g?

The *asymptotic complexity* is built into the kind of complexity theory we have been using so far. Time bounds describe how running time grows when input data size grows to an arbitrarily large amount. This is fair enough since our investigation about feasibility of problems concerns large inputs.

Let f and g be time bounds, i.e. functions of type $\mathbb{N} \to \mathbb{N}$. We have already encountered a definition (and notation) for f growing (much) faster than g, namely $g \in o(f)$ in Definition 13.5. It should be pointed out that $g \in o(f)$ does not make any statement about whether $f(n) > g(n)$ for small n. But if $g \in o(f)$ we know that if n gets really big (in other words, the size of the input of the problem at hand becomes really big) then $f(n)$ is significantly greater than $g(n)$.

We first consider $\mathtt{L} = \mathtt{WH^1LE}$.

Since we generalise from linear time bounds to arbitrary ones, there is one additionally condition in this version of the Hierarchy Theorem, namely that f and g are *time constructible*. What does that mean? Informally, f is time constructible means that we can "find out how much time $f(n)$ is available by a computation not taking more than the order of $f(n)$ steps" [5, Sect. 19.5]. This is an intuitive restriction and many familiar functions are time constructible (in particular linear functions, polynomials, exponentials, and arithmetic combinations of those, see Exercise 5).

Definition 15.3 A time bound $f : \mathbb{N} \to \mathbb{N}$ is called *time constructible* if there is a program p and a constant $c > 0$ such that for all $n \geq 0$ we have that:

$$\llbracket p \rrbracket \left(\ulcorner n \urcorner \right) = \ulcorner f(n) \urcorner \quad \text{and} \quad time_p(d) \leq c \times f(|d|) \quad \text{for all } d$$

where $\ulcorner n \urcorner$ denotes the encoding of number n in the datatype of the language used. This definition works for any programming language as every language should be able to encode natural numbers (thus the language superscripts have been omitted).

Non-computable functions (like the Busy-Beaver function) cannot be time constructible. Also, we need to have enough time to produce the result, thus sub-linear (small growing) functions are often not time constructible. In the linear Hierarchy Theorem 15.2 this condition was implicit as linear functions are automatically time constructible.

Theorem 15.3 (Asymptotic Hierarchy Theorem $\mathtt{WH^1LE}$) *If functions f and g are time constructible, $f(n) \geq n$, $g(n) \geq n$ and $g \in o(f)$, then it holds that:*

$$\mathbf{TIME}^{WH^1LE}(\mathscr{O}(f)) \setminus \mathbf{TIME}^{WH^1LE}(\mathscr{O}(g)) \neq \emptyset$$

Proof We proceed analogously to the Linear Time Hierarchy Theorem 15.2, i.e. we need to find a problem A such that:

1. A is decidable by a $\mathtt{WH^1LE}$ program $\mathtt{decA}$
2. $\mathtt{decA} \in \mathtt{WH^1LE}^{time(b \times f(n))}$, in other words $A \in \mathbf{TIME}^{WH^1LE}(b \times f(n))$ and thus $A \in \mathbf{TIME}^{WH^1LE}(\mathscr{O}(f))$
3. if $A \in \mathbf{TIME}^{WH^1LE}(\mathscr{O}(g))$ with a decision procedure $p \in \mathtt{WH^1LE}^{time(\mathscr{O}(g))}$, then we get a contradiction if we ask $p \in A$ using decision procedure p, thus asking actually $\llbracket p \rrbracket^{WH^1LE}(p) = \mathtt{true}$?

To obtain the appropriate $\mathtt{decA}$, we modify $\mathtt{diag}$ from Fig. 15.2 changing the assignment to variable $\mathtt{Timebound}$ to

```
Timebound := <f> n;
```

where f is the program that computes time bound f. This program exists since f is time constructible. Again, we can compile diag into a WH^{1}LE program decA and thus define:

$$A = \{\, d \mid [\![\mathtt{decA}]\!]^{\mathtt{WH^1LE}}(d) = \mathrm{true}\,\}$$

Exactly as for the Linear Time Hierarchy Theorem we can prove that decA $\in$ WH^{1}LE$^{\mathrm{time}(b \times f(n))}$ for some b, so condition (2) is met. For condition (3) that $A \notin$ **TIME**$^{\mathtt{WH^1LE}}(\mathcal{O}(g))$, as before, we carry out a proof by contradiction and assume $A \in$ **TIME**$^{\mathtt{WH^1LE}}(\mathcal{O}(g))$. This means that there is a WH^{1}LE program p that decides membership in A and also satisfies $time_p^{\mathtt{WH^1LE}}(d) \leq T(|d|)$ for all $d \in \mathbb{D}$ and a time bound T for which there is an $n_0 \in \mathbb{N}$ such that for all $n > n_0$ we have that $T(n) \leq c_T \times g(n)$. We now ask (as before): what is $[\![p]\!]^{\mathtt{WH^1LE}}(p)$? The crucial step here is to obtain:

$$time_p^{\mathtt{WH^1LE}}(p) \leq T(|p|) \implies time_p^{\mathtt{WH^1LE}}(p) \leq f(|p|) \qquad (15.1)$$

If we have that, then the same argument as in the Linear Time Hierarchy Theorem is applicable to derive a contradiction. As before, diag then allows the self interpreter enough time to interpret p and we get two semantically different deciders for A, a contradiction. It follows that $A \notin$ **TIME**$^{\mathtt{WH^1LE}}(\mathcal{O}(g))$ but by (2) $A \in$ **TIME**$^{\mathtt{WH^1LE}}(\mathcal{O}(f(n)))$

It remains to prove Eq. 15.1: It obviously suffices to show that $T(n) \leq f(n)$ where n is the size of p. Since $T(n) \leq c_T \times g(n)$ for $n > n_0$ it suffices to show that $c_T \times g(n) \leq f(n)$ for all n assuming $|p| > n_0$. This, however, we obtain from the assumption $g \in o(f)$, which implies

$$g(n) \leq \frac{1}{c_T} \times f(n) \quad \mathrm{for}\, n > n_1,$$

if $|p| > \max\{n_0, n_1\}$ (in which case we are done). But what do we do if $|p|$ is not large enough? Then we produce a program p_{new} from p that also meets the assumptions of condition (3), i.e. $p_{\mathrm{new}} \in$ WH^{1}LE$^{\mathrm{time}(\mathcal{O}(g))}$, just by padding p with code of sufficiently large size. That this can always be achieved will be discussed in Exercise 7.

The above theorem states that for two time bounds f and g (that are at least linear) such that f grows asymptotically faster than g there are more problems decidable in time (bounded by) $\mathcal{O}(f)$ than there are problems decidable in time (bounded by) $\mathcal{O}(g)$.

The Asymptotic Hierarchy Theorem can analogously be shown for languages WHILE, SRAM and even RAM (under the unit cost model [3]). The version for Turing Machines requires a stronger assumption as $g \in o(f)$ does not suffice but $g(n) \times \log g(n) \in o(f)$ does. The additional logarithmic factor is due to the extra cost for maintaining the program counter of the timed universal program. While maintaining the counter can be done in constant time in WHILE and RAM, it requires more moves on the tape of a TM.

It should be clear that the *Linear Time Hierarchy Theorem* is *not* a special case of the asymptotic one as $\lim_{n\to\infty} \frac{g(n)}{f(n)}$ would boil down to $\lim_{n\to\infty} \frac{a\times n}{a\times b\times n} = \frac{1}{b}$ which is not 0.

From the *Asymptotic Hierarchy Theorem* we can now, for instance, conclude that there must be problems that are not decidable in polynomial time but in exponential time:

Corollary 15.1 *There is a problem in **EXP** that is not in **P**, in other words the inclusion **P** $\subsetneq$ **EXP** is proper.*

Proof First we observe that $\mathbf{P} \subseteq \mathbf{TIME}(\mathcal{O}(2^n))$ as every polynomial is dominated by the exponential function.

By the Asymptotic Hierarchy Theorem now $\mathbf{TIME}(\mathcal{O}(2^n)) \subsetneq \mathbf{TIME}(\mathcal{O}(2^n \times 2^n))$ $= \mathbf{TIME}(\mathcal{O}(2^{2n})) \subsetneq \mathbf{EXP}$ from which immediately follows $\mathbf{P} \subsetneq \mathbf{EXP}$ by transitivity.

15.3 Gaps in the Hierarchy

It is worth pointing out that the hierarchy actually contains some gaps. What do we mean by "gaps"? For any function $h : \mathbb{N} \to \mathbb{N}$ for which $h(n) \geq n$ one can prove that there exists a time bound t such that any program with time bound $f(d) = h(t(d))$ has time bound t for all but finitely many inputs d. So the "extra time" described by h cannot help to decide more problems, i.e. $\mathbf{TIME}(\mathcal{O}(t(n))) = \mathbf{TIME}(\mathcal{O}(h(t(n))))$. But this means that there are arbitrarily large gaps represented by h for certain time bounds t. The question is: For which kind of t do those gaps exist? Those functions cannot be *time constructible* because otherwise this would contradict the Hierarchy Theorem. There is a corresponding theorem called *Gap Theorem* and more can be read about this in e.g. [5, Sect. 20.4] or other textbooks about complexity.

What Next?

The *Asymptotic Hierarchy Theorem* provides examples of problems that cannot be solved with a certain time bound, but since they have been constructed by *diagonalisation*, these are not very *natural* problems. As IT practitioners, we are of course particularly interested in *natural* (common) problems and we will have a closer look at those in the next two chapters.

Exercises

1. Prove Lemma 15.1, i.e. that the STEP macro in the $\mathrm{WH^1LE}$ timed universal program tu with input [p, d] is called at most $k \times n$ times for a program-independent constant k if $time_p^{\mathrm{WH^1LE}}(d) = n$.
   Hint: Analyse the changes of the code stack during execution of tu in terms of input program d.

2. As mentioned in the proof of Theorem 15.1, show that the STEP macro for the WH^1LE timed universal program can be executed in constant time independent of the size and shape of its input parameters.
3. Explain in detail why a timed version of the universal program (self-interpreter) for WHILE (for the untimed version see Sect. 7.2) would *not be efficient* in the sense of Definition 15.2.
4. Are the following statements true or false? Give a reason (*Hint: use robustness results and Hierarchy Theorems.*):

   a. There is a problem in $\mathbf{TIME}^{WH^1LE}(\mathcal{O}(n^2))$ that is not in $\mathbf{TIME}^{WH^1LE}(\mathcal{O}(5n))$.
   b. Any problem in $\mathbf{TIME}^{WH^1LE}(n^2)$ is in $\mathbf{P}^{GOTO}$.
   c. There is a $b > 0$ such that there is a problem in $\mathbf{TIME}^{WH^1LE}(b \times n)$ that is not in $\mathbf{TIME}^{WH^1LE}(n)$.
   d. Any problem in $\mathbf{TIME}^{WH^1LE}(n^2)$ is in $\mathbf{TIME}^{WHILE}(n^2)$.

5. Show that the following functions are time constructible (according to Definition 15.3):

   a. any constant function
   b. the identity function
   c. the (pointwise) addition of two time constructible functions
   d. the (pointwise) multiplication of two time constructible functions
   e. any linear function
   f. any polynomial function
   g. the exponential function $f(n) = 2^n$.

6. Show that $\mathbf{P}^{WH^1LE} \subsetneq \mathbf{EXP}^{WH^1LE}$, i.e. there is at least one problem in **EXP** that cannot be decided in polynomial time by a WH^1LE-program.
7. Show that for any WH^1LE-program p of size n there is a WH^1LE-program p' that has size $n + k$ such that $[\![p]\!]^{WH^1LE} = [\![p']\!]^{WH^1LE}$ and $p \in WH^1LE^{time(\mathcal{O}(g))}$ iff $p' \in WH^1LE^{time(\mathcal{O}(g))}$.
8. Let M be a Turing machine deciding a set A in time $f(n) \in o(n)$. Show that for any (!) $\varepsilon > 0$ there is a Turing machine M_{faster} deciding A in time g where $g(n) = \varepsilon \times f(n) + c_1 \times n + c_2$.
9. The Speedup Theorem for TMs from Exercise 8 is contradicting the Linear Time Hierarchy Theorem for WH^1LE (and our intuition for that matter). Explain the reason why the Speedup Theorem can be shown for TMs but not for WH^1LE or SRAM.

References

1. Ben-Amram, A.M., Jones, N.D.: Computational complexity via programming languages: constant factors do matter. Acta Inform. **37**, 83–120 (2000)
2. Cook, S.A.: An overview of computational complexity. Commun. ACM **26**(6), 400–408 (1983)

3. Cook, S.A., Reckhow, R.A.: Time-bounded random access machines. J. Comput. Syst. Sci. **7**, 354–375 (1973)
4. Dahl, C., Hesselund, M.: Determining the constant coefficients in a time hierarchy. Student report 94-2-2, DIKU (University of Copenhagen) (1994)
5. Jones, N.D.: Computability and Complexity: From a Programming Perspective. MIT Press, Cambridge (1997) (Also available online at http://www.diku.dk/neil/Comp2book.html)

Chapter 16
Famous Problems in P

> *What useful problems have polynomial time algorithms? What is their complexity?*

In order to fully grasp limits of computation, it is important to look at *both sides* of the limit. We will discuss problems that are not known to be feasibly computable in the next chapter, but first we look at the "right side" of the limit where we can actually compute efficiently. Programmers strive to make their systems run faster, fast programs is what companies make money with. Therefore, *all* computer scientists will have to think about the time complexity of their algorithms and programs eventually.

Programmers, of course, know that how one codes a solution has impact not only on correctness (bugs) but also on the performance of the resulting program. They often run into performance issues and computational complexity is the science lurking behind those issues. From a pragmatic point of view, it is pretty obvious that it makes a difference in runtime whether an algorithm uses two nested loops running in quadratic time or just one loop, running in linear time. But the devil is in the detail and sometimes quite ingenious algorithms can significantly reduce runtime complexity, and take a conceptual algorithmic idea to a practically usable piece of software. Here is an example:

The *Fast Fourier Transform* algorithm (FFT) is *"easily the most far-reaching algorithm in applied mathematics, the FFT revolutionized signal processing"*, states [9] which included FFT in a list of "top 10 algorithms of the 20th century." The FFT is a fast way to compute the *Discrete Fourier Transform* (DFT) *"used in many disciplines to obtain the spectrum or frequency content of a signal"* [14]. It is fast in the sense that it reduces the time complexity of computing the *Direct Fourier Transform* from quadratic runtime $\mathcal{O}(n^2)$ to a time bound that is $\mathcal{O}(n \log n)$ where n is the sample length of the signal. The reduction is achieved by using "divide-and-conquer" to split the sequence of samples into subsequences of odd and even indexed samples and carry out a DFT separately for each of them, thus halving the number of expensive multiplications one has to perform. This process can then be repeated

© Springer International Publishing Switzerland 2016
B. Reus, *Limits of Computation*, Undergraduate Topics in Computer Science,
DOI 10.1007/978-3-319-27889-6_16

until a sample length 2 is reached, which altogether leads to an exponential reduction in the number of multiplications.

Digital audio, digital storage, speech and image recognition are just a few examples that would not be possible without FFT. The emphasis here is on the first F "fast". The efficient algorithm was popularised in 1965 by *James Cooley*[1] and *John Tukey*,[2] although it is fair to say that it was heavily influenced by previous work. It turned out to have been already discovered by *Carl Friedrich Gauss*[3] about 160 years earlier. For a historic overview see [14].

Explaining the details of FFT requires a significant amount of mathematics and is beyond the scope of this book, so this chapter will take a closer look at some other natural problems that quite often occur in practice, and that are simpler to explain. For instance, the *Predecessor problem* (Sect. 16.2) is, one of the most studied problems in data structures. It is solved billions of times every day when a packet is forwarded from one router to another in an internet connection. The most popular packet forwarding policy is the *longest prefix match* forwarding rule which requires search for a longest address prefix in a forwarding table that matches the destination address of a packet. The problem, in general terms, is to find the largest element in a given set that is *smaller* than a given search value. In case of strings or addresses, smaller means being a prefix of the search value. A plethora of solutions have been proposed using various data structures to speed-up the search.[4] The *Predecessor problem* is thus also an excellent example for the trade-off between time and space complexity.

The good news is that *Predecessor* and all other problems we will discuss in this chapter have been proven to be in **P**. Although we do not really believe the Cobham–Edmonds/Cook–Karp Thesis (see Chap. 14), problems in **P** can be classified feasible if the polynomials appearing as their time bounds have a low degree. This will be the case for the problems in this chapter that are solved by algorithms with runtime bounds that are polynomials of degree not higher than 4.

The problems presented and their algorithmic solutions are quite famous, so they can be easily found by their names (online or in textbooks).[5] The problems included in this chapter are the Parsing problem (Sect. 16.3), Primality testing (Sect. 16.4), and various graph problems: Reachability (Sect. 16.5.1), Shortest Paths (Sect. 16.5.2), Maximal matchings (Sect. 16.5.3), Min-Cut & Max-Flow (Sect. 16.5.4), the Postman

---

[1]James William Cooley (born September 18, 1926) is an American (applied) mathematician who worked at the IBM Watson Research Centre.

[2]John Wilder Tukey (June 16, 1915–July 26, 2000) was an American mathematician and statistician who invented the box plot. He received the National Science Medal in 1973.

[3]Johann Carl Friedrich Gauss (30 April 1777–23 February 1855) is widely regarded as one of the greatest mathematicians of all time for his contributions to many different fields in mathematics, among them number theory, geometry, probability theory, planetary astronomy, ...

[4]For an overview see [6].

[5]The diligent Computer Science student might have already encountered some of them in their *Data Structures and Algorithms* or related modules.

problem (equivalently "Seven Bridges of Königsberg" in Sect. 16.5.5). The chapter concludes with a most versatile problem called Linear Programming (Sect. 16.6).

Before we start, let us first define the template for decision and optimisation problems, respectively.

16.1 Decision Versus Optimisation Problems

A decision problem is of the form "decide membership in A" (here A is a set of data) as discussed in Definition 2.13. A template for such a problem can be described by stating what the problem instance is and what the problem question is:

- (problem) **instance**: a particular data d (out of a given set or type)
- (problem) **question**: "is d in A"? (or short $d \in A$?)

Furthermore if we want to show that the problem is in **TIME**(f) we need to find an algorithm (procedure) p such that

- **Input**: p takes as input a data element d (of the set of allowable data = input type);
- **Output**: p decides $d \in A$ returning true or false as output accordingly; output type is therefore $\mathbb{B}$;
- **Time bound**: p has a runtime bound f.

An optimisation problem is of the form "find an optimal s in G" and a template can be described as follows:

- (problem) **instance**: some "scenario" G (often including a graph or a set of equations or numbers[6]
- (problem) **question**: find a solution s for G such that $m(s)$ is minimal/maximal where m is some (fixed) measure of the size of s.

If we want to show that the problem is in **TIME**(f) we need to find an algorithm (procedure) p with

- **Input**: G
- **Output**: solution s such that s is a minimal/maximal solution w.r.t. m;
- **Time bound**: p has a runtime bound f.

We can reformulate any *optimisation problem* as a decision problem of the following form:

- (problem) **instance**: a particular encoding of G and a positive natural number K
- (problem) **question**: is there a solution s for G such that $m(s) \leq K$ (or $m(s) \geq K$ for maximisation, respectively)?

---

[6]In any case something clearly defined.

Why, or in which sense, is it sufficient to consider the decision problem version of optimisation problems? Well, the decision problem is *not harder* than the optimisation problem. If we can solve the optimisation problem with time bound f then we can solve the decision problem version with time bound $g(n) = f(n) + p(n)$ where p is a polynomial. Why? Because if we have the optimal solution s we can check whether $m(s) \leq K$ (or $\geq$, respectively) in polynomial time in the size of s. So if the optimisation problem is in **P** so is the decision problem. Contrapositively, if the decision problem is not in **P**, the optimisation problem is not either. We are interested in problems that are not in **P** as they are more likely to be infeasible. And according to the above, it is sufficient to consider decision problem versions of optimisation problems.

16.2 Predecessor Problem

Search is a well-known problem and a simple linear search in a list is clearly $\mathcal{O}(n)$ where n is the length of the list. In this case, the length of the items is assumed to have a constant upper bound independent of n and thus any comparisons are $\mathcal{O}(1)$. Runtime can be improved using *binary search*, which gets (worst-case) complexity down to $\mathcal{O}(\log_2 n)$ but only works on *sorted* lists. If the list is not static, list insertion and deletion need to be considered as well, which are both $\mathcal{O}(n)$ for sorted lists (see Exercise 1).

Another less famous, but equally important, problem is searching a set of integer values for the largest value less than or equal to a given value. This is called the *Predecessor problem* and is the basis for string matching problems and network protocols.

Definition 16.1 (*Predecessor problem*)

- **instance**: a set $S \subseteq U$ containing n elements of (finite) type U and a search value $k \in U$. Type U supports a "less than or equal to" order relation, $\leq$; Let u denote the largest[7] element in U;
- **question**: what is the largest element in S that is smaller than k? In other words find $\max\{x \in S \mid x \leq k\}$.

Example 16.1 Let $\Sigma = \{a, b, c, \ldots, y, z\}$. We will look at two instances of the predecessor problem with U being a subset of the natural numbers and finite words over alphabet Σ, respectively.

1. Let $U = \{x \in \mathbb{N} \mid x \leq 31\}$ with the usual $\leq$-order on numbers and $u = 31$. Assume $S = \{0, 2, 5, 8, 24\}$ which implies $n = |S| = 5$. If the search value is $k = 4$ the answer to this instance of the Predecessor problem is 2, because 2 is largest number x in S for which $x \leq 4$ holds. If we set $k = 8$ the answer is 8, equally for $k = 11$, for $k = 30$ the answer is 24.

---

[7] w.r.t. $\leq$.

2. Let $U = \{ w \in \Sigma^* \mid |w| \leq 10 \}$ be the words over the standard alphabet (all lowercase) with the standard *lexicographic order*, i.e. the alphabetical order in which entries of a dictionary are sorted. To obtain a finite set, the length of words is restricted to 10. Let $S = \{ par, parade, part, pasta, port, portal, portcullis \}$ which implies $n = |S| = 7$ and $u = portcullis$ the largest string in S. If the search value is $k = pastor$, the answer to this instance of the Predecessor problem is *pasta*, because *pasta* is the, lexicographically speaking, largest word in S that is less or equal to *pastor*. If we set $k = tardis$ the answer is *portcullis*, equally for $k = potential$. For $k = apple$, the answer is *'not found'*.

The first example can actually be viewed as a word matching problem of the second type, setting $U = \{ x \in \{0, 1\}^* \mid |x| = 5 \}$ and $S = \{00000, 00010, 00101, 01000, 11000 \}$, i.e. using binary words to encode the numbers. In this case $u = 11111$ and its length is 5 (which corresponds to $\log_2 2^5$ in the number version of the problem). In this representation the search keys must also be represented in binary form, i.e. $k = 4$ becomes $k = 00100$ and so on.

The *Predecessor problem* has been widely researched and solutions are tradeoffs between time and space complexity. Space refers to the memory needed to store supporting data structures to speed up runtime. An additional tradeoff is the one between search and updates of the set U. The latter can be costly if the auxiliary data structure is complex. If we precompute and store all queries in an additional data structure, this structure needs $\mathcal{O}(u)$ space but the answer time is then just a constant lookup, i.e. search is $\mathcal{O}(1)$. The update of the auxiliary table is $\mathcal{O}(n)$. If n is large this may not be an ideal solution.

In practice, one strikes a compromise, using so-called *tries*.

Definition 16.2 A *trie* for a set S of words (over alphabet Σ) is a tree of prefixes in S. An edge from vertex s to vertex t exists if there is a letter c of the alphabet Σ such that $sc = t$. In this case the edge is labelled c. In cases where S is a set of numbers, one uses the binary representation of those numbers, then the prefixes are words over alphabet $\{0, 1\}$.

Example 16.2 The trie for the *Predecessor problem* $U = \{ x \in \{0, 1\}^* \mid |x| = 5 \}$ and $S = \{00000, 00010, 00101, 01000, 11000 \}$ from Example 16.1 is depicted in Fig. 16.1 as a subtree of a full binary tree of depth 5. The thick edges highlight the actual trie with its five leaves that represent the elements of S.

There are various algorithms to solve the Predecessor problem, offering different tradeoffs and using different data structures. A simple solution is to precompute and store the trie as a binary tree as outlined in Fig. 16.1, then perform the search by running along the tree starting from the root node (on the left) by selecting edges with labels that match the letters of the key (word). Either one travels along a path in the trie and the key is already in S, or at some point a letter does not match an edge label of the trie. In the latter case, one selects the nearest element in S, i.e. the nearest trie path. In Fig. 16.1 this is the first trie path *above* the (aborted) path of the key. Obviously, such a run through a binary tree representing a trie can be

Fig. 16.1 Trie for
Example 16.1

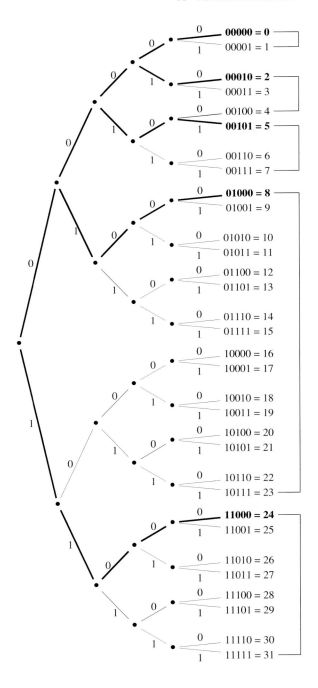

performed in the length of u (the maximum word), so search is in $\mathcal{O}(u)$. Note that u in this example is $\log_2 32$. The tree can be stored with space $\mathcal{O}(n \times u)$ where n is the number of elements in S. With so-called *fast tries* the search time can be cut down to $\mathcal{O}(\log_2 u)$ with $\mathcal{O}(n)$ space, and there are even more clever techniques in certain situations, see e.g. [6]. So search for prefixes (as search in general) can be done in logarithmic time if one uses the right data structure. The initialisation of the data structure will need polynomial time (or at least linear time) though. But it only has to be set up once if the set S does not change.[8]

16.3   Membership Test for a Context Free Language

We know that context-free languages (generated by context free grammars) are decidable. The decision procedure for such language is simply a parser for that language. But what is the time complexity of parsing?

Definition 16.3 (*Parsing problem*)

- **instance**: a context free grammar G over a (finite) alphabet Σ and a word s over alphabet Σ (i.e. $s \in \Sigma^*$)
- **question**: is s in the language generated by G?

A simple minded parser would just try all the possible derivations for the grammar. But due to branching there may be exponentially many such derivations. One has to do something a bit more clever to get a polynomial time algorithm. The clever trick is called *Dynamic Programming* which simply refers to the fact that one stores intermediate results of (sub)problems and (re-)uses them cleverly to compute further subproblems and eventually the desired result. This avoids unnecessary re-computation of known results which can speed up the process significantly. This technique had been discovered by *Richard Bellman*[9] in the 1950s [7] at a time *"when computer programming was an esoteric activity practiced by so few people as not to even merit a name. Back then programming meant "planning," and "dynamic programming" was conceived to optimally plan multistage processes"* [10, p. 174]. So Bellman was looking to solve an optimisation problem about discrete deterministic processes and their scheduling [7].

Probably the most famous algorithm for parsing is the Cocke–Younger–Kasami algorithm[10] (also called CYK, or CKY algorithm). It is named after the scientists who independently invented it in the late 1960s: John Cocke,[11] Daniel Younger,

---

[8]Otherwise one needs to consider insert and update operations.

[9]Richard Ernest Bellman (August 26, 1920–March 19, 1984) was an American mathematician famous for the discovery of dynamic programming. He won (among other awards) the John von Neumann Theory Prize in 1976.

[10]According to [16, Chap. 7] only Younger's version was ever published "conventionally" in [24].

[11]John Cocke (May 30, 1925–July 16, 2002) was an American computer scientist often called "the father of RISC architecture." He won the Turing Award in 1987.

and Tadao Kasami.[12] This algorithm performs "bottom-up" parsing constructing a so-called *parsing table* that stores which substrings can be generated from which non-terminals of the grammar. The most famous parsing algorithm runs in time $\mathcal{O}\big(|s|^3 \times |G|\big)$, i.e. in cubic time in terms of the size of the string times the size of the grammar.[13] Regarding asymptotic worst case complexity, this is the best algorithm for parsing.

16.4 Primality Test

This problem is a true decision problem and thus presented as such.

Definition 16.4 (*Primality Problem*)

- **instance**: a positive natural number n (the size of the number is measured logarithmically, i.e. as the number of digits of its binary representation)[14];
- **question**: is n a prime number?

That there is an algorithm with polynomial time bound (in the given strong sense regarding logarithmic size) has only be shown in 2002 in a famous award-winning[15] result by: *Manindra Agrawal, Neeraj Kayal, Nitin Saxena*[16] called *"PRIMES is in P"* [2].

"A remarkable aspect of the article is that the final exposition itself turns out to be rather simple. The text as published in Annals of Mathematics is a masterpiece in mathematical reasoning. It has a high density of tricks and techniques, but the arguments come in a brilliantly simple manner; they remain completely elementary ...if the number-theoretic assumptions hold, then the algorithm runs in nearly cubic time, which is amazing ...Thus "PRIMES is in P" is a perfect example for a new trend in derandomization" [5]. A good introduction to the problem of primality testing and its history is provided e.g. in [1].

The AKS algorithm[17] runs in $\mathcal{O}\big(\log_2^{12} n \times p(\log_2(\log_2 n))\big)$ where n is the problem instance (natural number). If we replace $\log_2 n$ by n as the input of the problem is the binary representation of the number, we obtain $\mathcal{O}\big(n^{12} \times p(\log_2 n)\big)$ which is clearly polynomial. Note that the exponent 12, under certain assumptions, can be greatly reduced.

The simple-minded primality test (also called *"trial division"*) that the reader might have learned already at school—namely to divide the given number K by

---

[12]Tadao Kasami (April 12, 1930–March 18, 2007) was a Japanese engineer and information theorist.
[13]If the grammar is fixed, one usually ignores the grammar coefficient.
[14]E.g. $|102| = |1100110| = 7$ whereas in the WHILE cost measure $|\ulcorner 102 \urcorner| = 103$ (a list with 102 nil entries).
[15]Actually they won several awards, among them the Gödel Prize and the Fulkerson Prize in 2006.
[16]All from the Indian Institute of Technology Kanpur.
[17]Usually called AKS-algorithm according to the initial letters of the surnames.

all odd numbers up to the square root of K—has a complexity of $\mathcal{O}\left(\sqrt{K}\right)$ (see Exercise 5). If we consider the input number K represented as a binary of n digits then trial division is $\mathcal{O}\left(\sqrt{2^n}\right) = \mathcal{O}\left(2^{\frac{n}{2}}\right)$. This is clearly not polynomial.

16.5 Graph Problems

Recall the formal definition of a graph:

Definition 16.5 (*(Directed) graph*) A directed graph consists of two sets: a set of vertices V and a set of edges E. Edges are simply pairs of vertices (s, t) the source and the target vertex of an edge. In undirected graphs one does not even distinguish source and target vertex.

Example 16.3 An example of an (undirected) graph with nodes labelled (a) to (g) is depicted in Fig. 16.2.

Moreover, (a, b)—or (b, a)—is an example of an edge in E, whereas (a, c)—or (c, a)—is not an edge in E.

Lemma 16.1 *Every undirected (weighted) graph can be represented as a directed (weighted graph).*

Proof Exercise 2.

16.5.1 Reachability in a Graph

Definition 16.6 (*Reachability Problem*)

- **instance**: a (directed or undirected) graph $G = (V, E)$, and two vertices $s, t \in V$;
- **question**: is there a path from s to t in G?

Fig. 16.2 Undirected graph $G = (V, E)$

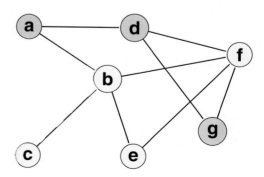

Example 16.4 In the graph depicted in Fig. 16.2 there is a path from node (a) to node (g). The path is $a \rightarrow d \rightarrow g$ and the corresponding nodes in the path are coloured red. There are other paths between (a) and (g) too, for instance $a \rightarrow b \rightarrow f \rightarrow g$ or $a \rightarrow d \rightarrow f \rightarrow g$ or even $a \rightarrow b \rightarrow e \rightarrow f \rightarrow g$.

Graph reachability can be solved by a simple breadth-first (or depth-first) search along the edges starting from vertex s. By visiting all edges (and remembering those visited edges) one can traverse the entire graph and find all reachable nodes.

16.5.2   Shortest Paths in a Graph

Definition 16.7 (*Weighted graph*) A *weighted graph* $G = (V, E, w)$ is a graph $G = (V, E)$ together with a function $w : E \rightarrow \mathbb{R}$ such that each edge has a weight expressed as a (real) number.

In the following, all graphs are undirected but the problems can also be expressed using directed graphs.

The graph in Fig. 16.3 is weighted. In this case all the weights are actually natural numbers in $\mathbb{N}$.

Definition 16.8 (*Shortest Path Problem*)

- **instance**: a weighted graph $G = (V, E, w)$
- **question**: What is the distance of the shortest path from v_1 to v_2 in G for each pair of vertices $v_1, v_2 \in V$? By shortest path we mean the path for which the sum of weights of all the edges on the path is lowest.

Example 16.5 The shortest path from node a to node g in the graph depicted in Fig. 16.3 is $a \rightarrow d \rightarrow g$ and the sum of weights of its edges, i.e. the required distance, is $3 + 2 = 5$.

Fig. 16.3 Weighted graph with shortest path from a to g

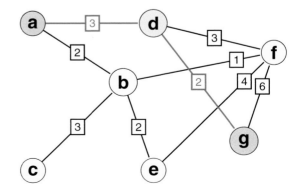

```
input:  directed weighted graph (V,E,weight);
output: array[][] of distances between nodes: dist;

for each v in V do
   for each w in V do
      if v=w
      then  dist[v][v] := 0       (* loops have no distance *)
      else  dist[v][w] := MAXINT  (* other distances very large *)
      end if
for each edge (u,v) in E
      dist[u][v] := weight(u,v)    (* set weight of edge (u,v) *)
for k = 1 to |V|
      for i = 1 to |V|
         for j = 1 to |V|
            if dist[i][j] > dist[i][k] + dist[k][j]
            then dist[i][j] := dist[i][k] + dist[k][j]
            end if
```

Fig. 16.4 The Floyd–Warshall algorithm solving the Shortest Path Problem

The Shortest Path Problem can be solved by the famous *Floyd*[18]*–Warshall*[19] *algorithm* that is presented (as pseudo-code) in Fig. 16.4 to demonstrate the *dynamic programming* technique.[20] The dynamic programming technique means that intermediate results (distances) are stored and re-used to improve complexity. Input is a weighted graph $G = (V, E, w)$.

Runtime of Floyd–Warshall

Let us analyse the algorithm's runtime behaviour. One can easily see that the first nested loop running through v and w is $\mathcal{O}(|V| \times |V|)$. The second loop runs through all edges, so it is $\mathcal{O}(|E|)$. The third nested loop running through k, i, and j is $\mathcal{O}(|V| \times |V| \times |V|)$. Since $E \subseteq V \times V$ we have that $|E| \leq |V| \times |V|$. So the dominating factor of the three summands is the last nested loop and thus the runtime of the algorithm is $\mathcal{O}(|V|^3)$.

The Floyd–Warshall algorithm in Fig. 16.4 computes the distances of shortest paths between any two vertices. This can be easily extended to compute a representation of the shortest paths as well without changing the complexity. The problem is called computing the *shortest path tree*. There is another famous algorithm, *Dijkstra's*[21] *algorithm* which only works for positive distances. It runs in $\mathcal{O}(|E| \times |V| + |V|^2 \times \log |V|)$. If the number of edges is roughly $|V|^2$ then

---

[18]Robert W. Floyd (June 8, 1936–September 25, 2001) was a famous American computer scientist who won the Turing award in 1978 for his work in the areas of program verification and algorithms.

[19]Stephen Warshall (November 15, 1935–December 11, 2006) was a renowned American computer scientist.

[20]Other algorithms we will not include in detail, as they can be easily found in the literature and online.

[21]Edsger Wybe Dijkstra (11 May 1930–6 August 2002) was a Dutch mathematician and physicist and eventually a pioneer of a then new branch of science: computing. He won the Turing

Fig. 16.5 A bipartite graph
with two different matchings

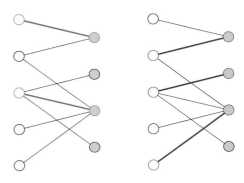

Dijkstra's algorithm and the *Floyd–Warshall algorithm* have the same order of complexity.

The shortest path problems have many applications, for instance in internet routing protocols. The *Open Shortest Path First* protocol [23] (OSPF), an interior gateway protocol for large enterprise networks, for instance, can deal with link failures of nodes by quickly computing a new shortest path for a route.

16.5.3 Maximal Matchings

A *matching* in a graph is a set of edges such that no two edges in the set share a common vertex. Often the graph is bipartite, that means its vertices can be divided into two disjoint sets of nodes A and B, such that nodes in A (and B, respectively) are not connected and the edges of the graph always connect a vertex from A with a vertex from B. If one wishes to match up pairs of competitors from two teams for a tournament, for instance, one would have edges between pairs of vertices that represent members of different teams. Since each competitor is supposed to have only one opponent, no two edges in the matching should share a vertex. Figure 16.5 shows two matchings for the same bipartite graph (where the different sets of vertices are coloured yellow and purple, respectively). The matching of size two on the left hand size is highlighted in red, and the matching of size three on the right hand side is highlighted in blue. Indeed, three is the largest size matching for the given graph.

Definition 16.9 (*Matching Problem*)

- **instance**: a (undirected) graph $G = (V, E)$;
- **question**: What is the largest matching in G?

(Footnote 21 continued)
award (among many prizes) in 1972 "for fundamental contributions to programming as a high, intellectual challenge; for eloquent insistence and practical demonstration that programs should be composed correctly, not just debugged into correctness; for illuminating perception of problems at the foundations of program design" [4].

A famous (first) algorithm to compute maximal matchings is the *Blossom algorithm* devised by *Jack Edmonds*[22] in 1965. This algorithm is $\mathcal{O}(|V|^4)$ but has been improved later to $\mathcal{O}(|V|^3)$ or also $\mathcal{O}(|V| \times |E|)$. *Norbert Blum* [8] found an algorithm in $\mathcal{O}(\sqrt{|V|} \times |E|)$ and some further improvements have been carried out since then [13].

16.5.4 Min-Cut and Max-Flow

A very popular problem in *Operations Research* is the widely studied optimisation problem called *Max-Flow* Problem. It is a specific network flow problem defined as follows:

Definition 16.10 (*Max-Flow Problem*)

- **instance**: a weighted directed graph $G = (V, E, w)$ (encoding a flow network where $w(u, v) \in \mathbb{Z}$ describes the maximum capacity of flow from node u to node v), a source node $s \in V$ and a sink node $t \in V$.
- **question**: What is the maximum flow from s to t in G?

This problem can be solved by the *Ford–Fulkerson algorithm*,[23] the complexity of this algorithm is $\mathcal{O}(|E| \times maxflow)$ where *maxflow* is the maximum flow of the network.

Definition 16.11 A *cut* in a flow network graph $G = (V, E, w)$ with source s and sink t (as defined in Definition 16.10) is a set of edges $S \subseteq E$ whose removal from E disconnects s and t. This can be equivalently viewed as a partition of the nodes V into two disjoint sets A and B such that $s \in A$ and $t \in B$. The *capacity* of a cut is the sum of the capacities of the edges in the cut.

The Min-Cut Problem requires to compute the minimum capacity of flow that one needs to disconnect in order to completely stop any flow from source to sink in the given network.

Definition 16.12 (*Min-Cut Problem*)

- **instance**: a weighted directed graph $G = (V, E, w)$ (encoding a flow network) where $w(u, v) \in \mathbb{Z}$ describes the maximum capacity of flow from node u to node v, a source node $s \in V$ and a sink node $t \in V$.
- **question**: What is the cut with minimum capacity for the given flow network?

---

[22]Jack R. Edmonds (born April 5, 1934) is an American computer scientist, famous for his contributions to the field of combinatorial optimization. He received the 1985 John von Neumann Theory Prize.

[23]Lester Randolph Ford, Jr. (born September 23, 1927) and Delbert Ray Fulkerson (August 14, 1924–January 10, 1976) are American mathematicians who are essentially famous for this algorithm.

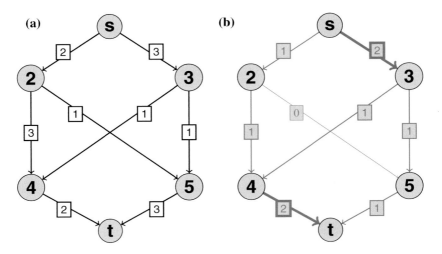

Fig. 16.6 **a** A flow network as graph **b** Max-Flow for the network on the *left*

Theorem 16.1 *Let G with source s and sink t be a flow network. The maximum flow is equal to the capacity of the minimum cut.*

Applications of this problem basically occur wherever networks occur: In a computer network one may wish to route as many packets as possible. In transportation one may wish to send as many trains as possible, where tracks have certain limits on the number of trains per any given time. Similar applications arise for road or phone networks. For more about this problem and its applications see [3].

Example 16.6 The graph in Fig. 16.6a shows a flow network as a graph. The labels on each directed edge denote capacities. The network's source node is *s* and its sink node is *t*, respectively. In Fig. 16.6b the same network is depicted but edges are now labelled with their maximum flow (in red) instead of their capacity. It is obvious from this picture that the maximum flow is 3. One can also see that the minimum cut is 3. To completely disrupt the flow from node *s* to *t* it suffices for instance to cut the edges $(s, 2)$ and $(s, 3)$, which together have a flow of 3. Alternatively, one could cut edges $(2, 4)$, $(3, 4)$, and $(3, 5)$ or the pair of edges $(4, t)$ and $(5, t)$. In either case the total flow of the edges is 3.

16.5.5 *The Seven Bridges of Königsberg*

Legend has it that *Frederick the Great of Prussia* once had to host a group of international dignitaries whom he wanted to show the beauty of the city, first and foremost the seven bridges across the river *Pregel*. He thus asked the mathematician *Leonhard*

(a)

(b)

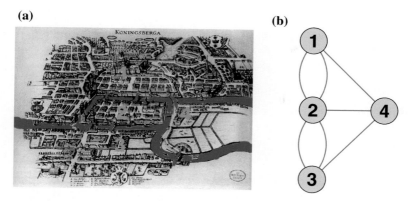

Fig. 16.7 The seven bridges of Königsberg. **a** Königsberg, river Pregel with bridges **b** …abstract as a graph

$Euler$[24] to compute a tour across all seven bridges that visits each bridge exactly once. A tour of course ends up where it started. The problem is depicted in Fig. 16.7a that shows the river's various arms in blue and the bridges in red.[25]

Euler abstracted away from the concrete layout of the river and bridges by using a graph where the vertices represent the various land masses (separated by the river) and the edges represent the bridges.[26] The green land masses 1–4 from Fig. 16.7a are the vertices of the graph in Fig. 16.7b that represent the abstraction from the problem.

Definition 16.13 (*Eulerian circuit Problem*)

- **instance**: an undirected (connected) graph $G = (V, E)$
- **question**: is there a circular path in G that visits each edge exactly once?

Euler could show that no tour exists that visits each bridge exactly once by proving a general result about circuits in graphs (now called *Eulerian circuits* in his honour). Such circuits are supposed to visit each edge exactly once and finish in the same vertex they started from. Euler circuits will be discussed in more detail in Exercise 7.

The standard algorithm for finding an Eulerian circuit is *Fleury's algorithm* [12], which is in $\mathcal{O}(|E|^2)$. There are algorithms with even better complexity as well, like *Hierholzer's algorithm* [15], which runs in $\mathcal{O}(|E|)$.[27]

A variation of this problem, where the length of the route is also important, is the so-called[28] *Postman Problem*: given a map of a (quarter of a) city with distances

---

[24]Leonhard Euler (15 April 1707–18 September 1783) was a great Swiss mathematician and physicist. His main mathematical contributions are in infinitesimal calculus and graph theory. He is also responsible for much of modern mathematical notation in analysis.

[25]The image of Köngisberg city centre is cut out of a larger engraving by Joachim Bering from 1613 and was printed on a memorial page of the 600th anniversary of the city in about 1813.

[26]Inventing the field of graph theory on the fly.

[27]Both algorithms are well-known and can be easily found online or in textbooks.

[28]*Kwan Mei-Ko* seems to have introduced the term *Postman problem* in [20].

(lengths) for all streets, find the *shortest* tour that visits each street at least once, and returns to the starting point. If we represent this problem graphically, streets correspond to edges and vertices to street corners. Weights are the distances. This is a generalisation of the 7-Bridges problem above.

Definition 16.14 (*Postman Problem*)

- **instance**: a weighted undirected graph $G = (V, E, w)$ where the weights correspond to (positive) distances.
- **question**: what is the shortest circuit in G that visits each edge ("street") at least once?

It is important that the graph in undirected. This means that the postman can walk along each street in *any direction*. There are no one way streets and also the cost of travelling in either direction is the same. We will see in the next chapter that this is crucial. Surprisingly maybe, if one mixes directed and undirected edges, or if the costs can depend on the direction travelled, one obtains much harder problems.

There are many real world applications. The Postman Problem is not only an issue for postmen's route planning, it can lead to fuel (and time) savings for rubbish collection, bus routing, road sweeping, snow-plough dispatch and so on. This Postman Problem is also often called *Route Inspection Problem*, as one is allowed to visit an edge more than once, so any kind of (pipe-)line inspection is thus another application scenario.

The first solution to the Postman Problem is due to *Jack Edmonds*[29] and *Ellis Johnson*[30] [11] and uses Eulerian circuits and Edmond's blossom algorithm. The runtime is $\mathcal{O}(|V|^3)$.

16.6 Linear Programming

Consider the following problem of solving linear inequalities.[31] It is so general that it has an unlimited number of applications. Formally it can be described as follows:

Definition 16.15 (*Linear Programming Problem*)

- **instance**: $\mathbf{x}$ is a n-vector of positive (real) variables, $\mathbf{x} \geq 0$ and $\mathbf{c}$ is column vector, such that $A\mathbf{x} \leq \mathbf{b}$ where A is an $m \times n$ matrix of integers,[32] and $\mathbf{b}$ is a row vector of integers of length m.
- **question**: maximize $\mathbf{c}^T \cdot \mathbf{x}$

---

[29]The author of the Blossom algorithm mentioned in Sect. 16.5.3.

[30]Ellis L. Johnson (born 26 July 1938) is an American mathematician famous for his work in combinatorial optimisation for which he received (among other awards) the John von Neumann Theory Prize in 2000 (with Manfred W Padberg).

[31]Some Computer Science students may know this problem from Discrete Mathematics or Operations Research.

[32]It has m rows and n columns.

Here is a simple example of an application of *Linear Programming* from [25]:

Example 16.7 Suppose that a farmer has a piece of land, say A km^2 large, to be planted with either wheat or barley or some combination of the two. The farmer has a limited permissible amount F of fertilizer and I of insecticide, each of which is required in different amounts per unit area for wheat (F_1, I_1) and barley (F_2, I_2). Let P_1 be the price of wheat, and P_2 the price of barley, each per km^2. The areas planted with wheat and barley are denoted x_1 and x_2 (in terms of km^2), respectively. The farmer would like to optimize the income, and needs to decide how much wheat or barley to plant. This can be expressed as a linear programming problem as follows:

$$
\begin{array}{lll}
P_1 \times x_1 + P_2 \times x_2 & & \text{optimize revenue} \\
x_1 + x_2 & \leq A & \text{total area constraint} \\
F_1 \times x_1 + F_2 \times x_2 & \leq F & \text{fertilizer constraint} \\
I_1 \times x_1 + I_2 \times x_2 & \leq I & \text{insecticide constraint}
\end{array}
$$

which gives rise to the following instance of the Linear Programming Problem as given in Definition 16.15:

$$
\mathbf{c} = (P_1 \; P_2) \qquad \mathbf{b} = \begin{pmatrix} A \\ F \\ I \end{pmatrix} \qquad A = \begin{pmatrix} 1 & 1 \\ F_1 & F_2 \\ I_1 & I_2 \end{pmatrix}
$$

The best performing algorithm on average for Linear Programming is the *simplex*[33] *algorithm* (also known as *simplex method*), discovered by *George Dantzig*, often cited as one of the top 10 algorithms of the 20th century [9] (see also discussion in Sect. 14.3). It does not run in polynomial time though [19]. *Leonid Khachiyan*[34] was the first to prove that Linear Programming is in **P** [18]. His algorithm was, however not very efficient. Karmarkar[35] improved the algorithm in 1984 [17]. His algorithm runs in $\mathscr{O}(n^{3.5} L^2 \ln L \, \ln L)$ where n is the number of variables and L is

---

[33]In geometry, a *simplex* is a generalization of the notion of a triangle or tetrahedron to arbitrary dimensions: a k-simplex is a k-dimensional polytope which is the convex hull of its $k+1$ vertices. The simplex algorithm uses a simplex to describe the solution space ("feasible region") and then runs along the edges to find the best corner.

[34]Leonid Genrikhovich Khachiyan (May 3, 1952–April 29, 2005) was a Soviet mathematician of Armenian descent who moved to the US in 1989. For proving the above mentioned result he was awarded the Fulkerson Prize of the American Mathematical Society and the Mathematical Programming Society. He even made the front page of the New York Times. An insightful and amusing account of this story and how journalists can misrepresent scientific results can be found in [21]. Khachiyan's work stimulated much research on the area, eventually leading to Karmarkar's breakthrough.

[35]Narendra Krishna Karmarkar (born 1957) is an Indian mathematician who was a Ph.D. student of Richard Karp. He developed an efficient algorithm for Linear Programming for which he received several awards.

the size of the entire input (in bits). This is $\mathscr{O}(n^{2.5})$ times better than Khachiyan's original algorithm.

Linear Programming is quite generic and can thus be applied to many real world problems. It plays an important role in many applications in transport, matching, and planning problems (economics). It can be used, for instance, to solve Min-Cut (Definition 16.12). Interestingly, and maybe surprisingly, a seemingly innocuous change to the Linear Programming Problem, namely restricting the solutions to be integers, leads to a much harder problem that we will discuss in the next chapter, the so-called *Integer (Linear) Programming* Problem.

What Next?

We have seen a number of useful problems, most of them optimisation problems, that can be all solved in polynomial time. Moreover, the worst-case asymptotic time bounds of those algorithms are polynomials of a small degree, usually 3 or 4. Next, we want to study some more common problems for which nobody so far has found a polynomial time algorithm.

Exercises

1. Consider the *binary search* algorithm of searching a value in a sorted list of size n. Explain why

   a. the time complexity of searching a value in the sorted list is $\mathscr{O}(\log_2 n)$.
   b. the time complexity of inserting a new element into the sorted list is $\mathscr{O}(n)$.
   c. the time complexity of deleting a new element from the sorted list is $\mathscr{O}(n)$.

   Time complexity always refers to asymptotic worst-case complexity.
2. Show Lemma 16.1.
3. Look up Dijkstra's algorithm and verify that it is $\mathscr{O}(|E| \times |V| + |V|^2 \times \log |V|)$ where (V, E) is the input graph.
4. Look up Edmonds' *Blossom* algorithm and verify that its complexity is $\mathscr{O}(|V|^4)$ where V is the set of vertices of the given graph.
5. Show in detail that the *trial division* primality test in Sect. 16.4 is $\mathscr{O}(\sqrt{2^n})$ where n denotes the number of (binary) digits of the input.
6. Look up the *Ford–Fulkerson algorithm* and show that it is $\mathscr{O}(|E| \times maxflow)$ where *maxflow* is the maximum flow of the network and E are the edges.
7. Consider the *Eulerian Circuit Problem* (7 Bridges Problem) from Definition 16.13.

   a. For which of the following six graphs does a required tour that visits each edge exactly once exist? It does not matter in which vertex the tour starts and ends, but it is important that the tour starts and ends in the same vertex.

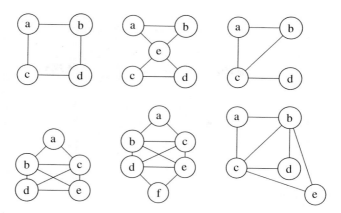

b. Consider the graphs from question (a) for which a tour as described above exists. What do they have in common?

c. Can you guess the the condition in *Euler*'s Theorem that states in which cases a tour (in connected graphs) exists?
Hint: Look at the edges! Note that in a more general variant one needs to first check whether the graph is connected, i.e. it does not have isolated vertices.

8. The number of edges a vertex v in an undirected graph $G = (V, E)$ is connected to, is called the *degree* of v. Show the following lemma:

The number of nodes of odd degree in an undirected graph G is always even.

Hint: The sum of the degrees of all the nodes in G must be an even number because each edge connects exactly two nodes.

9. Look up *Hierholzer's algorithm* and show that it has complexity $\mathcal{O}(|E|)$ where E is the set of edges in the graph.

10. Show that if an undirected weighted graph $G = (V, E, w)$ has an Eulerian circuit (in the sense of Definition 16.13) then this circuit must be the solution to the Postman Problem for the same graph G.

11. Provide a solution for the Linear Programming Problem of Example 16.7. In other words, tell the farmer how much of the area A should be used for wheat and how much for barley.

12. Convert the Max-Flow problem of Example 16.6 into a Linear Programming Problem. Then solve the Linear Programming problem and check that the solution is identical to that given in Example 16.6.

13. A *spanning tree* of an undirected graph $G = (V, E)$ is a subgraph (V, E_s)—where $E_s \subseteq E$—that is a tree itself and connects all the vertices V.
Consider the *Minimum spanning tree* Problem (MSTP) defined as follows:

- **instance**: an undirected and connected weighted graph $G = (V, E, w)$ with $w(u, v) \geq 0$ for every edge $(u, v) \in E$.

```
Input:   a connected, undirected, weighted graph G=(V,E,w).
Output:  a minimum spanning tree MST for G.

L  := E as list, sorted in increasing order by weight;
MST := empty tree;
for each edge e=(u,v) in L
        if u and v are disconnected in MST
        then add e to MST;
return MST;
```

Fig. 16.8 Kruskal's algorithm

Fig. 16.9 Graph for
Exercise 13b

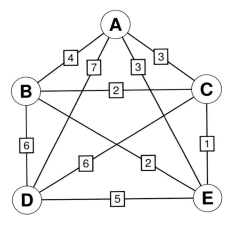

- **question**: What is a spanning tree of *G* with minimal total weight? The weight
  of a tree is the sum of the costs of its edges.[36]

  a. This problem has many practical applications in network design whether it
     is electrical,[37] telephone, road, or hydraulic networks. Explain why MSTP
     could be important for such networks.
  b. There have been suggested various algorithms to solve MSTP. One of them
     is *Kruskal's* Algorithm, which can be found in Fig. 16.8 (as pseudo code).
     Use Kurskal's algorithm to find a minimum spanning tree in the weighted
     graph of Fig. 16.9.
  c. Explain why *Kruskal's Algorithm* is called a *greedy algorithm*.

---

[36]There may be more than one minimum spanning tree for a given graph.

[37]The problem and a solution to it was first published under the name "Contribution to the solution of
a problem of economical construction of electrical networks" by Boråvka in 1926 to solve an engineering problem for the city of Moravia. For more details, see e.g. the historically interesting [22].

d. Determine the (order of) the complexity of *Kruskal's algorithm* in terms of $|V|$ and $|E|$ of the input graph G.
e. From the above conclude that MSTP is in **P**.

14. Convert the Postman Problem of Definition 16.14 into a Linear Programming Problem.

References

1. Aaronson, S.: The prime facts: from Euclid to AKS, available via DIALOG. http://www. scottaaronson.com/writings/prime.pdf (2003). Cited on 15 June 2015
2. Agrawal, M., Kayal, N., Saxena, N.: PRIMES is in P. Ann. Math. **160**(2), 781–793 (2004)
3. Ahuja, R., Magnanti, T., Orlin, J.: Network Flows: Theory, Algorithms, and Applications. Prentice-Hall, New Jersey (1993)
4. A.M. Turing Award Winners: Edsger Wybe Dijkstra. Available via DIALOG. http://amturing. acm.org/award_winners/dijkstra_1053701.cfm. Cited on 28 October 2015
5. Award committee: 2006 Gödel Prize, available via DIALOG. http://www.sigact.org/Prizes/Godel/2006.html (2006). Cited on 15 June 2015
6. Beame, P., Fich, F.E.: Optimal bounds for the predecessor problem and related problems. J. Comput. Syst. Sci. **65**(1), 38–72 (2002)
7. Bellman, R.: The theory of dynamic programming. Bull. Amer. Math. Soc. **60**(6), 503–515 (1954)
8. Blum, N.: A new approach to maximum matching in general graphs. Automata, Languages and Programming. Lecture Notes in Computer Science, vol. 443, pp. 586–597. Springer, New York (1990)
9. Cipra, B.A.: The best of the 20th century: editors name top 10 algorithms. SIAM News **33**(4) (2000)
10. Dasgupta, S., Papadimitriou, C.H., Vazirani, U.: Algorithms. McGraw-Hill, New York (2006)
11. Edmonds, J., Johnson, E.: Matching, Euler tours and the Chinese postman. J. Math. Progr. **5**(1), 88–124 (1973)
12. Fleury: Deux problèmes de Géométrie de situatio. Journal de mathématiques élémentaires 257–261 (1883)
13. Gabow, H., Tarjan, R.: Faster scaling algorithms for general graph-matching problems. J. ACM **38**, 815–853 (1991)
14. Heideman, M., Johnson, D.H., Burrus, C.S.: Gauss and the history of the fast fourier transform. ASSP Mag. **1**(4), 14–21 (1984)
15. Hierholzer, C.: Über die Möglichkeit, einen Linienzug ohne Wiederholung und ohne Unterbrechung zu umfahren. Mathematische Annalen **6**(1), 30–32 (1873)
16. Hopcroft, J.E., Motwani, R., Ullman, J.: Introduction to Automata Theory, Languages, and Computation, 3rd edn. Addison-Wesley, Boston (2006)
17. Karmarkar, N.: A new polynomial-time algorithm for linear programming. Combinatorica **4**(4), 373–395 (1984)
18. Khachiyan, L.G.: A polynomial algorithm in linear programming. Doklady Akademiia Nauk SSSR **244**, 1093–1096 (1979). English translation: Soviet Mathematics Doklady, 20(1):191–194
19. Klee, V., Minty, G.J.: How good is the simplex algorithm? Inequalities III. Shisha, O. (ed.). In: Proceedings of the Third Symposium on Inequalities held at the University of California, Los Angeles. pp. 159–175, Academic Press (1972)
20. Kwan, M.K.: Graphic programming using odd or even points. Chinese Math. **1**, 273–277 (1962)
21. Lawler, E.L.: The great mathematical Sputnik of 1979. Math. Intell. **2**(4), 191–198 (1980)

22. Nešetril, J., Nešetrilová, H.: The Origins of Minimal Spanning Tree Algorithms—Boråvka and Jarník. Documenta Mathematica: Extra Volume "Optimization Stories", pp. 127–141 (2012)
23. Pióro, M., Szentesi, A., Harmatos, J., Jüttner, A., Gajowniczek, P., Kozdrowski, S.: On open shortest path first related network optimisation problems. Perform. Eval. **48**(1), 201–223 (2002)
24. Younger, D.H.: Recognition and parsing of context-free languages in time n^3. Inf. Control **10**(2), 189–208 (1967)
25. Zisserman, A.: Linear Programming. Lecture 3, available via DIALOG. http://www.robots.ox. ac.uk/az/lectures/b1/lect3.pdf. Cited on 26 September 2015

Chapter 17
Common Problems Not Known to Be in P

For which common and useful problems are polynomial time algorithms (yet) unknown?

In this section we present another set of useful and famous problems. This time, however, nobody knows yet whether they are in **P**. No algorithms have been discovered yet that solve (decide) these problems in polynomial time (asymptotically). For simplicity, all problems in this chapter will be formulated as decision problems, even if they more naturally occur as optimisation problems.[1]

The problems included in this chapter are the Travelling Salesman Problem (Sect. 17.1), the Graph Colouring Problem (Sect. 17.2), the Max-Cut Problem (Sect. 17.3), which are all graph problems, the 0-1 Knapsack Problem (Sect. 17.4), and the Integer Programming Problem (Sect. 17.5). We also discuss the implications of the fact that no polynomial solutions are known (Sect. 17.6).

For all problems in this chapter one can still find a decision procedure, e.g. by implementing some kind of "brute force" generate-and-test strategy at the cost of generating exponentially many solution candidates (or more). Those simple-minded algorithms clearly cannot have a polynomial time bound. Spoiler-alert: in Chap. 18 it is argued that for all these (and many more optimisation) problems one can at least test in polynomial time, whether candidate solutions are actually solutions. Furthermore, Chap. 21 is all about how and to what extent solutions for such hard problems can be computed in practice in a reasonable amount of time.

---

[1] An advantage is that we do not have to care how long it takes to construct the solution.

© Springer International Publishing Switzerland 2016
B. Reus, *Limits of Computation*, Undergraduate Topics in Computer Science,
DOI 10.1007/978-3-319-27889-6_17

17.1 The Travelling Salesman Problem (TSP)

The *Travelling Salesman Problem* (TSP) has been mentioned in the introduction (Chap. 1) already. It is easy to understand and is therefore used heavily in research about optimisation algorithms and popular science publications alike. *"It belongs to the most seductive problems in combinatorial optimization, thanks to a blend of complexity, applicability, and appeal to imagination"* [4]. It is often used as a *benchmark* which allows different algorithms to be compared on their merits of finding solutions for "hard" problems.

A version of it was first defined by the Irish mathematician (physicist and astronomer) *W.R. Hamilton*[2] and the British mathematician *Thomas Kirkman*[3]. As a mathematical problem it was defined in 1930 by *Karl Menger*.[4] The name "Travelling Salesman Problem" was apparently introduced by American *Hassler Whitney*.[5] The origin of TSP lies with Hamilton's *Icosian Game*, which was a recreational puzzle based on finding a Hamiltonian cycle, a Sudoku of the 19th century, so to speak [5]. The problem is much liked by mathematicians, due to its simplicity and its numerous applications. A number of books[6] have been dedicated to it, requiring a varying degree of maths knowledge [2, 3, 10].

We now present TSP in decision problem form.

Definition 17.1 (*Travelling Salesman Problem*)

- **instance**: a road map of cities with distances, i.e. an undirected[7] weighted graph $G = (V, E, w)$ where vertices are cities and edges are roads such that weights $w(u, v)$ denote the distance between vertices (cities) u and v, a positive natural number k, a vertex a (the city to start in) and a vertex b (the city to end up in).
- **question**: is it possible to take a trip from vertex (city) a to b that passes through *all* cities such that the resulting trip has a length that is not greater than k?

This problem is important for haulage companies or any company that has to distribute goods to many cities (places). Saving miles means saving time and fuel, thus money.[8]

---

[2]Sir William Rowan Hamilton (3/4 August 1805–2 September 1865) was an Irish physicist, astronomer, and mathematician who made significant contributions to all those fields.

[3]Thomas Penyngton Kirkman (31 March 1806–3 February 1895) was a British mathematician.

[4]Karl Menger (January 13, 1902–October 5, 1985) was an Austrian-American mathematician and son of the famous Viennese economist Carl Menger. He is probably most renowned for the three-dimensional generalisation of the Cantor set and Sierpinski carpet (fractal curves) called Menger sponge.

[5]Hassler Whitney (23 March 1907–10 May 1989) was an American award-winning mathematician.

[6]In 2012, there has even been a movie released with this title [13] that won the "Best Movie" award at the Silicon Valley Film Festival.

[7]This is the so-called symmetric version of the TSP, there is also an asymmetric one where the cost of travelling from u to v does not equal the cost of travelling from v to u.

[8]Apparently bees can learn quickly to find the shortest tour from their bee hive to all the food sources (i.e. flowers) in a meadow. Researchers of Royal Holloway University, London, have

It should be pointed out that TSP is not just relevant for salesmen, haulage companies, and transport providers. The emerging trend of "Retail-as-a-Service" (RaaS)[9] may actually change our shopping experience in the future. It requires complex logistics to distribute goods to a large number of households. For such applications TSP can be generalised to multiple salesmen (or vehicles), the resulting problem is then known as the *Vehicle Routing Problem* (VRP). This is particularly important for RaaS and even more so the Vehicle Routing Problem with Time Windows (for delivery), Capacitated VRP, Multiple Depot VRP, VRP with Backhauls and VRP with Pick-up and Deliveries. For a survey of all these with details see [12].

TSP is furthermore relevant to microchip and other manufacturing processes. It pays off to find the shortest path for a tool, e.g. soldering equipment or a drill, to move during a production process. In the latter examples cities correspond to soldering points and edges to the potential moves of the soldering iron. As subproblem, TSP also appears in the DNA sequencing problem (cities are DNA fragments).

It is instructive to look how quickly the number of possible tours grows which prohibits a simple "generate-and-test" solution for TSP. For three cities there are only 12 possible tours, for 8 cities we have already 20,160 possible tours. Let us now increase the number of cities more steeply: for 20 cities we have over 60,000 *trillion* tours. And if we go beyond 70 cities, we have definitely more possible tours than there are atoms in the known universe.[10]

Note that ignoring the distances, a path through all vertices from A to B is also called a *Hamiltonian path*. Accordingly, a tour is called a *Hamiltonian circuit*. Again, as with Euler circuits, there is an abstract graph problem behind the more applied salesman-packaging of the problem.

It is worth pointing out the difference between the Travelling Salesman Problem and the Postman Problem (from Sect. 16.5.5), both request shortest tours, but in the first case every city needs to be visited once (i.e. vertex in the underlying graph), whereas in the second case every road (i.e. edge in the underlying graph) needs to be visited at least once.

(Footnote 8 continued)
published some results in 2010. However, it is questionable how big the route can be that they manage to solve.

[9]RaaS in the context of *Network Operating Systems* means "Routing-as-a-Service".

[10]One usually calculates between 10^{78} and 10^{80} atoms in the known universe. In the "J" series of the BBC panel game quiz show "QI" (Quite Interesting), host Stephen Fry mentioned a fact about 52!, the number of possibilities you can shuffle a pack of (poker) cards. The number is so big (around 8×10^{67}) that it is unlikely that anybody who shuffled a deck of cards before produced the same order of cards in the entire history of humanity! This requires, of course, that the deck is shuffled properly, but apparently 4 decent shuffles suffice. And of course, it just expresses an extremely low probability. So when Fry introduced his shuffling of a deck with the words "*I'm going to do something that has never been done by any human being since the beginning of time*", he is not quite right actually. He should have said "that is very unlikely (with virtually zero probability) to have been done by any human being before".

17.2   The Graph Colouring Problem

How many colours do we need to colour a map such that no adjacent counties (or countries) have the same colour?[11] This depends on the map, but in any case we never need more than 4 colours, even if some countries have many more neighbours. That four colours are always enough for a map is a famous result (The Four-Colour theorem) proved by *Kenneth Appel*[12] and *Wolfgang Haken*[13] in 1976 [1],[14] but has been already observed by Englishman Francis Guthrie in 1852.

Maps can be seen as special graphs (planar graphs) where vertices are countries and edges between countries exist if the two countries have a common piece of borderline. This is an abstraction process similar to the step Euler made from the bridges of Königsberg to a graph.

Therefore, we can formulate the above colouring problem for maps as a more general problem about colouring of graphs:

Definition 17.2 (*Colouring Problem*)

- **instance**: an undirected graph G and a number K of colours to be used where $K \geq 3$ (otherwise it is easy).
- **question**: using only K colours, can one colour graph G in such a way that each pair of adjacent nodes (connected by an edge) is coloured differently?

Example 17.1 An example of a graph with 6 vertices coloured in using only three colours (blue, red and green) is given in Fig. 17.1.

Although the Graph Colouring Problem's name may suggest that it were not applicable,[15] it actually occurs quite often in various disguises. Usually, when limited resources have to be managed that are in conflict with each other, this problem is somewhere not far away. In compiler construction, for instance, registers need to be allocated to variables. The question how many registers you need is a version of the Graph Colouring Problem. In air traffic or network traffic control one has a similar problem too: interfering planes or transmitting stations correspond to vertices that are connected by an edge. The number of colours needed corresponds to the maximal number of resources (registers to store variables in compilers, channels to deal with plane traffic or frequencies) needed at any given time to meet the requirements.

---

[11]In other words: countries that share a border need to be drawn in different colours.

[12]Kenneth Ira Appel (October 8, 1932–April 19, 2013) was an American mathematician. With Wolfgang Haken he solved the Four-Colour theorem, one of the most famous problems in mathematics. He was awarded the Fulkerson Prize.

[13]Wolfgang Haken (born June 21, 1928) is a German-American mathematician who works in topology. With Appel he proved the Four-Colour-Thereom and was awarded the Fulkerson Prize in 1979.

[14]The proof by Appel and Haken was the first proof of a major theorem that used computers to "compute" parts of the proof in order to deal with the many resulting different cases and combinatorial problems. It was therefore initially not trusted by everyone. In 2004 a fully machine-checked proof of the Four-Colour theorem has been carried out [7].

[15]When do you want to colour a graph in real life?

Fig. 17.1 A coloured
(non-planar) graph

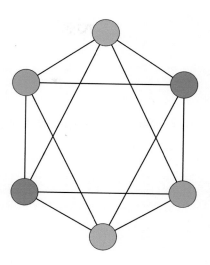

Note that the Colouring Problem for planar graphs, i.e. graphs that can be given
a layout such that no edges cross each other (and thus correspond to maps) is only
non-trivial for $K = 3$ due to the 4-Colour-Theorem mentioned above.

17.3   Max-Cut Problem

In Sect. 16.5.4 we encountered the Max-Flow and Min-Cut Problem. Let us consider
now a variation of the flow-cut problem, where we look for maximum capacity in
the cut between the edges of two subsets of vertices in the network (graph). This
problem is called Max-Cut. As a decision problem, Max-Cut can be described as
follows:

Definition 17.3 (*Max-Cut Problem*)

- **instance**: weighted graph $G = (V, E, w)$ where $w(u, v) \in \mathbb{Z}$ describes the maximum capacity of flow from u to v, and a number K.
- **question**: can we partition, i.e. cut, the vertices V into two sets S and T such that
  the weight of all edges E between a node in S and a node in T is greater or equal
  than K?

In a simplified version there is no weighting and one is interested in the partition with
the most edges between S and T. The Max-Cut Problem has important applications
in VLSI[16] circuit design. The partitioning of the circuit into subsystems is essential

---

[16]Very Large Scale Integration.

for optimizing connectivity of components and this partitioning can be optimised with the help of Max-Cut.

Given how similar the Max-Cut Problem is to Min-Cut / Max-Flow, it is surprising that only the latter are known to be in **P**. Max-Cut is *not* known to be in **P**.

17.4 The 0-1 Knapsack Problem

Consider the problem of packing a knapsack or a container with items such that a certain capacity of the container is not exceeded, optimising at the same time the value of the packed items. As a decision procedure we replace the optimum packing with one where the value of the packed items is above a certain threshold.

Definition 17.4 (*0-1 Knapsack Problem*)

- **instance**: a vector of n positive integers, representing the sizes, w_i $(1 \leq i \leq n)$ of the items, a natural number, representing the capacity or total weight of the container (knapsack), W, a vector of n positive integers, representing the profits (or values) of the items, p_i $(1 \leq i \leq n)$, and a natural number K
- **question**: Is there a way a subset of the n items can be packed (stored) in the container (knapsack) such that their profit adds up to at least K, without exceeding the total capacity W? This can be expressed by two inequalities and a boolean vector x_i $(1 \leq i \leq n)$, i.e. each x_i is either 0 or 1, indicating whether item i is in the container (1) or not (0):

$$\sum_{i=1}^{n} w_i \times x_i \leq W \qquad\qquad \sum_{i=1}^{n} p_i \times x_i \geq K$$

Again, this finds many applications in practice: consider truck loading, or any form of capacity planning. The so-called *Bin-packing* Problem is like *Knapsack* but it uses more than one container.

There are actually quite a number of similar problems known in the literature.[17] In [8] the problem is presented in an even simpler version without values and requiring that $\sum_{i=1}^{n} w_i \times x_i = W$. Also for this version no polynomial time algorithm is known.

The Knapsack Problem has a special status among the problems mentioned in this section. One might think that there is actually a polynomial time algorithm that solves it since one can prove the following result:

Proposition 17.1 *Any instance of the 0-1 Knapsack Problem can be solved in* $\mathcal{O}(n^2 \times p_{\max})$, *where (as in Definition 17.4) n is the number of items and $p_{\max}$ is the largest profit value.*

---

[17] Bounded Knapsack Problem, Multiple-Choice Knapsack Problem, Multiple Knapsack Problem, Subset-Sum Problem, Change-Making Problem and so on. In fact, the are even entire books [9, 11] dedicated to those problems alone.

Proof We again apply the concept of *dynamic programming* to produce a $n \times p_{\max}$ matrix V of integers where $V(i, v)$ denotes minimum weight of a selection among the first i items with a total profit that is at least v. This matrix can be computed by initialising $V(0, 0) := 0$ and $V(0, v) := \infty$ for all $0 \leq v$ and then computing in a nested loop: $V(i, v) := \min\{ V(i - 1, v), w_i + V(i - 1, v - p_i) \}$ for all $1 \leq n$ and $0 \leq v \leq \sum_i p_i$. The second term in the maximum is only defined if $p_i \leq v$. In cases where it is not defined $V(i, v)$ is simply set to $V(i - 1, v)$. Matrix V can be clearly computed in time $\mathcal{O}(n \times \sum_i p_i)$. The size of the optimal solution, v^*, is the largest v such that $V(n, v) \leq W$. The answer for the decision problem is "yes" if, and only if, $v^* \geq K$.

Is Proposition 17.1 not a contradiction to the claim that Knapsack is not known to be in **P**? The algorithm runs in time $\mathcal{O}(n^2 \times p_{\max})$ (see Exercise 5). We need to look at the size of the input altogether. We measure the size of integer numbers in terms of the digits needed to represent them. The input of the algorithm consists of $2n$ numbers for weight and profit, plus two more numbers W and K. One can assume that all the weights are smaller than W (see Exercise 2 in Chap. 21). Then the input size is $(\log_2 W) \times (n + 1) + (\log_2 p_{\max} \times n) + \log_2 K$ which is $\mathcal{O}(n \times (\log_2 p_{\max}))$. As runtime is $\mathcal{O}(n^2 \times p_{\max}) = \mathcal{O}(n^2 \times 2^{\log_2 p_{\max}})$, it follows that runtime is *exponential in the size of the input* and the above algorithm does *not* run in polynomial time. It would run in polynomial time if all input numbers of the problem were represented in the *unary* number system. This example shows how important it is to be precise about how the input size is measured. For reasonably small instances of numbers, however, the above mentioned algorithm based on dynamic programming does give good results. We get back to this in Chap. 21.

17.5   Integer Programming Problem

In Sect. 16.6 we met the *Linear Programming Problem* which referred to solving (finitely many) linear inequalities for a finite set of variables in $\mathbb{R}$. This problem is in **P**. If we just change one condition, namely requesting that the solutions for the variables be integers in $\mathbb{Z}$, things change significantly (and surprisingly maybe). We have to make this problem into a decision problem again, dropping therefore the optimisation function described in Definition 16.15 by vector **c**, just requiring the existence of a point satisfying the constraints:

Definition 17.5 (*Integer Programming Problem*)

- **instance**: A is an $m \times n$ matrix of real numbers, and **b** is a row vector of length m.
- **question**: is there a vector **x** of n positive *integers*, i.e. $\mathbf{x} \in \mathbb{Z}^n$ and $\mathbf{x} \geq 0$, such that $A\mathbf{x} \leq \mathbf{b}$?

Integrality restrictions often reflect natural indivisibilities of resources. For instance, one cannot divide a vehicle or a person in two or more parts, whereas the farmer from Example 16.7 was perfectly capable of dividing the area to grow wheat and barley into two arbitrary (real) amounts. Sometimes, the integrality constraints reflect the nature of the problem being a "decision-making" problem, when the only possible solutions allowed are 0 and 1.

The Integer Programming Problem is sufficiently general to be able to express other problems in this section, for instance the Knapsack problem (Definition 17.4) or the Max-Cut problem (Definition 17.3). This will be further explored in Chap. 20 and in Exercises 6 and 7.

As for all the other problems in this chapter, also for Integer Programming there is no known algorithm that solves it in polynomial time. This problem provides evidence that *restricting* a problem does not necessarily make it easier.

17.6 Does Not Being in P Matter?

There are, of course, many more problems which are unknown to be decidable in polynomial time. Garey and Johnson's book [6] is a classic text that lists many problems of that kind.

Although nobody has found yet any polynomial-time algorithms for the above problems, they need to be dealt with in practice on a case by case basis. If the input is small, a brute-force approach might work. If the input gets bigger, maybe some clever *branch and bound* techniques and other *heuristics*-based approaches do allow the solution of the problem for slightly bigger input.

There is some good news though, there is an upper bound for the problems presented, so they are not in a outrageously complicated complexity class. Alas, the class we can actually show they are in is **EXP**:

Proposition 17.2 *All the problems above are in **EXP**. It is important to point out that these are* decision problems, *not function (optimisation) problems. For some of the problems the optimisation problem version, where an actual solution or minimum has to be constructed, may not even be known to be in **EXP**.*

Proof This is relatively obvious for problems that have at most exponentially many possible solutions. This proposition will also follow more easily from the discussion in the next chapter.

One might think that not requiring the optimal solution but an *approximation* of the optimal solution makes the problem easier to solve, maybe to an extent that it becomes decidable in polynomial time. Alas, for most of the problems that yet have resisted polynomial time solutions, this is *not* the case, at least not unless we change the problem. We will discuss these and approaches of how to "ease the pain" of these hard problems in Chap. 21.

What next?

The problems in this chapter are not in **P**, but they have all in common that we can verify in polynomial time whether a given instance (candidate solution) is actually a solution. We will use this fact to define a new class of problems that more closely fits the problems we have seen in this chapter. This will us also give more of a handle to prove or disprove that the problems have polynomial time solutions.

Exercises

1. Consider the undirected weighted graph in Fig. 16.9 from Exercise 13, Chap. 16.
   a. The graph has five nodes, how many tours are there (starting from *A*)?
   b. Why does it not matter where the tour actually starts (and finishes)?
   c. In a "brute-force" way, find the shortest tour that visits every node (city) exactly once.

2. Look up the Held–Karp algorithm (also called Bellman–Held–Karp algorithm) that uses dynamic programming and solves the optimisation version of TSP (not the decision problem) and explain why its time complexity is $\mathcal{O}\big(|V|^2 \times 2^{|V|}\big)$.
3. Explain why the result of Exercise 2 proves that TSP (the decision problem) is in **EXP**.
4. Abstract away from the details of the map of South America and draw a graph that just represents the countries and their borderlines. Colour the twelve nodes representing the countries such that no two neighbouring countries have the same colour. We know that four colours should suffice.
5. Explain why the algorithm for 0-1 Knapsack given in the proof of Proposition 17.1 is $\mathcal{O}\big(n^2 \times p_{\max}\big)$.
6. Express the 0-1 Knapsack Problem (Definition 17.4) as an instance of the Integer (Linear) Programming Problem (Definition 17.5).
   Hint: The problem has already been given in a quite "linear programming" looking shape. However, one equation uses $\geq$ instead of $\leq$, and this needs to be changed using a basic algebraic transformation.
7. Express the Max-Cut Problem (Definition 17.3) as an instance of the Integer (Linear) Programming Problem (Definition 17.5).
   Hint: Introduce a variable for each vertex that encodes which partition the vertex is in, and a variable for each edge that encodes whether the edge is cut.
8. Consider the following problem, called *Vertex Cover Problem*:

   • **instance**: an undirected graph $G = (V, E)$ and a positive number k.
   • **question**: is there a subset $V_s \subseteq V$ of vertices, such that $|V_s| \leq k$ and for each edge $(u, v) \in E$ either $u \in V_s$ or $v \in V_s$ (or both)?

   The set V_s then "covers" all edges of G.
9. Show that *Vertex Cover* can be expressed as a special case of an Integer Linear Programming Problem (Definition 17.5).
   Hint: Introduce a variable for each vertex v of the graph that encodes whether v is in the cover set V_s.

10. Assume a county has five major cities which all need hospitals. It needs to be decided where to build (or close in a more realistic modern day scenario) hospitals. The minimum number of hospitals should be built, but there is one condition (C):

> From each city at least one hospital should be not less than a 10 minutes drive away.

A table T with 5×5 entries is given whose entries $T(u, v)$ state in minutes how long it takes to drive (an ambulance) from city u to city v. A group of local politicians would like to know whether 3 hospitals are sufficient to meet the condition (C). Explain how this question can be formulated as an Integer (Linear) Programming Problem (Definition 17.5).

References

1. Appel, K., Haken, W.: Every planar map is 4-colourable. Part I: discharging & Part II: reducibility. Ill. J. Math. **21**(3), 429–567 (1977)
2. Applegate, D.L., Bixby, R.E., Chvátal, V., Cook, W.J.: The Traveling Salesman Problem: A Computational Study. Princeton University Press, Princeton (2007)
3. Cook, W.J.: In Pursuit of the Traveling Salesman: Mathematics at the Limits of Computation. Princeton University Press, Princeton (2012)
4. Cook, W.J., Cunningham, W.H., Pulleyblank, W.R., Schrijver, A.: Combinatorial Optimization. Wiley, New York (1998). Reprinted 2011
5. Cummings, N.: A brief history of the travelling salesman problem. Available via DIALOG http://www.theorsociety.com/Pages/ORMethods/Heuristics/articles/HistoryTSP.aspx (2000). Accessed 26 Nov 2015
6. Garey, M.R., Johnson, D.S.: Computers and Intractability: A Guide to the Theory of NP-Completeness. W.H. Freeman, San Francisco (1979)
7. Gonthier, G.: Formal proof-the four-color theorem. Not. AMS **55**(11), 1382–1393 (2008)
8. Karp, R.M.: Reducibility among combinatorial problems. In: Miller, R.E., Thatcher, J.W. (eds.) Complexity of Computer Computations, pp. 85–103. Springer, New York (1972)
9. Kellerer, H., Pferschy, U., Pisinger, D.: Knapsack Problems. Springer, Berlin (2004)
10. Lawler, E.L., Lenstra, J.K., Rinnooy Kan, A.H.G., Shmoys, D.B. (eds.): The Traveling Salesman Problem: A Guided Tour of Combinatorial Optimization. Wiley, New York (1985)
11. Martello, S., Toth, P.: Knapsack Problems: Algorithms and Computer Implementations. Wiley, New York (1990)
12. Toth, P., Vigo, D. (eds.): The Vehicle Routing Problem. Society for Industrial and Applied Mathematics (2001)
13. Travelling Salesman Problem: Official Movie Site. Available via DIALOG. http://www.travellingsalesmanmovie.com (2012). Accessed 19 June 2015

Chapter 18
The One-Million-Dollar Question

In which complexity class are our hard problems if they cannot be shown to be in P? If this class was different from P what impact would that have?

The following quote addresses a phenomenon that we can observe also for the problems encountered in Chap. 17, for which we do not know whether they are in **P**:

> In theory it is much harder to *find* a solution to a problem than to *recognize* one when it is presented.
> Stephen A. Cook [3]

Although it is difficult to find efficient solution to the problems of Chap. 17 (even as decision problems), it turns out to be easy to check whether a certain candidate solution actually *is* a solution. As this appears to be such a common pattern, we will define a complexity class for it, called **NP**. The question whether this new class and **P** describe the same class of problems is a big open question, with a one-million-dollar price awarded for it. Whatever the answer is, whether the classes are equal or not equal, this will have dramatic impact on the IT landscape.

Cook gives another example of the enormous difference between finding a solution and checking whether a presented candidate is a solution [3]: *Rubik's cube*[1]: An unscrambled cube can be quickly identified as each face has a solid colour. The unscrambling action, however, can take a very long time, at least to the untrained person[2] that does twist the faces randomly. After all, there are 43,252,003,274,489,856,000 different permutations to go through [6].

---

[1]Rubik's cube is a three-dimensional combination puzzle invented by Ernő Rubik, a professor of interior architecture and design, to help explain three-dimensional geometry in 1974. The original cube consists of $3 \times 3 \times 3$ smaller cubes that rotate around 3 axes. It is considered one of the top-selling puzzle games of all time with allegedly more than 350 million cubes sold worldwide [6].

[2]Currently, the fastest ever time to solve the Cube is 5.25 seconds recorded by Collin Burns, an American teenager, in April 2015. This is according to the "World Cube Association" [11], and one may be surprised to hear that such an organisation exists.

© Springer International Publishing Switzerland 2016
B. Reus, *Limits of Computation*, Undergraduate Topics in Computer Science,
DOI 10.1007/978-3-319-27889-6_18

The quick verification will only need to check a given candidate solution. Such a candidate solution can be constructed by nondeterministically guessing its parts. Alternatively, a certificate for the solution can be used by the verifier. Using these ideas we will define a new class **NP** of problems (Sect. 18.1) for which verifiers exists that run in polynomial time on input of polynomial size. This class **NP** is contained in **EXP** and contains all our problems from Chap. 17 (see Sect. 18.4), which are therefore in **EXP**. The fact that they are in **EXP** (the class of problems decided by programs with exponential[3] time bound) just means that only decision procedures with very bad asymptotic complexity are *easy* to find. We will also briefly point out that neither the class **NP** nor **EXP** are the *worst* with respect to asymptotic complexity. There are much worse classes and problems that can only be decided in double, treble (and even more) exponential time.

The class **NP** can also be understood as class of problems "accepted" by a non-deterministic program. This is discussed in Sect. 18.2. It turns out that **NP** is again a robust class (Sect. 18.3) and that it gives rise to the biggest open problem in (theoretical) Computer Science (Sect. 18.5).

18.1   The Complexity Class NP

First of all, let us define the new complexity class. We make use of the fact (we already observed) that the problems for which no polynomial time decision procedure is known, have something in common: namely that one can *check* whether a given candidate solution is actually a solution in polynomial time. A program that can perform such a check is a "verifier". We thus define:

Definition 18.1 (*Verifier*) An L-*verifier* for a problem $A \subseteq \{0, 1\}^*$ is a L-program p (that always terminates) such that

$$A = \{ d \in \{0, 1\}^* \mid [\![p]\!]^{\mathrm{L}} (d, c) = \text{true for some } c \in \{0, 1\}^* \}$$

The extra input c is called the *certificate* used to verify that d is in A and c may be different for different instances d. The runtime of a verifier is measured only w.r.t. d and not the certificate c.

In analogy with the definition of **TIME** in Definition 13.2 we can now define a class of problems based on their verifiers:

Definition 18.2   1. The class of *problems L-verifiable in time f* is:
   $\mathbf{NTIME}^{\mathrm{L}}(f) = \{A \subseteq \{0, 1\}^* \mid A \text{ has an L-verifier } p \in \mathrm{L}^{time(f)}\}$
   2. $\mathbf{NP}^{\mathrm{L}}$ is the class of problems that have polynomial time L-verifiers, or in other words, $\mathbf{NP}^{\mathrm{L}} = \{A \subseteq \{0, 1\}^* \mid A \text{ has an L-verifier } p \in \mathrm{L}^{ptime}\}$

---
[3]Recall that the exponent can be a polynomial in n.

3. **NLIN$^\text{L}$** $= \{A \subseteq \{0, 1\}^* \mid A$ has an L-verifier $p \in \text{L}^{lintime}\}$
4. **NEXP$^\text{L}$** $= \{A \subseteq \{0, 1\}^* \mid A$ has an L-verifier $p \in \text{L}^{exptime}\}$

As usual, we drop the superscript if the language in question is clear from the context or irrelevant.

Now, why are those classes called **NTIME**, **NP**, **NLIN**, and **NEXP**, respectively? The "**P**" of course stands for "Polynomial" (time) but what about the "**N**"? The reason for the "**N**" is that the above mentioned classes can be defined differently. Traditionally, the definition relies on nondeterministic Turing machines. If Turing machines are given in the classic style as state transformers (extensions of automata), such a nondeterministic Turing machine is one where the transition function is actually a transition *relation*, so for any state and tape symbol there may be *several different* instructions. For the machine program style presentation given earlier, one simply adds a nondeterministic program instruction. This will be explained in the next section.

18.2   Nondeterministic Programs

We will now extend the semantics of our various definitions of languages (and machines) from Chap. 11 with a nondeterministic feature. For the machine-like models of computation, where programs are lists of labelled instructions, we define a new *nondeterministic* instruction:

$$\ell: \quad \texttt{goto} \quad \ell_1 \quad \texttt{or} \quad \ell_2$$

with semantics as follows:

$$(\ell, \sigma) \to (\ell_1, \sigma) \quad \text{if } I_\ell = \quad \texttt{goto} \quad \ell_1 \quad \texttt{or} \quad \ell_2$$
$$(\ell, \sigma) \to (\ell_2, \sigma) \quad \text{if } I_\ell = \quad \texttt{goto} \quad \ell_1 \quad \texttt{or} \quad \ell_2$$

This means that there are now *two* possible successor configurations when executing $\texttt{goto } \ell_1 \texttt{ or } \ell_2$ and not just one as before. Instead of a sequence of configurations we obtain a branching tree of configurations and Fig. 18.1 depicts an example of a tree of configurations describing the execution of a nondeterministic program. Whenever there is a $\texttt{goto } \ell_1 \texttt{ or } \ell_2$ instruction executed we obtain two new branches where the computation continues either at instruction ℓ_1 or instruction ℓ_2. Of course, both branches might lead to completely different computations and outcomes. Some might not even terminate. In Fig. 18.1 the paths ending in three vertical dots could be such nonterminating paths, for instance.

In order to introduce nondeterminism into WHILE and WH^{1}LE, we add a further command that has the same effect as the nondeterministic jump above:

$$\texttt{choose} \quad C_1 \quad \texttt{or} \quad C_2$$

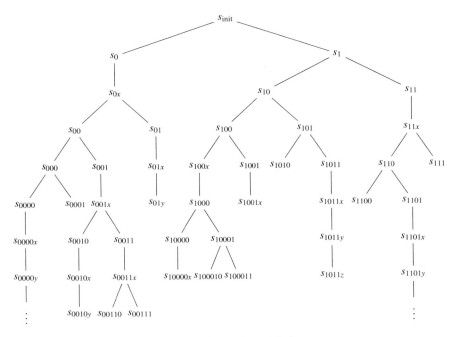

Fig. 18.1 Configurations of a sample run of a nondeterministic program

with the following semantics

$$\texttt{choose}\ C_1\ \texttt{or}\ C_2 \vdash \sigma \to \sigma' \quad \text{if}\ C_1 \vdash \sigma \to \sigma'\ \text{or}\ C_2 \vdash \sigma \to \sigma'$$

And this means that for a certain C and a certain store σ we may now have more than just one σ' for which $C \vdash \sigma \to \sigma'$ holds. If we have nondeterministic programs, then we cannot write any longer

$$[\![p]\!]^{\mathrm{L}}(d) = e.$$

In other words, the semantics of computation models has *not any longer a functional type*

$$[\![-]\!]^{\mathrm{L}} : \text{L-data} \to \text{L-data}_\bot$$

since there is no unique result any longer. Instead the semantics is now a relation between input and output:

$$[\![-]\!]^{\mathrm{L}} \subseteq \text{L-data} \times \text{L-data}_\bot$$

We will therefore have to redefine the class of problems decidable by nondeterministic programs within a given time bound. Instead of "decided by" we use the term "accepted by" (pointing thus out that a nondeterministic program is used).

Definition 18.3 (*Accepted set of a nondeterministic program*) A nondeterministic L-program p *accepts* input d if $(d, \text{true}) \in [\![p]\!]^{\text{L}}$. The set $Acc(p) \subseteq$ L-data, called the set of data *accepted by* p, is defined as follows:

$$Acc(p) = \{\, d \in \text{L-data} \mid p \text{ accepts } d \,\}$$

Note that it is sufficient to have *one* accepting execution sequence, this kind of non-determinism is therefore called *"angelic"*.[4] Due to the nature of our nondeterministic commands in programs we do not need to specify how such accepting sequences can be obtained. The right sequence is, so to speak, "guessed" any time there is a nondeterministic instruction executed. Since we do not have any influence on the guesses, we can as well view them as determined by flipping a coin. For acceptance of input we only require the existence of one accepting sequence. We can thus consider the nondeterministic choices be made in a "magic" way such that we obtain an accepting sequence. Next we will define a time measure for nondeterministic programs in a way such that we can assume that the choices produce a minimal accepting sequence with respect to runtime.

18.2.1 Time Measure of Nondeterministic Programs

The next problem that occurs is that we need to change the definition of runtime bounds for nondeterministic programs since we now have more than one execution sequence. Analogously to the input/output behaviour, one generalises the runtime usage measure *time* from a function to a relation.

Definition 18.4 Let p be a nondeterministic L-program. The time usage relation for p is of type:

$$time\_nd_p^{\text{L}} \subseteq \text{L-data} \times \mathbb{N}_\bot$$

which measures the time for each linear execution sequence (for a fixed number of outcomes for any choice instruction) according to the old Definitions 12.1 and 12.4, respectively. Then a time measure function for nondeterministic programs can be defined for the above relation as follows:

$$time_p^{\text{L}}(d) = \min\{\, t \mid (d, t) \in time\_nd_p^{\text{L}} \text{ such that } p \text{ accepts input } d\}$$

---

[4]The kind of nondeterminism that corresponds to requiring that all computation sequences need to be accepting is called *"demonic"* instead.

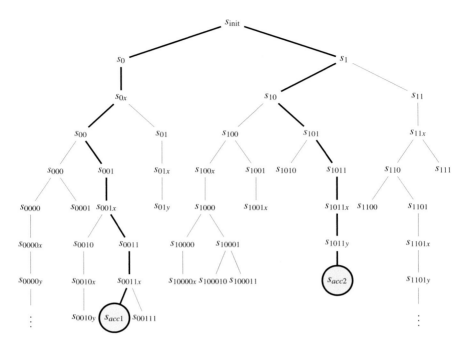

Fig. 18.2 Configurations tree with accepting paths in bold

The runtime of a nondeterministic program p is therefore the *shortest length of an accepting path in the tree of configurations*, i.e. the minimal possible time it takes to accept the input. In Fig. 18.2 some (assumed) accepting paths in the configuration tree from Fig. 18.1 have been highlighted. The shortest accepting path[5] is the one moving right through state s_1 and ending in (accepting) state s_{acc2}.

Using the above definition gives one another way to define the class $\mathbf{NP^L}$, namely as the class of problems accepted by a nondeterministic L-program with a polynomial time bound (according to the above).

The reader might wonder what happens if there are *no accepting sequences* whatsoever for nondeterministic program p with input d. Then the above definition yields $\text{time\_nd}_p^{\text{L}}(d) = \min \emptyset$ which usually is defined as $+\infty$. Since our time functions have to have results in $\mathbb{N}_\perp$, we instead define $\text{time\_nd}_p^{\text{L}}(d) = \perp$ if p does not have any accepting sequences for input d. If $\text{time\_nd}_p^{\text{L}}(d) = \perp$ then all sequences are *either* non-terminating or produce a result that is different from true.

---

[5] With length 7.

18.2.2   Some Basic Facts About NP

The first observation is that **P** is contained in **NP**:

Proposition 18.1 *P ⊆ NP.*

Proof Let $A \in$ **P**. Just use the program p that decides A in polynomial time as verifier. It does not need an extra certificate, so the certificate is the empty string. We already know that p runs in polynomial time so it is also a polynomial verifier.

As mentioned at the beginning of this chapter, the reason for the "N" in **NP** is the following Theorem:

Theorem 18.1 $A \in \mathbf{NP}^{TM}$ *if, and only if, A is accepted by a nondeterministic Turing machine (NTM) with polynomial time bound. Similarly, $A \in \mathbf{NLIN}^{TM}$ if, and only if, A is accepted by a nondeterministic Turing machine (NTM) with linear time bound and $A \in \mathbf{NEXP}^{TM}$ if, and only if, A is accepted by a nondeterministic Turing machine (NTM) with exponential time bound*

Proof Let $A \in \mathbf{NP}^{TM}$. We have to construct a nondeterministic Turing machine (NTM). This machine simulates the verifier by guessing the certificate nondeterministically. It is important here that the certificate has a size polynomial in the size of the input. This ensures that the runtime of the NTM is polynomial according to Definition 18.4. For the reverse direction, assume M is a NTM that accepts A. One constructs a deterministic verifier that simulates the execution path of the NTM determined by the certificate (which is an additional parameter to the verifier) describing corresponding guesses of M. Since the certificate has a length polynomial in the size of the input, the runtime of the verifier is polynomial.

Similar theorems can be shown for other notions of computations we discussed earlier.

So "NP" stands here for **N**ondeterministic **P**olynomial, as a problem in this class is *accepted by a nondeterministic polynomially bounded program* (e.g. Turing machine program).

It should be pointed out, first of all, that without any further time bounds the problems (languages) accepted by a nondeterministic Turing Machine p, $Acc(p)$, is just *recursively enumerable*, in other words *semi-decidable*, in the previously used sense. So the additional time bound is necessary to justify the term "decidable" for problems of the form $Acc(p)$. One might wonder therefore, why it should be acceptable to just consider the runtime of accepting sequences, and why we do not require the nondeterministic program to terminate and, while we're at it, even more require the program to return false when the input is not accepted.

The answer for both relies on the following fact: for any nondeterministic program that accepts a set A in time bounded by f, we can write another nondeterministic program that accepts A but cuts off all non-accepting computations sequences after $f(|d|)$ steps (like the efficient timed universal program did) and returns false. This program will need some extra steps but overall the runtime of the new program will still be polynomially bounded if f was a polynomial. So as far as **NP** is concerned, it does not matter what happens in non-accepting paths of the nondeterministic program.

18.3   Robustness of NP

We answer two questions here. Firstly, can nondeterministic programs solve problems that deterministic programs can't solve? Secondly, does it matter which language we enrich by nondeterministic commands (so can we solve more problems with nondeterministic Turing machine programs than we can solve with, say, nondeterministic While-programs)?

The answer to the first question is no. The Church–Turing thesis still holds also for nondeterministic machines and models. To see this, one only needs to compile a nondeterministic program into a deterministic one. This can be done by the concept called "dove-tailing".[6] We run all the different possible computation sequences in a sequential way. One can e.g. write an interpreter for nondeterministic programs that simulates the execution of all possible branches concurrently. Since there can be many nondeterministic commands in a program, the tree of possible transition steps can branch quite enormously, in the worst case after n steps there could be 2^n different execution branches. This means that the runtime of this interpreter will be exponentially slow (in the number of nondeterministic jumps taken). By partially applying the interpreter to the program in question we get a new program that is now deterministic. Note that the exponential slow-down in the simulating deterministic program is actually the intended motivation for using nondeterministic programs in the first place.

The answer to the second question is also *no*. Analogously to the deterministic case (**P**) we get the following:

Theorem 18.2   *Aside from data-encoding issues*

$$NP^{TM} = NP^{SRAM} = NP^{GOTO} = NP^{WHILE}$$

The proof is similar to the **P** version in the deterministic case, the only difference is that for **NP** we have to compile the verifier rather than the decision procedure.

It is worth pointing out that all problems in **NP** are also automatically in **EXP**.

Proposition 18.2

$$NP^{TM} \subseteq EXP^{TM}$$

---

[6]We have used "dove-tailing" in Chap. 9, Exercise 1(d) to show that if a set and its complement are both semi-decidable then the set is already decidable. The "dove tail" is split and the deterministic computation we do needs to be split up into all those potential computations (in a lock step way in order to avoid running into a non-terminating computation).

Proof Assume we have a problem $A \in$ **NP**. We know by Theorem 18.1 that A is accepted by a NTM in polynomial time in the sense of Definition 18.4. Assume this polynomial is p_1. We have to produce a deterministic decision procedure for A that runs in exponential time. We do this "brute-force" by simulating sequentially all the potential paths the NTM may take with input x in a breadth-first style in order to avoid getting caught in a non-terminating path. Whenever one such simulated path returns true the deterministic program returns true as well. Otherwise, one has to wait until all paths (they are all of finite length) have executed fully and returns false. Each choice doubles the number of paths the length of which is bounded by $p_1(|x|)$. Since each step of a Turing machine costs one unit, the deterministic decision procedure is $\mathcal{O}\left(2^{p_1(n)}\right)$. This proves that A is in **EXP**.

Note that this result holds also for other models of computations than TM by the Cook–Karp thesis.

18.4 Problems in NP

Let us look at some examples of problems that are in this class **NP**. All the common problems we encountered in Chap. 17 are actually in **NP**. This is not surprising, as this was our goal.

Theorem 18.3 *The* Traveling Salesman *Problem, the* Colouring[7] *the* Max-Cut *Problem, the* 0–1 Knapsack *Problem, and the* Integer Programming *Problem are all in* NP.

Proof The definition of **NP** makes it relatively easy to define a deterministic program that with the help of a certificate verifies the existence of a solution in polynomial time. The certificate is simply the (encoding of the) solution itself:

1. The certificate encodes a candidate for the solution (i.e. a tour, colouring, cut, etc.).
2. Verify that the encoded candidate solution has the required property (i.e. the tour visits each city once and has total length $\leq K$, the colouring is correct and uses not more than K colours, etc.).
3. The verification needs to be possible in polynomial time in terms of the size of the problem instance (e.g. a graph with a number as "quality measure"). For the given problems, it should be again obvious that the verification can be done polynomially in the size of the problem instance. For instance, for the TSP problem—where the input of the verifier is a graph $G = (V, E)$ and a number K—one needs to check that a given candidate for a tour visits all vertices which is done in time $\mathcal{O}(|V|^2)$, that start and end are the same vertex, which can be done in time $\mathcal{O}(1)$, that for each itinerary from v_1 to v_2 there is actually an edge, which can be done

---

[7] With $k \geq 3$.

in time $\mathcal{O}(|V| \times |E|)$, and that the sum of distances travelled is smaller or equal K, which can be done in time $\mathcal{O}(|V|)$. Putting all these checks together, the verification runs in polynomial time in the size of the input which for a graph G with number K is $|V| + |E| + K$.[8] Details of other problems will be discussed in exercises (see e.g. Exercise 1).

Corollary 18.1 *From Theorems 18.2 and 18.3 follows immediately that all the problems in Chap. 17 are in* ***EXP***.

It may now appear as if **EXP** contained all the "kings and queens of hard problems", but it turns out there are problems that cannot be decided even in exponential time. An example how to construct such a problem will be discussed in Exercise 8. There is one very famous decision problem for so-called *Presburger*[9] *arithmetic* who introduced it in 1929 [8]. Now, hang on, we know that the *Entscheidungsproblem* and thus (Peano) arithmetic is *undecidable* from Sect. 9.5, so how can there be a decision procedure for arithmetic? It can be explained by the fact that Presburger arithmetic is much weaker than (Peano) arithmetic, as it only includes addition and equality, and *no* multiplication. This makes it decidable. *Michael Fischer*[10] and *Michael Rabin*[11] [7] then proved in 1974 that there is a constant $c > 0$ such that any decision procedure for Presburger arithmetic must be *at least* in $\mathcal{O}(2^{2^{cn}})$ which is called "double exponential". But this is not the end of the road. In fact there are problems that are not even in

$$\mathcal{O}\left(2^{2^{\cdot^{\cdot^{\cdot^{2^{cn}}}}}}\right).$$

The problems in the cumulative hierarchy above, first defined in [10], are called *elementary*.[12] A problem that is not even elementary in this sense is, for instance, the problem whether two given regular expressions that may use the complement operator (i.e. ¬) are equivalent [9]. As far as this book goes, this will be the hardest decidable problem we mention.

---

[8] If the number is represented in binary then use $\log_2 K$ instead of K.

[9] Mojżesz Presburger (27 December 1904–1943?) was a Polish mathematician, logician, and philosopher and a student of Alfred Tarski. In his memory the European Association for Theoretical Computer Science created the annual Presburger Award for young scientists in 2010.

[10] Michael J. Fischer (born April 20, 1942) is an American (applied) mathematician and computer scientist and an ACM Fellow. He is known for his work in concurrency and he is a pioneer of electronic voting systems.

[11] Michael Oser Rabin (born September 1, 1931) is an Israeli computer scientist. Together with Dana Scott he won the Turing Award for their work on finite nondeterministic automata in 1976.

[12] Elementary? Not really, my dear Watson. The name "elementary" was chosen for that class of problems as they are still decidable and thus "elementary" in that sense (and only in that sense).

18.5 The Biggest Open Problem in (Theoretical) Computer Science

The big open problem actually is whether **NP** is the same as **P**.

$$\mathbf{P} \stackrel{?}{=} \mathbf{NP}$$

If **P** and **NP** were the same class then for all the hard problems we discussed in Chap. 17—for which nobody has found a polynomial time algorithm to date—there would actually have to exist such a polynomial time algorithm. This open problem has attracted much attention. The *Clay Mathematics Institute* that offers prize money of one million dollar for a (proven correct) answer to seven "Millennium Problems" [4], has included **P = NP**? as one of those seven[13] problems [5] although, strictly speaking, it is not (just) a mathematical problem but one of computation and thus theoretical computer science.

Every few years somebody makes a valid effort and claims to have solved **P = NP**? (usually attempting to prove **P ≠ NP**) but so far all those attempts have been quickly refuted by the experts in the field.[14]

Fact is that still nobody knows whether **P = NP**. We will see shortly that the answer to this question hugely affects computer science as we know it. In a way, we've already noticed this when we observed that for none of the problems in Chap. 17 we know any polynomial time algorithms. Basically, most experts in the field believe that **P** is not equal to **NP**. *Scott Aaronson* puts it like this:

> *If NP problems were feasible, then mathematical creativity could be automated. The ability to check a proof would entail the ability to find one. Every Apple II, every Commodore, would have the reasoning power of Archimedes or Gauss* [2].

Moreover, the other *Millennium Problems* could then most likely be solved by a program too.

The questions whether **P = NP** appears to have even wider impact outside computer science. If **P ≠ NP** was true, it may imply that time travel is impossible and that we can never have "godlike powers" in physical reality [1]. We will get back to the connection to physics in Chap. 23.

A small number of researchers think that maybe **P = NP** is true but that the polynomial time algorithms required are so complicated that no-one can (ever) understand and thus find them. However it may be, it remains the most famous open problem in complexity theory, but it turns out that there are many, many more open problems (and some of them we meet in later chapters). One could certainly argue that we know a relatively small amount about computational complexity so far. There appear to be

---

[13] Note that one of them, the *Poincaré Conjecture* has been proved in 2002 by Russian mathematician *Grigoriy Perelman* who was eventually awarded the Millennium Prize from the *Clay Institute* in 2010 but declined the prize, as he declined many other prestigious awards before.

[14] The most recent such proof attempt that received wide press coverage was submitted in August 2010 by *Vinar Deolalikar* from HP Labs.

more *open* problems than there are solved ones. Researchers usually argue that the reason for this is that we have not found yet the right mathematical tools to tackle them. To add just quickly another (open) problem we can already formulate: it is unknown whether $\mathbf{NP} = \mathbf{EXP}$, we only know half of this result (see Theorem 18.1). Therefore, if we are looking for a problem that is not even in $\mathbf{NP}$ then $\mathbf{EXP}$ can't help for now. We will discuss in Exercise 7 the existence of decidable problems, that are not even in $\mathbf{NP}$.

What Next?

We have seen that our hard problems fall into the class $\mathbf{NP}$, but we do not know whether $\mathbf{P} = \mathbf{NP}$. However, we would like to distinguish the problems that are hard, and lack efficient decision procedures so far, from the ones that we can prove are in $\mathbf{P}$. In the next chapter we define a class of problems that is contained in $\mathbf{NP}$ and characterizes exactly those hard problems. We will show that this class can serve as "lower bound" in the sense that any problem in this new class of hard problems is feasible if, and only if, $\mathbf{P} = \mathbf{NP}$.

Exercises

1. Show that the 0-1 Knapsack problem is in $\mathbf{NP}$.
2. Show that the *Vertex Cover* problem from Exercise 8 in Chap. 17 is in $\mathbf{NP}$.
3. Let $\mathbf{NP}_T$ be the class of problems accepted by a nondeterministic program such that *all* its possible execution sequences terminate in polynomial time, not just the minimal accepting one. By definition we have that $\mathbf{NP}_T \subseteq \mathbf{NP}$. Explain why also the other inclusion $\mathbf{NP} \subseteq \mathbf{NP}_T$ holds. From this it follows trivially that $\mathbf{NP}_T = \mathbf{NP}$. This result means that it does not matter whether we just consider the time the shortest accepting run takes or whether we consider the runtime of all accepting runs.
4. Show that $\mathbf{NP}$ is closed under union and intersection.
5. Show part of Theorem 18.2, namely that $\mathbf{NP}^{\mathrm{GOTO}} = \mathbf{NP}^{\mathrm{WHILE}}$.
6. It is unknown whether $\mathbf{NP} = \mathbf{EXP}$. But if it were, what important result would immediately follow?
7. Are there decidable problems that are *definitely* (provably) not even in $\mathbf{NP}$? We already know that the problems in $\mathbf{EXP}$ won't do the job as we don't know whether $\mathbf{NP} = \mathbf{EXP}$. Therefore, proceed as follows: Use the class $\mathbf{NEXP}$ instead and argue that $\mathbf{NP} \subsetneq \mathbf{NEXP}$.[15]
8. Exercise 7 was about the existence of problems that are not in $\mathbf{NP}$. Clearly, a problem that is not in $\mathbf{EXP}$ cannot be in $\mathbf{NP}$. Show that there must be a problem that is not in $\mathbf{EXP}$.
   Hint: Use the Hierarchy Theorem of Sect. 15.2.

---

[15]In Chap. 20 Exercise 15 we will see a concrete problem that is not in $\mathbf{NP}$.

References

1. Aaronson, S.: Guest column: NP-complete problems and physical reality. ACM Sigact News **36**(1), 30–52 (2005)
2. Aaronson, S.: Quantum Computing Since Democritus. Cambridge University Press, Cambridge (2013)
3. Cook, S.A.: Can computers routinely discover mathematical proofs? Proc. Am. Philos. Soc. **128**(1), 40–43 (1984)
4. Clay Mathematics Institute: Millennium problems, available via DIALOG. http://www.claymath.org/millennium-problems. Cited on 16 June 2015
5. Clay Mathematics Institute: P versus NP problem, available via DIALOG. http://www.claymath.org/millennium-problems/p-vs-np-problem. Cited on 16 June 2015
6. Cube Facts: Rubik's®—the home of Rubik's cube, available via DIALOG. https://uk.rubiks.com/about/cube-facts/. Cited on 30 June 2015
7. Fischer, M.J., Rabin, M.O.: Super-exponential complexity of Presburger arithmetic. Proc. SIAM-AMS Symp. Appl. Math. **7**, 27–41 (1974)
8. Presburger, M.: Über die Vollständigkeit eines gewissen Systems der Arithmetik ganzer Zahlen, in welchem die Addition als einzige Operation hervortritt (English translation: Stansifer, R.: Presburger's Article on Integer Arithmetic: Remarks and Translation (Technical Report). TR84-639. Ithaca/NY. Department of Computer Science, Cornell University)). Comptes Rendus du I congrès de Mathématiciens des Pays Slaves, 1929, pp. 92–101 (1930)
9. Stockmeyer, L.J.: The Complexity of Decision Problems in Automata Theory and Logic, PhD Thesis. MIT, Cambridge (1974)
10. Stockmeyer, L.J., Meyer, A.R.: Word problems requiring exponential time: Preliminary report. In: Proceedings of the Conference on Record of 5th Annual ACM Symposium on Theory of Computing, pp. 1–9. ACM (1973)
11. World Cube Association: Results, available via DIALOG. https://www.worldcubeassociation.org/results/regions.php. Cited on 30 June 2015

Chapter 19
How Hard Is a Problem?

> *How can we characterise and formalise that problems are
> harder than others in a class?*

We have seen that it is unknown whether $\mathbf{P} = \mathbf{NP}$. Thus, we don't know whether the hard problems in Chap. 17 are actually in $\mathbf{P}$ (and thus most likely feasible) or not. We do not have any non-trivial lower bounds for them either.

Since we don't know whether there are any problems in $\mathbf{NP}$ that are not in $\mathbf{P}$, we want to at least identify the hardest problems in $\mathbf{NP}$ in some precise sense. After all, they are the most likely candidates not to be in $\mathbf{P}$. But how can we identify such problems?

The idea is to consider problems that are at least as difficult (hard) as all the other problems in the same class. We already know a technique to compare problems regarding their difficulty: reduction (see Sect. 9.4 which we refresh in Sect. 19.1). *Richard Karp*,[1] in his seminal paper [4], lists 21 problems that are in a very precise sense "hardest in $\mathbf{NP}$" and a specific form of reduction is suggested and used to prove that fact, called polynomial time reduction introduced in Sect. 19.2.

Those "hardest" problems (introduced in Sect. 19.3) do give us a handle to attack the $\mathbf{P} = \mathbf{NP}$ problem. In fact, we can show that it suffices to find a single polynomial algorithm that solves just one of those hardest problems in $\mathbf{NP}$ in order to conclude that $\mathbf{P} = \mathbf{NP}$. If we assume that $\mathbf{P}$ does *not* equal $\mathbf{NP}$ the previous observation provides us actually with a lower bound for $\mathbf{NP}$-complete problems: An $\mathbf{NP}$-complete problem is not in $\mathbf{P}$ unless $\mathbf{P} = \mathbf{NP}$. For this reason $\mathbf{NP}$-complete problems are one of the big success stories of complexity theory.

Moreover, in this chapter we will also learn that most optimisation problems are known to be either in $\mathbf{P}$ or hardest w.r.t. $\mathbf{NP}$. There are, however, a few notable exceptions: stubborn problems of which we do not know much more than that they are in $\mathbf{NP}$.

---

[1] Richard Manning Karp (born January 3, 1935) is an American computer scientist, most known for his work in complexity and computability theory for which he received, among many other awards, the Turing Award in 1985.

© Springer International Publishing Switzerland 2016
B. Reus, *Limits of Computation*, Undergraduate Topics in Computer Science,
DOI 10.1007/978-3-319-27889-6_19

19.1   Reminder: Effective Reductions

In Definition 9.4 we wrote $A \leq_{rec} B$ to express that A can be effectively reduced to B. This means that there is a computable and total reduction function f such that $x \in A$ if, and only if, $f(x) \in B$. We can reduce the question (problem) "is x in A" to the question "is $f(x)$ in B?" in an effective way (since f is computable by an "effective procedure"). Consequently, B is at least as hard as A. If we show this for any A in a certain class of problems, then B is at least as hard as all other problems in the class. It is hardest w.r.t. the given class.

A simple example for a reduction is the reduction from the *Travelling Salesman Problem* (TSP) as discussed in the introduction presented in Sect. 17.1. The version introduced there in Definition 17.1 allows for the tour to start and finish in different cities. If one wishes to come back to the starting point one could define a version of TSP, let's call it "TSPRound", that uses just one city as part of the instance of the problem.

Definition 19.1 (*Roundtrip Travelling Salesman Problem* (*TSPRound*))

- **instance**: a road map of cities with distances, i.e. a weighted undirected graph $G = (V, E, w)$ where vertices are cities and edges are roads with weights $w(u, v)$ denoting the length (distance) between vertices (cities) u and v, a positive natural number k and the start and finish (city) a.
- **question**: is it possible to take a round trip that start and finishes in city a and passes through *all* cities such that the resulting trip has a length that is not greater than k?

We can define a reduction function f that maps instances of TSP to instances of TSPRound as follows:

$$f(\langle G, K, a \rangle) = \langle G, K, a, a \rangle$$

where G is the graph of the map, K is the limit for the length of the tour and a is the city where the tour is to start and end. Of course, this is a particularly trivial reduction function that is clearly computable. If the two problems in the reduction are not so similar then it can be really difficult to find an appropriate reduction function.

19.2   Polynomial Time Reduction

If we consider reductions between problems within one (complexity) class that considers runtime bounds (as **P** and **NP** do) then we want the reduction not to lead out of this complexity class. Therefore, we have to add an additional condition on the reduction function. Not only must it be computable and total, also the additional time its execution takes, i.e. the time the reduction function takes to be executed, must not go beyond the bounds prescribed by the complexity class. As we consider complexity classes **P** and **NP**, this means the reduction function must be computable in *polynomial* time itself.

Definition 19.2 (*Polynomial time reduction*) The relation $A \leq_P B$ defines *polynomial time reduction* between problems A and B. In other words, it states the existence of an effective reduction from A to B in the sense of Definition 9.4 where the reduction function can be implemented by a program with polynomial time bound.

The chosen type of reduction is called "many-one" reduction because several values in A can be mapped to the same value in B. The test "is in B?" can therefore be used just one time (at the end).[2] This kind of many-one polynomial time reduction was introduced by *Emil Post*[3] and in the context of computational complexity championed by *Richard Karp*. The reduction between the two versions of TSP is clearly a polynomial time reduction as the reduction function can be implemented by a (deterministic) program that runs in constant and thus polynomial time.

Next we will look at some useful results about polynomial time reduction that we will use—implicitly and explicitly—in the following. They are very much related to Theorem 9.4 that helped us deduce (un-)decidability results with the help of effective reducibility $\leq_{\mathrm{rec}}$. We can prove an analogous result to Theorem 9.4 about $\leq_{\mathrm{rec}}$, but now we use polynomial time reduction $\leq_P$ and **NP** (or **P**) instead of the class of decidable problems.

Theorem 19.1 (Downward closure of **(N)P**) *If* $A \leq_P B$ *and* B *is in* **NP** *then* A *is also in* **NP**. *Similarly, If* $A \leq_P B$ *and* B *is in* **P** *then* A *is also in* **P**.

Proof We first show this for **NP**. So assume $A \leq_P B$ and B is in **NP**. From the latter it follows that there is a polynomial time verifier p for B. We have to construct a polynomial time verifier for A, let's call that one q. If we could achieve

$$[\![q]\!](x, c) = [\![p]\!](f(x), c) \quad (\dagger)$$

then we'd know that

$$[\![q]\!](x, c) = \text{true for a certificate } c \text{ iff } [\![p]\!](f(x), c) = \text{true for a certificate } c$$
$$\text{iff } f(x) \in B$$
$$\text{iff } x \in A$$

Note that the last step is a consequence of the fact that $A \leq_P B$. We thus know that q is a verifier for A. But how do we define q and why is it polynomial in the size of x? First of all we assume that all verifiers and other programs are given in WHILE. This is not a limitation due to Cook's thesis discussed in Chap. 14. Let us assume that p looks like that given in Fig. 19.1 and has polynomial time bound, say $p_2(n)$.

---

[2]There are other variations of reduction where e.g. the question "is in B?" can be asked several times.

[3]Emil L. Post (February 11, 1897–April 21, 1954) was an American mathematician and logician, also famous for his "Post System" that can be used as a model of computation.

```
p read xc {
  x := hd xc;
  c := hd tl xc;
  B
}
write y
```

Fig. 19.1 Verifier p showing $B \in \mathbf{NP}$

```
q read xc {
  x := hd xc;
  c := hd tl xc;
  x := <r> x;        // x := f(x)
  B                  // run p on input f(x)
}
write y
```

Fig. 19.2 Program q made from bits of p and r

We know that function f is computable by (deterministic) program r that runs in polynomial time, say $p_1(n)$, such that $x \in A$ iff $f(x) \in B$. We therefore define the program q as given in Fig. 19.2 where the macro call <r>x computes $f(x)$. This new program q fulfils condition (†) by construction. It remains to show that q runs in polynomial time. We have that

$$\begin{aligned} time_q(d, c) &= 2 + 3 + 4 + \text{"time for x :=< r \ \ x > "} + p_2(|f(d)|) \\ &= 9 + (1 + p_1(|d|)) + p_2(|f(d)|) \\ &\le 10 + p_1(|d|) + p_2(p_1(|d|)) \end{aligned}$$

We have used in the last inequality that $p_2(|f(d)|) \le p_2(p_1(|d|))$ which follows from the fact that p_2 is monotonic and $|f(d)| \le p_1(|d|)$ since the size of the output of program p that computes f cannot be larger than the time it takes to run p and produce this output.[4] So we know that q has a polynomial time bound since $10 + p_1(n) + p_2(p_1(n))$ is a polynomial in n which follows from the fact that the composition of two polynomials is a polynomial again, so $p_2(p_1(n))$ is a polynomial.

For the second statement the proof is analogous as the composition of the deterministic programs p and r in this case gives us a deterministic q.

Another result similar to the $\le_{rec}$ case is that $\le_P$ is *transitive*. Recall that this means that if $A \le_P B$ and $B \le_P C$ then $A \le_P C$.

---

[4]This follows from the definition of time measures as the construction of large expressions is accounted for in the time measure. For each time unit consumed one can at most produce one "bit" of the output.

```
q read x {
  y := <r1> x;      // y = f1(x)
  z := <r2> y       // z = f2(f1(x))
}
write y
```

Fig. 19.3 Program q that computes $f_2(f_1(x))$ in WHILE-syntax

Theorem 19.2 (Transitivity of $\leq_P$) *If $A \leq_P B$ and $B \leq_P C$ then $A \leq_P C$.*

Proof From $A \leq_P B$ follows that there is an f_1 computable by a program r_1 in polynomial time (say $p_1(n)$) such that $x \in A$ iff $f_1(x) \in B$. From $B \leq_P C$ follows that there is an f_2 computable by a program r_2 in polynomial time (say $p_2(n)$) such that $x \in B$ iff $f_2(x) \in C$. From both follows immediately that $x \in A$ iff $f_2(f_1(x)) \in C$. So the reduction function for $A \leq_P C$ is the composition of f_1 and f_2. Clearly this composition is again computable by first running r_1 and then r_2. We can now write a program q that implements the composition of f_1 and f_2 as outlined in Fig. 19.3.

It remains to check that q runs in polynomial time. We can calculate

$$time_q(d) = 2 + (1 + p_1(|d|)) + (1 + p_2(|f_1(d)|))$$

We don't know exactly what the size of $f_1(d)$ will be, but we can give an upper bound for it just like in the proof of the previous theorem and therefore we get that

$$time_q(d) \leq 2 + (1 + p_1(|d|)) + (1 + p_2(p_1(|d|)))$$

and therefore

$$time_q(d) \leq p_1(|d|) + p_2(p_1(|d|)) + 4$$

Now the right hand side is a polynomial again as the addition and the composition of two polynomials gives a polynomial.

19.3 Hard Problems

Definition 19.3 (*Complete Problem*) For any complexity class $\mathscr{C}$ a problem H is called *complete* for the class $\mathscr{C}$ if problem H is itself in $\mathscr{C}$ and is "hardest" in $\mathscr{C}$ in the sense that for all other problems $A \in \mathscr{C}$ we have that $A \leq H$ (for an *appropriate* instantiation of reduction).

We are interested in particular in the instance of $\mathscr{C}$ being **NP**, so we are interested in **NP**-complete problems for reasons that become clearer soon:

> a problem H is called *NP-complete* if problem H is itself in **NP** and is "hardest" in **NP** in the sense that for all other problems $A \in$ **NP** we have that $A \leq_P H$

In Karp's paper [4] such problems were still called "(polynomial) complete" with an obvious emphasis on the polynomial time reduction. Knuth reports in [6] how the switch of focus on the respective problem class came about.[5],[6]

A problem class can have more than just one "hardest" problem.[7]

Definition 19.4 (*Hard Problem*) We call a problem $\mathscr{C}$-hard if it is at least as hard as any other problem in $\mathscr{C}$. In this case the problem does not have to be in the class itself.

Obviously then the following holds:

Corollary 19.1 *A problem is NP-complete iff it is NP-hard and itself in NP.*

We can now use polynomial time reduction $\leq_P$ to prove results for **NP**-complete (and **P**-complete) similar to the ones above for **NP** and **P**:

Theorem 19.3 *If $A \leq_P B$ and A is NP-hard then B is also NP-hard.*

Proof To show that B is **NP**-hard we have to take any problem H in **NP**, and show that it can be reduced to B in polynomial time. Since A is **NP**-hard we know that it must be true that $H \leq_P A$. Since we also have that $A \leq_P B$ we get $H \leq_P A \leq_P B$ and—since $\leq_P$ is transitive—we obtain $H \leq_P B$ as required.

Corollary 19.2 (Upward-closure of completeness) *If $A \leq_P B$ and A is NP-complete and B is in NP then B is also NP-complete.*

Proof Follows from the above theorem and the fact that a problem is by definition **NP**-complete if it is **NP**-hard and itself in **NP**.

With the help of **NP**-complete problems we can show a very useful result that characterises the big problem we have in not knowing whether **P** = **NP**. It provides us with an angle to attack this big open problem:

---

[5]*"Another last minute terminology substitution happened when Aho, Hopcroft, and Ullman substituted "NP-complete" for "Polynomially-complete" in their text on Algorithms even though they had already gotten galley proofs using the original name. The name was changed at that late date as the result of a poll conducted throughout the Theoretical Computer Science community (suggested names were NP-Hard, Herculean Problem, and Augean Problem)"* (see footnote 6) [6]. It was actually Knuth himself who triggered the discussion about the terminology in [5].

[6]In Greek mythology Heracles (Hercules) was known for his strength. As a penance for having killed his family, he had to perform ten labours, the fifth of which was to clean the stables of Augeas in one day. It should be noted that Augeas was known for having the most cattle in the country. Heracles solved this task by rerouting the local rivers into the stables (which was then considered cheating, but this is another story…).

[7]For those who know about the properties of order relations: relation $\leq_P$ is not totally ordered, it is not even quite an order. It is just a so-called *preorder* (or *quasi-order*). It is reflexive and transitive but not anti-symmetric, i.e. it does not hold that $A \leq B$ and $B \leq A$ implies $A = B$. We can have, for instance two different "hardest" problems such that each must be reducible to the other but they are different problems still.

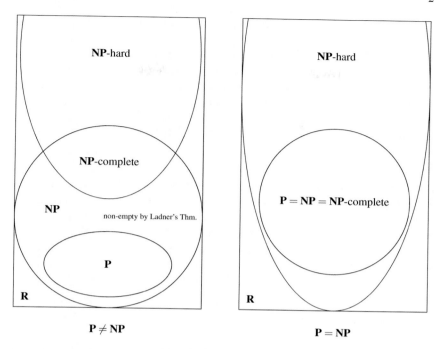

Fig. 19.4 "Two possible worlds view" on **P** = **NP**

Theorem 19.4 *If any **NP**-complete problem is already in **P** then **P** = **NP** (and the biggest open problem in theoretical computer science is solved).*

Proof Let H be the **NP**-complete problem that, by assumption, is also in **P**. We only have to show that **NP** $\subseteq$ **P** since we know that **P** $\subseteq$ **NP** holds anyway (see previous chapter) and from both we immediately can infer that **P** = **NP**.[8] To show **NP** $\subseteq$ **P** we take any problem $A \in$ **NP** and show that $A \in$ **P** holds as well (using the assumptions we have about H; of course this does not hold in general!). H is **NP**-complete and $A \in$ **NP** means by definition of **NP**-completeness that $A \leq_P H$. We now know also that H is in **P**. By the "Downward-closure of **P**", Theorem 19.1, we can conclude that $A \in$ **P**.

We don't know yet whether **P** = **NP**, and therefore we have two possible scenarios as highlighted in Fig. 19.4 which shows the two cases. Recall that **R** from Definition 9.2 denotes the class of decidable problems.

If **P** is not equal to **NP**, then by Ladner's Theorem [7] there must be at least one problem that is in **NP** but that is not **NP**-complete nor in **P**. These problems are called **NP**-intermediate problems. We omit the details here but point out that existence of such a problem is shown by diagonalisation yielding a very unnatural

---

[8] See Proposition 2.1.

problem. There are some more natural *candidates* for such intermediate problems which will be discussed in Sect. 20.6.

There are other interesting results that indicate that standard techniques won't work to prove whether $\mathbf{P} = \mathbf{NP}$. We won't go into details here, but the interested reader is referred to "relativized classes" where programs (or Turing machines) are allowed to ask an oracle a (complex) question and this costs just one unit of time. In that respect, the so-called *Baker-Gill-Solovay Theorem* [1] has been groundbreaking as it shows that there can be no relativizing proof for $\mathbf{P} = \mathbf{NP}$.

What Next?

We have seen how to compare problems within complexity classes and defined in a precise way the hard ones. We have also seen some results that provide a recipe how to show that a given problem is hard w.r.t. a class if we already know another such hard problem via reduction. We are mainly interested in the class $\mathbf{NP}$, so in the next chapter we will actually meet a first $\mathbf{NP}$-complete problem. Then we will also show that all the problems of Chap. 17 are $\mathbf{NP}$-complete and we will look into games and puzzles as problems and confirm that some famous ones are hard in the formal sense explained in this chapter.

Exercises

1. Recall the problems *Max-Cut* and *Integer Programming* from Chap. 17. Show that Max-Cut $\leq_P$ Integer Programming.
   Hint: Exercise 7 in Chap. 17.
2. Recall the problems *0-1 Knapsack* and *Integer Programming* from Chap. 17. Show that 0-1 Knapsack $\leq_P$ Integer Programming.
   Hint: Exercise 6 in Chap. 17.
3. Show Corollary 19.2 in detail.
4. Show that the concept of $\mathbf{R}$-completeness, based on effective reducibility *reduce*, is not useful (and thus never mentioned anywhere) by showing that every non-trivial problem in $\mathbf{R}$ is already $\mathbf{R}$-complete.
5. Describe the program q required for the proof of Theorem 20.11. It can use the semi-decision procedure p for A and the input d and it must hold that $[\![p]\!](d) =$ true iff $(q,q) \in \text{HALT}$.
6. Independently of the validity of $\mathbf{P} = \mathbf{NP}$ one can actually give two very simple problems that clearly are in $\mathbf{NP}$, but cannot be $\mathbf{NP}$-complete as there cannot be any reduction from non-trivial problems to them.

   a. What are these two very simple problems?
   b. Why do they not count as $\mathbf{NP}$-intermediate problems?

7. Let us show that the class of $\mathbf{NP}$-complete problems lacks wider closure properties by giving two $\mathbf{NP}$-complete problems such that the resulting problem under the given operation cannot be $\mathbf{NP}$-complete:

   a. Show that the class of $\mathbf{NP}$-complete problems is not closed under intersection

b. Show that the class of **NP**-complete problems is not closed under union.

Hint: Make clever use of Exercise 6.

8. What techniques do we have to prove that a given problem is *not* in **P**? Which depend on the assumption that $\mathbf{P} \neq \mathbf{NP}$?

9. Show that the Halting Problem HALT from Definition 8.3 is **NP**-hard but not **NP**-complete.

References

1. Baker, T.P., Gill, J., Solovay, R.: Relativizations of the P =? NP question. SIAM J. Comput. **4**(4), 431–442 (1975)
2. Greenlaw, R., Hoover, H.J., Ruzzo, W.L.: Limits to Parallel Computation: P-Completeness Theory. Oxford University Press, Oxford (1995)
3. Jones, N.D., Laaser, W.T.: Complete problems for deterministic polynomial time. Theor. Comput. Sci. **3**(1), 105–117 (1976)
4. Karp, R.M.: Reducibility among combinatorial problems. In: Miller, R.E., Thatcher, J.W. (eds.) Complexity of Computer Computations, pp. 85–103. Springer, New York (1972)
5. Knuth, D.E.: Terminological proposal. SIGACT News **6**(1), 12–18 (1974)
6. Knuth, D., Larrabee, T., Roberts, P.M.: Mathematical Writing No. 14. Cambridge University Press, Cambridge (1989)
7. Ladner, R.: On the structure of polynomial time reducibility. J. ACM (JACM) **22**(1), 155–171 (1975)

Chapter 20
Complete Problems

Which problems are NP-complete? How common are
NP-complete problems? What problems are complete in other
classes?

We have seen how to define hard problems in a class in terms of polynomial reduction. In particular, we have defined a notion of an **NP**-complete problem which is itself in **NP** and also **NP**-hard. In this chapter we will look at some *examples* of **NP**-complete problems and discuss how widespread these problems actually are. A famous result by Cook and Levin [3, 17] will provide us with a very first **NP**-complete problem in Sect. 20.1. To this problem one can then apply (an appropriate notion of) reduction to show that many others are **NP**-hard as well (Sect. 20.2). For instance, we will argue that all the decision problems from Chap. 17 are also all **NP**-complete, following another milestone paper [16].

There is a plethora of interesting **NP**-complete problems. In Sect. 20.3 we discuss some games and puzzles that we already find "hard" in a colloquial sense. We will present results that appropriately defined decision problem versions of those games are also "hard" in the sense of **NP**-hard. Chess, Sudoku, and tile-matching games will be covered. However, such games must first be generalised to allow for arbitrary large instances, because otherwise asymptotic time complexity is not applicable. This changes the nature of the games (some like chess more than others).

The complexity of evaluating database queries is the topic of Sect. 20.4. We will learn that even restricted queries may be **NP**-complete w.r.t. the size of the query expressions. Restricting to so-called acyclic queries with acyclic joins, however, gives rise to polynomial query expression complexity.

Reachability in networks using policy-based routing with general *packet filters* is also shown to be **NP**-complete in Sect. 20.5. The restricted problem, where routing is merely done by forwarding based on matching longest prefixes of IP addresses, is polynomially decidable though.

In Sect. 20.6 we briefly discuss the existence of so-called **NP**-intermediate problems that are not known to be in **P** nor to be **NP**-complete. Finally, we consider the concept of completeness for other complexity classes: **P** and **RE** (Sect. 20.7).

© Springer International Publishing Switzerland 2016
B. Reus, *Limits of Computation*, Undergraduate Topics in Computer Science,
DOI 10.1007/978-3-319-27889-6_20

The diligent reader will find that the classic book by Garey & Johnson [6] contains a large catalogue of **NP**-complete problems. We will refer to their list of **NP**-complete problems regularly in this chapter.

20.1   A First NP-complete Problem

So far we have seen results about **NP**-complete problems, but as far as we know, there could still be no such **NP**-complete problem in the first place. In this section we present such a problem. Then we can apply the previous results to it in order to prove that other problems are **NP**-complete as well. Note that we have carried out a similar task in the Computability section. It was first shown that the Halting Problem is undecidable and then reduction was used to show that other problems are undecidable as well.

The first concrete **NP**-complete problem we will encounter is about Boolean expressions (or formulae). Recall that a Boolean expression, or formula, is made up of variables, binary conjunction operator $\wedge$ ("and"), disjunction operator $\vee$ ("or") and unary negation operator $\neg$. An example is $(x \wedge y) \vee \neg z$.

Recall that a boolean expression $\mathscr{F}$ is *closed* if it has no variables but variables have been replaced by truth values "true" and "false". A closed expression $\mathscr{F}$ can be evaluated by the familiar rules of Boolean algebra, such as *true* $\wedge$ *false* = *false* or *true* $\vee$ *false* = *true*.[1]

Definition 20.1 (*Truth assignment and evaluation*) A *truth assignment* for $\mathscr{F}$ is a function θ mapping variables (of $\mathscr{F}$) to truth values (i.e. true and false). This mapping is similar to the stores we used to execute WHILE-programs where each program variable was mapped to a binary tree. If we have truth assignment θ for a boolean expression $\mathscr{F}$ then we can apply the truth assignment to the expression, obtain a closed formula and then evaluate this. For this we briefly write $\theta(\mathscr{F})$.

Example 20.1 For instance, if θ maps x to *true*, y and z to false, and $\mathscr{F} = (x \wedge y) \vee \neg z$ then $\theta(\mathscr{F}) = $ (true $\wedge$ *false*) $\vee \neg$*false* which can be evaluated to true.

Often boolean expressions are brought into normal forms:

Definition 20.2 (*conjunctive normal form (CNF)*) A boolean expression is in *conjunctive normal form* iff it is a finite conjunction of finite disjunctions of literals:

$$(A_{11} \vee A_{12} \ldots \vee A_{1n_1}) \wedge \cdots \wedge (A_{m1} \vee A_{m2} \vee \ldots \vee A_{mn_m})$$

where each A_{ij} is a literal, i.e. either a variable or a negated variable (x or $\neg x$). The disjunctive formulae $(A_{i1} \vee A_{i2} \ldots \vee A_{in_i})$ are called *clauses*.

It is well known that any boolean expression can be brought into *conjunctive normal form* (see Exercise 1).

---

[1] There are so-called "truth tables" for each Boolean operator that fully explain its semantics.

Definition 20.3 (*Satisfiability*) A boolean expression $\mathscr{F}$ is called *satisfiable* if it evaluates to true for some truth assignment θ.

Example 20.2 An example for a Boolean expression in CNF is:

$$(p \vee \neg q) \wedge \neg q \wedge (\neg p \vee p \vee q)$$

The example demonstrates that clauses can have different numbers of literals (here 2, 1, and 3) and can consist of just one literal (clause 2). It also demonstrates that a variable can appear positively and negatively in the same clause (here in clause 3 we have p and $\neg p$).

The above expression in CNF is satisfiable. Simply choose q to be *false*. This already makes the first and second clause *true*. The third clause is *true* whatever p and q are since $\neg p \vee p$ is a *tautology*, meaning that it is true independently of the value of p.

Definition 20.4 (*SAT*) The *Satisfiability problem*,[2] short SAT, is defined as follows:

$$\text{SAT} = \{\mathscr{F} \mid \mathscr{F} \text{ is a satisfiable boolean CNF expression }\}$$

In other words, the SAT problem can be presented like this:

- **instance**: a boolean expression $\mathscr{F}$ in CNF (conjunctive normal form)
- **question**: is $\mathscr{F}$ satisfiable?

SAT is clearly decidable, as we can simply try all possible truth assignments for $\mathscr{F}$. Since there can only be finitely many variables in $\mathscr{F}$, we can test all truth assignments and return true if we found one that lets $\mathscr{F}$ evaluate to true. If $\mathscr{F}$ contains n variable occurrences[3], the size of the formula is $\mathscr{O}(n \times \log_2 n)$ if we represent variable names in binary notation. In the worst case, there exist 2^n possible truth assignments to check (as for each variable there are two choices: true and false). So a brute-force search must take at least $2^n \times \mathscr{O}(n \times \log_2 n)$ steps and thus SAT is in **EXP**.

But we know more, thanks to a famous theorem by Cook (whom we have already encountered) and Leonid Levin[4]:

Theorem 20.1 (Cook–Levin-Theorem [3, 17]) *SAT is NP-complete.*

Proof (Sketch) We need to show two facts:

1. that SAT is in **NP** and
2. that all other **NP** problems can be reduced to it.

---

[2]This version of the problem is also sometimes called CNFSAT, when SAT is used to denote the problem for unrestricted propositional formulae.

[3]The actual number of different variables is then less than or equal n.

[4]Leonid Levin (born November 2, 1948) is a Soviet-American computer scientist who has apparently proved the same theorem as Stephen Cook independently behind the (then still existing) "iron curtain".

Item 1 is easy. Let n be the number of different variable occurrences. The verifier takes as certificate the right truth assignment θ which is linear in the size of n and evaluation is then linear in the size of the formula (and thus polynomial). To show item 2 we need to prove that any problem $A \in \mathbf{NP}$ can be reduced to SAT. This is difficult as we don't know what A is exactly. We therefore apply Theorem 18.1 which states that a problem is in $\mathbf{NP}$ iff it is accepted by a nondeterministic Turing Machine M in polynomial time where p denotes the concrete polynomial bound. The proof idea is now as follows: the reduction function must map input x to a formula $\mathscr{F}$ that is satisfiable if, an only if, $x \in A$, i.e. if running M on input x gives rise to an accepting path. Therefore, $\mathscr{F}$ must describe all possible paths (sequences of states) of Turing machine M and be satisfiable iff there is an accepting path. We briefly present the main concepts[5]: use boolean variable $C_{i,j,s}$ that states whether at time i cell j contains the symbol s. Since we have polynomially bounded runs, i.e. $0 \le i \le p(|x|)$, we know that only a certain finite part of the tape can be visited (since each move takes one unit of time) so only finitely many $C_{i,j,s}$ will be needed. Similarly, introduce variables $Q_{i,q}$ that state whether the machine at time i is in state q (or executes the instruction labelled q), and variables $H_{i,j}$ that state whether at time i the head of the machine scans cell j.

Using finitely many of those variables, one can encode a correct run of a Turing machine as Boolean expression (Exercise 4). The expression will be carefully constructed in CNF such that it is satisfiable only if the run it encodes is possible on the given machine with the given input, *and* accepting. One also needs to prove that the construction of the formula can be carried out in time polynomial in the size of the input, but this follows from the fact that the run of the machine is polynomially bounded and the constructed formula is therefore also polynomially bounded in size.

As usual precision is key: if one formulated SAT for formulae in *disjunctive normal form* the problem would be decidable in constant time (!) and thus actually be almost "the opposite" of $\mathbf{NP}$-complete.[6] This is the topic of Exercise 3.

The following slightly restricted version of (CNF)SAT was also shown to be $\mathbf{NP}$-complete by [3] via a reduction from SAT. The simpler shape of the clauses makes it easier to work with. It will be mentioned in the next section so we introduce it here:

Definition 20.5 (*3SAT* [3])

- **instance**: a boolean expression $\mathscr{F}$ in CNF such that every clause has *exactly* three literals.
- **question**: is $\mathscr{F}$ satisfiable?

---

[5] A more detailed proof can be found in many textbooks.

[6] There are many examples of this: consider Max-Cut versus Min-Cut or Integer Programming versus Linear Programming, TSP versus Postman Problem, and so on.

Corollary 20.1 ([3]) *3SAT is **NP**-complete.*

Proof This follows from a polynomial time reduction to SAT and Theorem 20.1. See Exercise 5.

20.2   More NP-complete Problems

The group of problems we encountered in Chap. 17 do not have any polynomial time solution yet, but we have seen that they are in **NP** in Theorem 18.3. As we have introduced the concept of **NP**-completeness, we can now even show that these problems are actually not only "hard" but they are "hardest" in **NP**.[7]

Theorem 20.2 ([16]) *The* Travelling Salesman *Problem, the* Graph Colouring *Problem, the* 0-1 Knapsack *Problem, the* Max-Cut Problem, *and the* Integer Programming *Problem are all **NP**-complete.*

Proof Richard Karp already provided various chains of reductions between those problems, starting with SAT, in his seminal paper [16]. He used a simpler version of Integer Programming, called *0-1 Integer Programming* where all the solution variables are only allowed to have either value 0 or 1. Also the TSP problem does appear in [16] only in the shape of the *Hamiltonian Circuit Problem* (so no weights for edges, i.e. distances between cities). Karp proved the following polynomial time reductions between all the **NP**-complete problems in his paper, including the following ones for the problems referred to in this theorem:

- SAT $\leq_P$ TSP[8]
- 3SAT $\leq_P$ Graph Colouring[9] (note that 3SAT is **NP**-complete according to Theorem 20.1.)
- Graph Colouring $\leq_P$ 0-1 Knapsack[10]
- 0-1 Knapsack $\leq_P$ Max-Cut[11]
- SAT $\leq_P$ 0-1 Integer Programming

Where Karp is not the original source of the reduction he provides the source in [16], which is an instructive read presenting very many reductions very concisely. A few of those reductions will be discussed in the exercises. As seen above, all problems listed in the theorem are **NP**-hard. Since we also know that all the listed problems are in **NP** (see Theorem 18.3) we can apply the "upwards-closure of **NP**-complete" result from Corollary 19.2 and conclude that the problems listed are also **NP**-complete.

---

[7]Recall that a consequence of this is that if any of those problems were in **P** we could deduce **P** = **NP**.

[8]Karp [16] actually showed SAT $\leq_P$ Clique $\leq_P$ Node Cover $\leq_P$ Directed Hamiltonian Circuit $\leq_P$ Undirected Hamiltonian Circuit.

[9]Karp [16] named this problem "Chromatic Number".

[10]Karp [16] actually showed that Graph Colouring $\leq_P$ Exact Cover $\leq_P$ 0-1Knapsack.

[11]Karp [16] actually showed that 0-1 Knapsack $\leq_P$ Partition $\leq_P$ Max-Cut.

Table 20.1 Problem Classification according to [6]

Problem name	Reference number
SAT	[L01]
3SAT	[L08]
TSP	[ND22]
Graph Colouring	[OPEN5]
0-1 Knapsack	[MP9]
Max-Cut	[ND16]
Integer Programming	[MP5]

When [6] was published, it was still an open problem whether Graph Colouring (for $k \geq 3$) was actually **NP**-complete

Analogously to the reductions in the proof of Theorem 20.2, one can show that many ("several hundred" according to [12, Chap. 25.3]) interesting (and useful) problems are **NP**-complete. The classic book [6] actually lists "some 300" [12]. It also introduced a classification system using specific codes that are often used for reference in order to be able to uniquely identify **NP**-complete problems. It should be pointed out though that the problems in [6] are presented as (optimisation) function problems and not as decision problems like here. For reference we provide the classification codes in Table 20.1.

Note that Cook and Levin's result did not find too much attention until Karp [16] showed that many *interesting* problems are actually **NP**-complete[12] via the concept of polynomial time reduction.

20.3 Puzzles and Games

Games are of natural interest to most people, particularly computer scientists as games are about problem solving "for fun". So a natural question to ask is this:

How difficult is it to find a winning strategy for game XYZ or solve a puzzle XYZ?

Or in the words of computational complexity:

In what complexity class is the decision problem: can I win game XYZ from a given position? In what complexity class is the decision problem of solving puzzle XYZ from a given position?

One can also generalise the above questions, abstracting away from the aspect of winning:

Can one get from position A to position B in game or puzzle XYZ?

The latter is commonly called the *Reachability Problem*.

---

[12][14, Chap. 25.6].

Fig. 20.1 Chess board

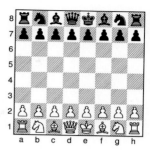

Fig. 20.2 Sudoku

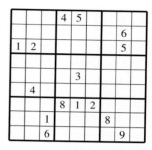

Fig. 20.3 A generic "match-three" game screen

In order to be able to apply what we have learned about our asymptotic worst-case complexity measures, we need to ensure that the problem instances can grow in size. Unfortunately, this is *unrealistic* for playing games as we know them. A chess board is 8×8 squares and similarly all other games have a certain fixed-size board or game state. But it does not make sense to ask whether the "normal" 8×8 version of chess is **NP**-complete. For our considerations, we can only consider games with *arbitrarily large descriptions*, as we do asymptotic complexity after all, so the questions above only make sense for a *generalised $n \times n$ board*. Such a generalisation is possible, but for games commonly known in their fixed-size version like chess, it may result in a rather different kind of game, and playing it may have little in common with the original game.

We will present the results for three types of games: *Chess* (Fig. 20.1), *Sudoku* (Fig. 20.2), and "match-three" based games like *Bejeweled*[TM] or *Candy Crush*[TM] (Fig. 20.3)[13].

"*The match-3 puzzle game market is a cluttered space. ...PopCap's Bejeweled sits at the top of the list alongside similar-looking titles that share similar names. There's Jewel World—a match-3 gem game, Jewel Mania—a match-3 gem game, Candy Crush Saga—a match-3 candy game, Candy Blast Mania—a match-3 candy game, Matching With Friends—a match-3 matching game and, well, some 2,000 more.*" [24]

The intuition that interesting games are "hard" to play will be confirmed by the fact that all those games are **NP**-hard or, even worse, **EXP**-hard.

20.3.1   Chess

Chess is a popular two-player strategy board game which is probably 1,500 years old and originated in Asia. It became very popular in the 19th century across the world and it was in the second half of that century when tournaments started. The rules are relatively complicated, so we will not repeat them here (they can be found e.g. in [29]).[14]

First, the game has to be generalised to arbitrary size for our purposes which is explained in [5]:

"*We let generalized chess be any game of a class of chess-type-games*[15] *with one king per side played on an $n \times n$ chessboard. The pieces of every game in the class are subject to the same movement rules as in 8×8 chess, and the number of White and Black pawns, rooks, bishops and queens each increases as some fractional power of n. Beyond this growth condition, the initial position is immaterial, since we analyze the problem of winning for an arbitrary board position.*"

Definition 20.6 (*Chess problem*)

- **instance**: an arbitrary position of a generalized chess-game on an $n \times n$ chessboard
- **question**: can White (or Black) win from that position?

The number of possible configurations in $n \times n$ chess is obviously bounded by 13^{n^2} as there are six different pieces for each player (and there can be no piece on a square). Therefore,[16] the Chess Problem is in **EXP**. But one can show more about this version of chess:

---

[13]*Bejeweled*[TM] is a trademark of PopCap Games and *Candy Crush*[TM] is a trademark of King.com Ltd.

[14]Non-chess players are encouraged to look them up and play this classic game.

[15]Without the 50 move rule that says that if no pawn was moved and no piece captured in the last 50 moves then it's a draw.

[16]Of course, not all these positions are legal positions but this can be ignored for this approximation.

Theorem 20.3 ([5]) *The Chess Problem (as defined in Definition 20.6) is* **EXP-**
complete.

Proof One uses polynomial time reduction of the problem G_3 (as defined in [23]) to
Chess. More details can be found in [5].

It is worth remembering what this means for classic chess: *"Thus, while we may
have said very little if anything about* 8×8 *chess, we have, in fact, said as much
about the complexity of deciding winning positions in chess as the tools of reduction
and completeness in computational complexity allow us to say."* [5]

20.3.2 Sudoku

*"The game in its current form was invented by American Howard Garns in 1979
and published by Dell Magazines as "Numbers in Place." In 1984, Maki Kaji of
Japan published it in the magazine of his puzzle company Nikoli. He gave the game
its modern name of Sudoku, which means "Single Numbers." The puzzle became
popular in Japan and was discovered there by New Zealander Wayne Gould. ...He
was able to get some puzzles printed in the London newspaper The Times beginning
in 2004. Soon after, Sudoku-fever swept England"* [19] and the rest of the world in
the following year.

The standard version of Sudoku uses a 9×9 grid with 81 *cells*. The grid is
divided into nine 3×3 *blocks* (i.e. blocks of size 3). Some of the 81 cells are already
containing numbers from 1 to 9. These filled-in cells are called *givens*. The goal
of the game is to fill in the remaining empty cells of the entire grid with digits
$\{1, 2, 3, 4, 5, 6, 7, 8, 9\}$ such that each row, each column, and each block contain
each of the nine digits exactly once.

Again, one has to generalise the game, for which the block size comes in handy.
The block size is 3 (with 3×3 cells) in the standard version described above and
visualised in Fig. 20.2. One can now generalise the size, which is often also called
rank, from 3 to n. "A Sudoku of *rank* n is an $n^2 \times n^2$ square grid, subdivided into n^2
blocks, each of size $n \times n$." [19]. In this generalised game the numbers used to fill
the cells range now from 1 to n^2 so they may not be single digits any longer. After
these generalisations we can now define a decision problem:

Definition 20.7 (*Sudoku problem*)

- **instance**: a Sudoku board of rank n partially filled with numbers in $\{1, 2, \ldots, n^2\}$.
- **question**: can one fill the blank cells with numbers in $\{1, 2, \ldots, n^2\}$, such that
  each row, each column, and each of the blocks contain each number exactly once?

So the Sudoku Problem as defined in Definition 20.7 is not actually about the solver
(producing the solution would not give us a decision problem) but rather an important
problem for the setter, namely whether the given starting position is actually solvable.

Theorem 20.4 *The Sudoku Problem (as defined in Definition 20.7) is **NP**-complete.*

Proof A stronger result was proved in [15], namely that Sudoku is **ASN**-complete, where **ASN** denotes the class of problems with a particular form of question: namely whether there are **a**lternative **s**olutions of the given problem with respect to certain already given solutions. Since SAT can be easily shown to be **ASP**-complete, it follows that Sudoku is also **NP**-hard. It can be easily seen that Sudoku is in **NP**, as one can write a verifier that takes a solution candidate as certificate, and then verifies that this candidate is correct by checking for n^2 columns, n^2 rows, and n^2 blocks, whether each number in $\{1, \ldots, n^2\}$ occurs exactly once. This can clearly be done in $\mathcal{O}(n^4)$.

20.3.3   Tile-Matching Games

With the advent of smartphones and tablets, a certain class of single player puzzle games rose to immense popularity that are "easy to learn but hard to master" [22]. These games were already available as browser-based or video games before. Some of the most successful such games are *Bejeweled*^TM and *Candy Crush Saga*^TM. *"Candy Crush is played by 93 million people every day, and it accrues an estimated $800,000 daily through players purchasing new lives and boosters that help them to conquer new levels. All told, half a billion people have downloaded the free app"* [22]. These are so-called tile-matching games, where the player manipulates tiles in order to make them disappear depending on the surrounding tiles (in *Bejeweled*^TM tiles are actually gems and in *Candy Crush*^TM tiles are sweets). The objective is to score highly rather than win or solve, which probably constitutes a big factor in the addictive nature those games appear to have. The key strategy is *"known as a variable ratio schedule of reinforcement and is the same tactic used in slot machines; you can never predict when you're going to win, but you win just often enough to keep you coming back for more."* [22]

Since there is no winning the game as such, one needs to define a slightly different set of problems, for instance (see [10]):

1. Is there a sequence of moves that allows the player to pop a specific tile?
2. Can the player get a score of at least x?
3. Can the player get a score of at least x in less than k moves?
4. Can the player cause at least x gems to pop?

Note that Problems (2–4) can be derived from Problem 1 which is the one we focus on below. Again, for complexity analysis one has to "generalize these games by only parameterizing the size of the board, while all the other elements (such as the rules or the number of tiles) remain unchanged." [10]

Theorem 20.5 ([10]) *Given an instance of Bejeweled*^TM*, the problem of deciding whether there exists a sequence of swaps that allows a specific gem to be popped*

*(Problem 1) is **NP**-hard. The decision problems (2–5) are **NP**-hard as well. The same result holds for many other "match-three" games, including Candy Crush^{TM}.*

Proof For Problem 1 one uses a reduction from a variation of SAT, called 1in3PSAT. In this version, one is given n variables $x_1, \ldots x_n$ and m clauses, each having at most 3 variables. The question is whether we can choose a subset of variables to set to true such that every clause has *exactly one* true variable. Then one can reduce from Problem 1. The details are very technical and can be found in [10].

Note that Problem 1 allows an unbounded number of moves, but in Candy Crush^{TM} the number of moves is actually limited and so Problem 3 is sufficient. Another reduction to show **NP**-hardness can be found in [27] using reduction from standard 3SAT.[17] In this bounded case, any guessed move sequence must be of polynomial length (since the length of the sequence is bounded by a constant) and thus Candy Crush is also in **NP** and therefore **NP**-complete.

20.4 Database Queries

Databases play an important role in the modern IT landscape and most Computer Science students are familiar with relational databases. The (worldwide) market for relational database management systems alone was worth about \$30 billion in 2014 [28]. Relational databases usually use SQL [2], a declarative query language, to formulate "questions" about the data stored and the database engine then retrieves and processes the data as required. Since large software systems execute huge volumes of queries every second, the obvious question in the context of complexity theory is:

What is the time complexity of running such queries?

First of all, we need to define exactly what the problem here is and in particular the problem instance. Is the database or the query expression the instance, or both? *Moshe Vardi*[18] introduced a taxonomy [25] that distinguishes between *data complexity*, *expression complexity*, and *combined complexity*, and showed that there is an exponential gap between data and expression complexity. For a fixed query the dependency on the size of the data (data complexity) is polynomial (and it better be really fast as this is one of the success stories of database implementation). Expression complexity refers to the time to evaluate a query in terms of the size of the query itself. Combined complexity considers evaluation in terms of size and data, but is usually dominated by expression complexity. One may wonder why the query

---

[17]In conclusion, [27] raises another interesting question: Since hundreds of millions of hours globally are spent playing Candy Crush^{TM} and similar games, could one profit from this by using the manpower to solve other **NP**-complete problems hidden in the games?

[18]Moshe Ya'akov Vardi is an Israeli computer scientist who teaches at Rice University in the US. His research spans many fields of computer science, including database systems, complexity theory, and program specification and verification. He has won numerous awards.

evaluation problem is so hard. For a really large query, an exponential number of intermediate joins may need to be computed that are much larger than the input, and this may prove to be inefficient[19] [26].

In order to formulate a problem, we need to fix the query language. There is a rich body of work in this field of query complexity. Let us look into one particular kind of query language that vaguely resembles SQL: *conjunctive queries*. The logical representation suits the fact that we need to define a decision problem and the restriction to conjunctive queries means that those roughly correspond to select-project-join queries which "constitute the most basic, most important, and best-studied type of database queries" [7].

Definition 20.8 A *relational database* consists of a finite set D, the domain (for simplification there is just one domain, but it could easily encode generally all types in use) and finitely many relations $R_i \subseteq D^{r_i}$ where r_i is called the rank (or degree) of the relation R_i and $1 \le i \le s$.

We follow [1], the first paper on the subject, to define conjunctive queries on relational databases:

Definition 20.9 A *conjunctive query* for a relational database $(D, R_1, \ldots, R_s)$ is a boolean database query expression that asks whether there exist $y_1, \ldots, y_n$ that fulfil a conjunction of relations. More precisely, they are of the form:

$$\exists y_1 \ldots \exists y_n . \alpha_1 \wedge \ldots \wedge \alpha_m$$

where $\exists$ stands for "there exists", the α_i are atomic formulae that are of the form $R_j(t_1, \ldots, t_k)$ for $1 \le j \le s$, and each t_l $(1 \le l \le k)$ is either a variable y_j $(j \in \{1, \ldots, n\})$ or a constant.

Note that every such conjunctive query can be expressed as SQL query. Let us look at an example.

Example 20.3 Assume we have the following three schemas

```
ORDER(oid,account)
PRODUCT(pid,seller)
INFO(oid,pid,quantity)
```

Assume further (for simplicity) the domain D are strings (with a maximum length of some sort) and, for reason of brevity, let us abbreviate the ORDER relation O, the PRODUCT relation P and the INFO relation I. We can now write, for instance, a conjunctive query that checks whether there is an order for account number *A123* that includes a product in quantity of 100 that is sold by *Acme*.

---

[19]In a relational database system there is an optimizer which tries to avoid exactly this. And of course, real-life queries are—most of the time—not very large.

$$\exists x, p.\ O(x, \text{`A123'}) \land I(x, p, 100) \land P(p, \text{`Acme'})$$

The variable x relates the first two relations and variable p the last two. This can be expressed as SQL query in the following way:

```
SELECT DISTINCT 1
FROM INFO I JOIN ORDER O ON O.oid = I.oid
        JOIN PRODUCT P ON P.pid = I.pid
WHERE  O.account = 'A123' AND
        P.seller  = 'Acme'
```

If there are more relations in the conjunction then the join condition of the SQL query will become more complicated. The query's SELECT uses a constant 1 which in SQL represents true and DISTINCT ensures that we only see one instance of 1 in case there are more than one such orders that satisfy the query. An empty result table corresponds to the conjunctive query being false.

Definition 20.10 (*Query Evaluation Problem*)

- **instance**: given a relational database $(D, R_1, \ldots, R_s)$ and a boolean conjunctive query ϕ
- **question**: does ϕ evaluate to true in database $(D, R_1, \ldots, R_s)$?

Theorem 20.6 ([1]) *The Query Evaluation Problem is NP-complete when applying the combined complexity measure, i.e. measure time usage w.r.t. database and query expression size.*

This well-known result was first proved in a seminal paper by Chandra and Merlin in 1977. "*However, the main factor responsible for the high complexity of the problem is the length of the input formula and not the size of the database. This does not fit the practical situation very well, because usually one evaluates a short query against a large database.*" [9]. One way to try to remedy this issue is "parameterised complexity" (see [4]). But this does not really lead to (parameterised) tractability either (see [21]) and a detailed exposition is beyond the scope of this book. However, [31] was the first to show that simplifying the conjunctive queries helps. The simplification uses a notion of "acyclicity" defined next.

Definition 20.11 (*join tree*) Let Q be a Boolean query:

$$\exists \mathbf{y}. R_1(\mathbf{z}_1) \land R_2(\mathbf{z}_2) \land \ldots \land R_m(\mathbf{z}_m)$$

where $R_i(\mathbf{z}_i)$ are the atoms of Q. We use vectors $\mathbf{y}$ for finite lists of variables and $\mathbf{z}$ for a finite lists of expressions that can be either constants or variables. A *join tree* for Q is a tree T defined as follows:

- The nodes of T are the atomic formulae $R_i(\mathbf{z}_i)$ for $1 \le i \le m$.
- If a variable x occurs in two different atomic formulae $R_i(\mathbf{z}_i)$ and $R_k(\mathbf{z}_k)$ of Q, then it occurs in *each* atomic formula on the unique path of tree T between $R_i(\mathbf{z}_i)$ and $R_k(\mathbf{z}_k)$.

This leads to a natural description of how the different atoms are joined via variables in an ordered way. Note that not every query actually has a join tree.

Definition 20.12   A conjunctive query is called *acyclic* if it has a *join tree*.

Example 20.4   The query of Example 20.3 is not cyclic as one can define a join tree T for it as follows: The nodes of T are the atomic formulae $O(x, \text{‘A123’})$, $I(x, p, 100)$, $P(p, \text{‘Acme’})$. Variable x appears in two atomic formulae and variable p appears in two. In either case we must make sure that the corresponding atomic formulae are connected in the tree which looks as follows:

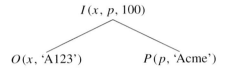

Further examples, also for *cyclic queries*, will be discussed in Exercise 9.
    The evaluation of *acyclic* conjunctive queries is tractable:

Theorem 20.7   ([31]) *When restricted to* acyclic *boolean conjunctive queries, the* Query Evaluation Problem *(Definition 20.10) is in* **P**.

The result has been refined further in [7, 9] and others but these interesting results are beyond the scope of this book.
    Adding union and negation to conjunctive queries one obtains *relational algebra* which is the semantic basis for relational database theory, but Theorem 20.7 won't hold any longer, as the extension introduces too many different orders to evaluate joins.
    As a consequence, there is large scope for optimization of large queries. One "high-level" technique is rewriting the query into a simpler (smaller) but equivalent[20] one. This has been suggested in [1] who also showed that near optimal rewrites of queries always exist.

20.5   Policy Based Routing

An important property of computer networks is reachability: can packets be successfully sent from one network device to another? As devices, including end-points (computer) and switches, are linked via connections, the problem can (once

---

[20]i.e. returning the same result as the original query.

again) be abstracted into a (network) graph problem. We already have encountered reachability for graphs in Sect. 16.5.1 but now we have additional forwarding policies to deal with, which makes the problem harder.

Definition 20.13 (*Network*) A *network* is a tuple $G = (V, E, \mathscr{P})$ such that $G = (V, E)$ is a directed graph where the set of vertices V corresponds to network devices and the set of edges E corresponds to (unidirectional) connections between them. The *policy function* $\mathscr{P}$ maps each edge $(u, v) \in E$ and each *symbolic packet* p to a boolean formula describing the forwarding policy for packet p travelling from u to v.[21] Therefore, $\mathscr{P}(e, p)$ evaluates to true for edge e and packet p, if, and only if, p is forwarded along e.

The policies expressible include general policies like forwarding and filtering. Packet transformation is more complicated and involves a shift from representing symbolic packets to a history array of packets with each entry representing the packet's content after a "hop" along a connection. Details can be found in [18].

Example 20.5 Consider a network $G = (V, E, \mathscr{P})$ with two vertices s and t that represent network routers. We now express examples of the above mentioned standard policies. We use *p.dst* and *p.src* to denote the destination IP address and the source IP address of packet p, respectively. The function $prefix_n(adr)$ selects and returns the first n bits of adr.[22] The addresses in *Internet Protocol version 4* (IPv4)[23] are four bytes long.

Forwarding   Assume packets are forwarded from s to t iff their destination address
        has prefix 139.184.0. Then $\mathscr{P}((s, t), p)$ is the formula

$$prefix_{24}(p.dst) = 139.184.0$$

Filtering   Assume router s blocks packets from travelling to router t iff their des-
        tination address has prefix 139.184. Then $\mathscr{P}((s, t), p)$ is the formula

$$prefix_{16}(p.dst) \neq 139.184$$

Two policies represented as formulas P_1 and P_2 can be conjoined in the formula $P_1 \wedge P_2$.

Definition 20.14 (*Network Reachability Problem* [30]) The *Network Reachability Problem* is defined as follows:

---

[21] So $\mathscr{P}((u, v), p)$ can be interpreted as the set of rules of network device (router) u for destination v depending on the packet p to be forwarded. Usually, only (parts of) the header of packet p is considered, but not its payload.

[22] We won't go into the details of *Classless Inter-Domain Routing* (CIDR) address notation taught in every *Network* course.

[23] This protocol still routes most traffic today despite the advent of IPv6.

- **instance**: a network $G = (V, E, \mathscr{P})$ as defined in Definition 20.13 and two vertices $s \in V$ and $t \in V$
- **question**: is there a packet that can be sent from s to arrive at t?

In other words, the question is whether there exists a symbolic packet p and a path π from s to t in G such that p satisfies the constraints $\mathscr{P}(e, p)$ for all edges e along the path, i.e. for all $e \in \pi$.

Theorem 20.8 ([18]) *The* Network Reachability Problem *of Definition 20.14 for networks that only use forwarding via the* longest prefix match *rule (and no filtering or packet transformations) is in* **P**.

Proof This can be shown by adapting the dynamic-programming based Floyd–Warshall algorithm from Fig. 16.4 in Sect. 16.5.2 such that it produces logical formulae the satisfiability of which expresses reachability instead of shortest distances. It is important that these formulae are in *disjunctive normal form* with literals being instances of the *Predecessor Problem*. We know from Sect. 16.2 that the latter is decidable in constant time. In other words, we establish a polynomial time reduction NetworkReachability $\leq_P$ DNFSAT (with DNFSAT from Exercise 3) and then apply Theorem 19.1. The details will be discussed in Exercise 13.

The following theorem states that the *Network reachability Problem* in presence of packet filtering is much harder:

Theorem 20.9 ([18]) *The* Network Reachability Problem *of Definition 20.14 that only use forwarding and packet filtering (and no packet transformations) is already* **NP**-*complete*.

Proof Employ Corollary 19.2 with a reduction SAT $\leq_P$ NetworkReachability and the result follows as we already know SAT is **NP**-complete (Theorem 20.1). The reduction function can be defined as each filter predicate can be arbitrarily complicated. The details are discussed in in Exercise 14.

It is proved in [18] that the problem is **NP**-complete even if the filter can only use one bit in the packet header if one allows parallel edges in graphs, more precisely multigraphs.

20.6   "Limbo" Problems

Most (optimisation) problems can be shown to be either in **P** (see Chap. 16) or they can be shown to be **NP**-complete problems like the problems from Chap. 17 or problems mentioned earlier in this chapter. There are, however, a few problems that are in limbo,[24] i.e. we don't know (yet) whether they are decidable in polynomial

---

[24]Limbo denotes an uncertain period of awaiting a decision or resolution. The name origins in theology where it refers to a certain part of "hell" but is now used generally in the aforementioned sense.

time or **NP**-complete. The most two famous "limbo" problems from **NP** are listed below.

The first problem is about prime factors. The prime factors of a positive integer are the prime numbers that are needed to produce the original number by multiplication. For instance, $12 = 2^2 \times 3$ so the prime factors are 2 and 3. If we consider $30 = 2 \times 3 \times 5$ then the prime factors are 2, 3, and 5.

Definition 20.15 (*Factorization Problem*)

- **instance**: an integer number $n > 1$
- **question**: what are n's prime factors?

This can be made into a decision problem as well:

- **instance**: integer numbers n and m such that $1 < m < n$
- **question**: does n have a factor $a \le m$?

This problem is of particular importance and we will come back to it in Sect. 21.4.2. As an exercise the reader is invited to show that it is in **NP**. We might now think, it may be **NP**-complete. Well it may be, but this is also unknown.

The second famous problem not known to be either in **P** or **NP**-complete is the following:

Definition 20.16 (*Graph Isomorphism Problem*)

- **instance**: two undirected graphs $G_1 = (V_1, E_1)$ and $G_2 = (V_2, E_2)$.
- **question**: are the two graphs G_1 and G_2 isomorphic? In other words is there a bijection f between V_1 and V_2 that extends to a bijection on edges E_1 and E_2, i.e. such that $(v, w) \in E_1$ if, and only if, $(f(v), f(w)) \in E_2$.

The problem has been shown to be in **P** for restricted classes of graphs like trees or planar graphs [11]. It finds some applications, for instance in chemistry, when looking for matching molecule structures in a database, and in circuit design. In those cases it can be solved efficiently, so it is widely believed that it is in **P**. Just as the manuscript of this book is about to be completed, there is news that *László Babai*[25] appears to have proved that the *Graph Isomorphism Problem* is quasi-polynomial.[26] If this was indeed true it would be *very close* to being in **P**. *Quasi-polynomial* refers to time complexity $\mathcal{O}\left(2^{(\log_2 n)^c}\right)$ for a constant c, whereas polynomial runtime is $\mathcal{O}\left(2^{c \log_2 n}\right)$ for constant c in comparison.[27]

---

[25]László Babai (born July 20, 1950) is an award-winning Hungarian mathematician and computer scientist from the University of Chicago.

[26]But the proof was not peer-reviewed yet at the time of writing.

[27]$2^{c \log_2 n} = (2^{\log_2 n})^c = n^c$.

20.7 Complete Problems in Other Classes

We can extend the notion of completeness to other classes we discussed. For instance, we can ask whether there are problems that are **EXP**-complete, **P**-complete, or **RE**-complete. An **EXP**-complete problem is discussed in Exercise 15. In the following we will briefly address **P**- and **RE**-completeness.

20.7.1 P-complete

For the definition of **P**-completeness polynomial time reduction is *not appropriate* as there are such reductions between all non-trivial **P** problems. This follows from the fact the reduction function has enough time to implement the required decision procedure. Consider two problems A and B, both in **P**. If B is not trivial, we know there is at least one $t \in B$ and $s \notin B$. Then we can simply define a polynomial time computable reduction function f as follows:

$$f(x) = \begin{cases} t & \text{if } x \in A \\ s & \text{if } x \notin A \end{cases}$$

Since A is in **P**, $x \in A$ can be decided by a program that runs in polynomial time and therefore f can be computed in polynomial time.

So for useful reductions within **P** one needs a different notion of effective reduction, most commonly *logarithmic space reduction*. Space complexity is beyond the scope of this book, but it should be clear by the name that the reduction function in this case can be effectively computed with space (memory) the size of which is logarithmic in the size of the input. This kind of reduction is usually denoted $\leq_{logs}$.

Definition 20.17 (*P-complete*) A problem H is called **P**-*complete* if problem H is itself in **P** and is "hardest" in **P** in the sense that for all other problems $A \in$ **P** we have that $A \leq_{logs} H$.

Theorem 20.10 *Among the list of the many P-complete problems are the Linear Programming Problem (Definition 16.15) and $CF^{\neq\emptyset}$, the problem whether a context-free grammar generates the empty language.*[28]

A long list of **P**-complete problems can be found in [13, Appendix A]. A few problems are also discussed in [8]. Furthermore, we can also consider complete problems in other classes like **RE** (see Definition 9.2), the class of semi-decidable problems.

---

[28]The empty language is the empty set of words, $\emptyset$, so it does not contain any words at all. Admittedly, not a very useful language, but that makes it even more important to be able to check efficiently that one's grammar does not describe this language.

20.7.2 RE-complete

Theorem 20.11 *The Halting problem, HALT, is* **RE***-complete w.r.t.* $\leq_{rec}$.

Proof We have to show that a given semi-decidable problem A can be effectively reduced to HALT. From the fact that A is semi-decidable we obtain a program p such that $[\![p]\!] (d) =$ true iff $d \in A$. Now we define the reduction function $f(d) = (q, q)$ such that $[\![p]\!] (d) =$ true iff $(q, q) \in$ HALT. It is left as Exercise 5 to find out what q must look like (it must obviously depend on p and d).

What Next?

We have seen that many combinatorial optimisation problems are **NP**-complete but appear in all kinds of applications in logic, graphs, networks, logistics, databases, compiler optimization and so on. The concept proved to be very versatile and an "intellectual export of computer science to other disciplines" [20]. But if so many common and useful problems are **NP**-complete, and we do not know any polynomial time algorithms to solve them, how are we dealing with them in practice? This seems to be slightly contradictory from a practical point of view. We will resolve this apparent contradiction in the next chapter.

Exercises

1. Show that every boolean expression (built from variables, negation, conjunction, disjunction, true, false) can be translated into conjunctive normal form.
   Hint: Show this by structural induction on the shape of the formula and use Boolean algebra laws like e.g. De Morgan's laws and distributive laws.
2. Let us encode a CNF formula in WHILE's datatype as a list of clauses that are, in turn, lists of literals. A literal is encoded as list of two elements. For positive literals the list contains the encoding of 1 followed by the encoded variable name as a number (in unary as usual). For negative literals the list contains the encoding of 0 followed by the encoded variable name as a number (in unary).

   a. Implement the evaluation operation in WHILE that takes a list containing the encoded formula and an encoding of a truth assignment and returns (the encoding of) 0 or 1. A truth assignment is a list of truth values encoded as 0 and 1. The value for the nth variable is in the nth position.
   b. Analyse the time complexity of the evaluation operation.
   c. Assume binary encoding of the variable numbers was used. How would that change the algorithm and its complexity?

3. A boolean expression is in *disjunctive normal form* iff it is a finite disjunction of finite conjunction of literals:

$$(A_{11} \wedge A_{12} \ldots \wedge A_{1n_1}) \vee \cdots \vee (A_{m1} \wedge A_{m2} \wedge \ldots \wedge A_{mn_m})$$

where each A_{ij} is a literal, i.e. either a variable or a negated variable (x or $\neg x$). The conjunctive formulae $(A_{i1} \wedge A_{i2} \ldots \wedge A_{in_i})$ are called *clauses*.

Assume the SAT problem was formulated using *disjunctive normal form* and not conjunctive normal form, i.e. as DNFSAT:

- **instance**: a boolean expression $\mathscr{F}$ in DNF (disjunctive normal form), i.e. a formula of the form $C_1 \vee C_2 \vee \ldots \vee C_n$ where each C_i is itself a *conjunction* of literals $l_1 \wedge l_2 \wedge \ldots \wedge l_n$.
- **question**: is $\mathscr{F}$ satisfiable?

 a. Show that DNFSAT is in **LIN**.
  Hint: Reformulate what satisfiability of a formula in DNF means exactly to find an algorithm that checks satisfiability in time linear in the size of the formula.
 b. It is well known that any Boolean expression can be brought into *disjunctive normal form*. Show that every Boolean expression (built from variables, negation, conjunction, disjunction) can be translated into disjunctive normal form.
  Hint: Show this by structural induction on the shape of the formula and use Boolean algebra laws like e.g. De Morgan's laws and distributive laws.
 c. From the fact that (CNF)SAT is **NP**-complete and DNFSAT is in **LIN**, what can be concluded about the complexity of transforming an arbitrary Boolean expression into DNF? Analyse the complexity of the algorithm that does the transformation discussed in Exercise 3b.

4. Encode an arbitrary run of a (one tape) Turing machine M as Boolean expression in conjunctive normal form. This requires the introduction of a number of variables: $Q_{i,k}$ states that M is in state k at time i; $H_{i,j}$ states that the head of M at time i is reading cell no. j; $C_{i,j,k}$ states that at time i on tape cell no. j we find symbol k. One needs to describe the run of M on initial word w in terms of many clauses (disjunctions) that are then "bolted together" via conjunctions to obtain a CNF expression. This is quite tedious but possible (see e.g. [6]).

5. The 3SAT problem from Definition 20.5 is like SAT with the additional restriction that all clauses (conjuncts) must have *exactly* three literals.

 a. Show that 3SAT is **NP**-hard by a polynomial time reduction SAT $\leq_P$ 3SAT.
  Hint: transform each clause of length n into n−2 clauses of length 3 that are not necessary logically equivalent to the original clause, but are satisfiable iff the original clause was satisfiable.
 b. Show that the 3SAT problem is **NP**-complete.

6. In Exercise 5 we restricted clauses of SAT to three literals. Let us go one step further and restrict to two literals. So let 2SAT be the problem that is like SAT with the additional restriction that all clauses (conjuncts) must have *exactly* two literals. Show that this even simpler problem 2SAT is actually in **P**.
 Hint: Consider each clause $l_1 \vee l_2$ as implication $\neg l_1 \Rightarrow l_2$ (reasonably adapted for negative literals l_1 and l_2, respectively). Develop an algorithm ("resolution") that investigates for each variable whether it has to be true or false to satisfy the

formula. In case a variable turns out to have to be both – true and *false – the formula is clearly not satisfiable.*

7. The *Vertex Cover Problem* from Exercise 8 in Chap. 17 was shown to be in **NP** in Exercise 2 from Chap. 19. Show that it is actually **NP**-complete by defining an appropriate polynomial time reduction: 3SAT $\leq_P$ Vertex Cover.

8. Consider the following problem SAT*:

   - **instance**: a boolean expression $\mathscr{F}$ in DNF, i.e. *disjunctive normal form* (see Exercise 3).
   - **question**: is $\mathscr{F}$ a *tautology*? In other words, does $\mathscr{F}$ evaluate to true *for all* possible truth assignments?

   In [3] SAT is, somewhat confusingly, actually defined as SAT*.[29] Considering Exercise 3, one might think that this is wrong but in actual fact SAT* only defines a "logically dual" problem to SAT. To see this proceed as follows:

   a. Show that if $\mathscr{F}$ is in CNF then $\neg\mathscr{F}$ is in DNF. And, vice versa, if $\mathscr{F}$ is in DNF, then $\neg\mathscr{F}$ is in CNF.
   b. Show that $\mathscr{F} \in$ SAT iff $\neg\mathscr{F} \notin$ SAT* and thus that SAT and SAT* are in the given sense *dual* to each other.
   c. Moreover, conclude that SAT $\leq_P$ SAT* and SAT* $\leq_P$ SAT.

9. Assume the following relational database schemas (from ([7]) are given :

   ```
   WORKS(empNo,projNo,task)
   MANAGES(empNo,projNo,task)
   RELATIVE(empNo,empNo)
   ```

   a. Write a boolean conjunctive query that checks whether there exists an employee *e* who works for a project that is managed by a relative of *e*.
   b. Write a boolean conjunctive query that checks whether there exists a manager who has a relative working in the company.
   c. Try to build the join tree for both queries. For one of the queries above you will not be able to build a join tree, as one of the queries above is actually *cyclic*.

10. Consider the following boolean conjunctive acyclic query (from [7])

$$\exists x, y, z, u, v, w.\ R(y,z) \wedge G(x,y) \wedge S(y,z,u) \wedge S(z,u,w) \wedge T(y,z) \wedge T(z,u)$$

    Compute its join tree.

11. Explain how a boolean conjunctive query can be evaluated in polynomial time, using its join tree.

12. A *k*-clique in an undirected graph $G = (V, E)$ is a set of *k* vertices (nodes) $S \subseteq V$ such that G has an edge between every pair of nodes in this set S, or in

[29]This had already been observed by [16] who uses some of the results in [3].

other words, for all $v, w \in S$ with $v \neq w$ we have that $(v, w) \in E$.
Consider the following *Clique Problem*:

- **instance**: an undirected graph $G = (V, E)$ and a positive natural number k
- **question**: does G contain a k-clique?

  a. Draw a graph with a 3-clique.
  b. Show that Clique is in **NP**.
  c. Show that Clique is **NP**-hard by a reduction SAT $\leq_P$ Clique. Define the reduction function as follows: Given a Boolean formula in conjunctive normal form $\mathcal{F} = C_1 \wedge C_2 \wedge \ldots \wedge C_n$, let $f(\mathcal{F}) = (G, n)$ where $G = (V, E)$ is an undirected graph such that

$$V = \{ \text{ occurrences of literals in } \mathcal{F} \}$$
$$E = \{ (v, w) \mid v \text{ and } w \text{ are not in the same conjunct of } \mathcal{F} \text{ and } v \text{ is not} $$
$$\text{the negation of } w \text{ and vice versa} \}$$

Let us consider a concrete example. For the CNF expression

$$\mathcal{F} = (A \vee \neg B) \wedge (B \vee C) \wedge (\neg A \vee C)$$

we therefore obtain $f(\mathcal{F}) = ((V, E), 3)$ with

$$V = \{ A, \neg B, B, C_1, \neg A, C_2 \}$$
$$E = \{ (A, C_1), (A, C_2), (A, B), (C_1, \neg B), (C_2, B), (C_2, \neg B)$$
$$(\neg A, \neg C_1), (\neg A, \neg B), (\neg A, B), (C_2, C_1) \}$$

Observe the two occurrences of C that need to be distinguished as C_1 (second literal) and C_2 (third literal).
  i. Draw the graph G.
  ii. Check that there is indeed a 3-clique in G if, and only if, there is a truth assignment θ of $\mathcal{F}$ such that $\theta(\mathcal{F}) = \text{true}$.
  Generalising the above result, one can then infer that $\mathcal{F}$ with k conjuncts is satisfiable iff $f(\mathcal{F}) = (G, k)$ and G has a k-clique.
  iii. Explain why the reduction function f from above is computable in polynomial time.

13. Assuming that policy functions only encode longest prefix forwarding, give the details of the reduction NetworkReachability $\leq_P$ DNFSAT from Theorem 20.8.
14. Assuming that policy functions only encode longest prefix forwarding and unrestricted packet filters, give the details of the reduction SAT $\leq_P$ NetworkReachability from Theorem 20.9.
15. Consider the following problem DAC:

- **instance**: a deterministic Turing machine M, some input word x, and a positive natural number k represented in *binary* (!)

- **question**: does M accept x (i.e. return true) in at most k steps?

  a. Show that DAC is in **EXP**.
  b. Show that DAC is **EXP**-hard, i.e. any other problem in **EXP** can be reduced to it and the reduction can be performed in polynomial time.
  c. Conclude that DAC is **EXP**-complete.

16. Show that the class of **P**-complete problems is closed under complement. In other words, show that if a problem is **P**-complete then its complement must be also **P**-complete. Note that it is unknown whether **NP**-complete problems are closed under complement or not.

References

1. Chandra, A.K., Merlin, P.M.: Optimal implementation of conjunctive queries in relational data bases. Proc. STOC'77. pp. 77–90, ACM (1977)
2. Connolly, T., Begg, C.: Database Systems: A Practical Approach to Design, Implementation, and Management. 5th edition, Pearson (2009)
3. Cook, S.A.: The complexity of theorem proving procedures. Proc. STOC '71, pp. 151–158, ACM (1971)
4. Downey, R.G., Fellows, M.R.: Parameterized Complexity. Springer, Berlin (1999)
5. Fraenkel, A.S., Lichtenstein, D.: Computing a perfect strategy for $n * n$ chess requires time exponential in n. J. Comb. Th. A **31**, 199–214 (1981)
6. Garey, M.R., Johnson, D.S.: Computers and Intractability: A Guide to the Theory of NP-Completeness. W.H. Freeman, New York (1979)
7. Gottlob, G., Leone, N., Scarcello, F.: The complexity of acyclic conjunctive queries. J. ACM **48**(3), 431–498 (2001)
8. Greenlaw, R., Hoover, H.J., Ruzzo, W.L.: Limits to Parallel Computation: P-Completeness Theory. Oxford University Press, Oxford (1995)
9. Grohe, M., Schwentick, T., Segoufin, L.: When is the evaluation of conjunctive queries tractable? In Aharanov, D., Ambainis, A., Kempe, J., Vazirani, U. (eds): Proc. STOC'01, pp. 657–666, ACM (2001)
10. Gualà, L., Leucci, S., Natale, E.: Bejeweled, Candy Crush and other match-three games are (NP-)hard. In: IEEE Conference on Computational Intelligence and Games, CIG, pp. 1–8, IEEE (2014)
11. Hopcroft, J., Wong, J.: Linear time algorithm for isomorphism of planar graphs. In: Proceedings of the Sixth Annual ACM Symposium on Theory of Computing, pp. 172–184 (1974)
12. Johnson, D.S.: A brief history of NP-completeness 1954–2012. In Götschel (ed.): Optimization Stories, Book Series, Vol. 6, pp. 359–376, Documenta Mathematica (2012)
13. Jones, N.D., Laaser, W.T.: Complete problems for deterministic polynomial time. Theor. Comput. Sci. **3**(1), 105–117 (1976)
14. Jones, N.D.: Computability and complexity: From a Programming Perspective[31]. MIT Press, Cambridge (Also available online at http://www.diku.dk/~neil/Comp2book.html) (1997)
15. Jones, N.D., Laaser, W.T.: Complete problems for deterministic polynomial time. Theor. Comput. Sci. **3**(1), 105–117 (1976)
16. Karp, R.M.: Reducibility Among Combinatorial Problems. In: Miller, R.E., Thatcher, J.W. (eds.) Complexity of Computer Computations, pp. 85–103. Springer, New York (1972)
17. Levin, L.A.: Universal Sequential Search Problems[32](Russian original appeared in: Problemy Peredachi Informatsii **9** (3), 115–116). Problems of Information Transmission, 9:3, 265–266 (1973)

18. Mai, H., Khurshid, A., Agarwal, R., Caesar, M., Godfrey, P., King, S.T.: Debugging the data plane with Anteater. ACM SIGCOMM Comput. Commun. Rev. **41**(4), 290–301 (2011)
19. Math Explorers' Club, Cornell Department of Mathematics: The Maths Behind Sudoku. Available via DIALOG. http://www.math.cornell.edu/mec/Summer2009/Mahmood/Intro. html Cited on 17 June 2015
20. Papadimitriou, C.H.: Lecture Notes in Computer Science. In: Degano, P., Gorrieri, R. (eds.) NP-completeness: A Retrospective. Marchetti-Spaccamela, A.: ICALP '97, pp. 2–6. Springer, Heidelberg (1997)
21. Papadimitriou, C.H., Yannakakis, M.: On the complexity of database queries. Proc. PODS '97, pp. 12–19, ACM (1997)
22. Smith, D.: This is what Candy Crush Saga does to your brain. Neuroscience, Notes & Theories in The Guardian, 1 April 2014. Available via DIALOG. http://www.theguardian.com/science/blog/2014/apr/01/candy-crush-saga-app-brain Cited on 18 June 2015 (2014)
23. Stockmeyer, L.J., Chandra, A.K.: Provably difficult combinatorial games. SIAM J. Comput. **8**, 151–174 (1979)
24. TraceyLien: From Bejeweled to Candy Crush: Finding the key to match-3. Polygon article, Feb 26, 2014. Available via DIALOG. http://www.polygon.com/2014/2/26/5428104/from-bejeweled-to-candy-crush-finding-the-key-to-match-3 Cited on 30 June 2015 (2014)
25. Vardi, M.: The complexity of relational query languages (Extended Abstract). In: STOC '82, pp. 137–146, ACM (1982)
26. Vardi, M.: On the complexity of bounded-variable queries (extended abstract). Proc. PODS '95, pp. 266–276, ACM (1995)
27. Walsh, T: Candy Crush is NP-hard. arXiv preprint arXiv:1403.1911 (2014)
28. Woodie, A.: RDBMs: The Hot New Technology of 2014? Datanami News Portal, December 12, 2013. Available via DIALOG. http://www.datanami.com/2013/12/12/rdbms_the_hot_new_technology_of_2014_/ Cited on 30 June 2015 (2013)
29. World Chess Federation: Laws of Chess. Available via DIALOG. https://www.fide.com/component/handbook/?id=124&view=article Cited 25 June 2015
30. Xie, G.G., Zhan, J., Maltz, D., Zhang, H., Greenberg, A., Hjalmtysson, G., Rexford, J.: On static reachability analysis of IP networks. In: Proceedings of the IEEE INFOCOM 2005, Vol. 3, pp. 2170–2183(2005)
31. Yannakakis, M.: Algorithms for acyclic database schemes. In: Proceeding of the 7th Very Large Data Bases '81 Vol. 7, pp. 82–94 VLDB Endowment IEEE (1981)

Chapter 21
How to Solve NP-Complete Problems

If there are no known feasible algorithms for NP-complete problems how do we deal with them in practice?

Hundreds of well-known and useful problems have been shown to be **NP**-complete. Some of them we have discussed in Chap. 20. We do not yet know whether solutions to these problems can be computed in polynomial time, and thus we don't know whether these problems are *feasible*.[1] We only know that despite all efforts for the last six decades nobody has found such an algorithm for any of them. Nor have any attempts to prove $\mathbf{NP} = \mathbf{P}$ or $\mathbf{NP} \neq \mathbf{P}$ been successful. So what does the working programmer do when faced with such a problem? At first, the results we have seen seem to imply that software developers are in serious trouble. In some sense they are, but there is often a way out of the trouble—to a degree.

This is what we will discuss in this chapter. The following questions will be answered: what do software developers do in practice to *"ease the pain"*[2] so that they can produce some sort of acceptable solution for any of those **NP**-complete problems? And is there also a benefit in certain problems not being known to be decidable in polynomial time? New and yet to be fully exploited notions of computability, like *Quantum computing* and *Molecular (DNA) computing* will be discussed in the final part of the book.

More precisely, in this chapter we will discuss:

- exact algorithms (Sect. 21.1):
  We know exact algorithms with polynomial runtime are unknown, but how far can one get with them?
- approximation algorithms (Sect. 21.2):
  will giving up the expectation to find optimal solutions help? What if we allow the solution to be a certain percentage worse than the optimum?

---

[1] In the literature one also finds the synonym *tractable*.

[2] This terminology has been use in Chap. 5 of the popular science book "Computers LTD" [20].

© Springer International Publishing Switzerland 2016
B. Reus, *Limits of Computation*, Undergraduate Topics in Computer Science,
DOI 10.1007/978-3-319-27889-6_21

- parallelism (Sect. 21.3):
  will using more computing devices help?
- randomization (Sect. 21.4):
  will using probability theory and randomised algorithms help?

In Sect. 21.5 we then take the very popular Travelling Salesman Problem as an example and see what results are known for it, in particular based on the previously mentioned approaches. Finally, we will investigate in which sense a problem for which polynomial time solutions are yet unknown can be a good thing by looking at the Factorisation Problem (Sect. 21.6).

In this chapter we will not deal with the decision version of the problems mentioned, but with the function problems that ask for an optimal solution.[3]

21.1   Exact Algorithms

Usually, exact algorithms only work for relatively small input. In the case of the Travelling Salesman Problem, for instance, for up to 60–80 cities. One uses branch-and-bound and branch-and-cut techniques to reduce the search space (without losing any potential optimal solutions in this case). For some applications this may already be sufficient. Most of the time, however, larger input is required to be dealt with.

Some clever algorithms in co-operation with using several CPUs (see Parallelism below) actually manage to find optimal TSP solutions for thousands of cities. For example, Concorde TSP Solver [10] is a freely available C library to solve TSPs. Concorde achieved an optimal solution for a TSP with over 85,000 locations (for VSLI chip design), however using several CPUs and 136 CPU years. Concorde uses the Dantzig, Fulkerson, and Johnson's method [13] which is based on the well known *simplex method* [36] discussed already in Sect. 16.6.

21.2   Approximation Algorithms

Instead of finding an optimal solution—e.g. computing the shortest route for the Travelling Salesman Problem (TSP), or graph colouring with fewest colours—one can also attempt to find a decent solution that is not optimal, but can be computed in polynomial time. Of course, the solution must not be arbitrarily poor. To be useful, one needs some form of guarantee about its quality. How close the computed solution is to the optimal solution is called the (worst-case) *approximation factor*. One has to distinguish minimisation and maximisation:

---

[3]The class of function problems corresponding to **NP** is usually called **NPO** for **NP**-optimisation problem.

Definition 21.1 (*α-approximation algorithm*) An algorithm A for a maximisation problem is called *α-approximation algorithm* if for any instance x of the problem we have that

$$A(x) \leq OPT(x) \leq \alpha \times A(x)$$

where $OPT(x)$ is the optimal value for instance x of the problem. Analogously, an algorithm A for a minimisation problem is called *α-approximation algorithm* if for any instance x of the problem we have that

$$OPT(x) \leq A(x) \leq \alpha \times OPT(x)$$

We call α the *approximation factor*. In either case of minimisation or maximisation it always holds that $\alpha \geq 1$.

There are two types of approximation: either one is satisfied with a constant factor, or one demands something stronger, namely that the optimal solution can be approximated with a ratio that is arbitrarily close to 1. Note that if α is 1, then the problem can be solved exactly. For instance, if the solution shall be no more that 50 % worse than the optimal minimal solution, then one has a constant factor of 1.5. Approximations only make sense for the function problem version of the **NP**-complete problems we have seen and not the decision version.

There are actually complexity classes for problems that can be approximated either way. The first class guarantees only constant approximation factors:

Definition 21.2 The complexity class **APX** (short for "approximable") is the set of optimization problems in **NP** that admit polynomial-time *approximation algorithms*[4] where the approximation factor is *bounded by a constant*.

If one has an algorithm that can approximate arbitrarily close to the optimum and it runs in polynomial time, the problem in question has a *polynomial time approximation scheme*.

Definition 21.3 (*PTAS*) The problem class **PTAS** is defined as the class of function optimisation problems that admit a **p**olynomial **t**ime **a**pproximation **s**cheme: this means that for any $\varepsilon > 0$, there is an algorithm that is guaranteed to find a solution which is at most ε times worse than the optimum solution in time polynomial in the problem instance (ignoring ε). This means that the approximation factor is $1 + \varepsilon$.

Clearly, the definition of **PTAS** is stronger than that of **APX**, as we can approximate arbitrarily close, so we have that **PTAS** $\subseteq$ **APX** (just fix one ε of choice).

Sometimes, changing the problem slightly *does* help as we will see below for TSP. Consider a modified TSP, called *MetricTSP*. This problem is like TSP but with an extra assumption about the distances between cities (weights of edges of the graph). The restriction is that the so-called *triangle inequality* holds for distances. We discuss this in more detail in Sect. 21.5.2.

---

[4]The term "approximation algorithm" has been introduced in [24] (according to [25]).

It is much easier to come up with algorithms that run in polynomial time and produce approximative solutions *without any quality bound* (so they might produce some really poor results for certain inputs), or that only run in polynomial time for certain inputs. Heuristic Search can be applied to reduce the (exponentially large) search space in a clever way. With approximative solutions it may happen that we cut off the optimal solution, giving up on optimality. Methods taught usually in *Artificial Intelligence* can be used as well, for instance *genetic algorithms* or *ant colony algorithms*.

Limits of Approximation Algorithms

In the last four decades or so, computational complexity theory has very much shifted towards analysis of approximation problem versions of **NP**-hard problems. There has been tremendous success on many fronts, suggesting clever approximation algorithms with various degrees of approximation ratios and worst-case upper runtimes. For a short survey, see e.g. the very readable [26], which also gives a short history of **NP**-completeness, and for an in-depth treatise see [47, 50].[5]

Surprisingly, one might say, there are quite complicated, iron-cast *limits* for approximating many of the **NP**-complete problems. It turns out that for many **NP**-complete problems the version that asks for an approximative solution (of a certain quality) is also **NP**-complete. This is a disappointing fact. In general, it has to be said that the landscape of **NP**-complete problems is not uniform in this respect. Some problems might have good approximations algorithms, others have not.

The modern study of inapproximability is a large field which progressed leaps and bounds triggered mainly by two events:

- the discovery that probabilistic proof systems could give a model for **NP** that could be used to show the inapproximability of the problem *Independent Set*[6] due to [17],
- and proof of the PCP Theorem which characterizes **NP** in terms of (interactive) probabilistic proof systems.

The latter is to approximation theory what Cook–Levin's Theorem 20.1 is to **NP**-completeness. Various versions were proved by [5, 6] in 1991–92, using many previous results that are beyond the scope of this introductory book (see [26]).

PCP stands for "**p**robabilistic **c**hecking of **p**roofs". This technique originated in cryptography in order to convince someone that one has a certain knowledge (proof) without revealing all the information.[7] The prover writes down a proof in an agreed format, and the verifier checks random bits of the proof for (local) correctness, so

---

[5] Vazirani opens with this quote [50, Preface]:

> Although this may seem a paradox, all exact science is dominated by the idea of approximation.
> Bertrand Russell (1872–1970)

[6] Independent Set is explained in Exercise 1.

[7] More details, also about the history of this theorem can be found e.g. in [4].

to speak. By repeating this process, due to the randomness, various parts of the proof will be checked, and if all these checks are affirmative, one can assume with a reasonably high probability that the proof is actually correct. The proof can, for instance, verify that a candidate solution to one of our **NP**-complete problems is a correct solution, thus verifying that for a given problem instance the answer is 'true'.

So one generalises the concept of a *verifier* in the original definition of **NP** (Definition 18.2). A verifier is a deterministic algorithm proving that some input x is correct with the help of a certificate c of which one can use only information that is of polynomial size in x as the verifier must run in polynomial time. For a *probabilistic polynomial time verifier* the input x is now accompanied by some further limited ("oracle") access to the certificate c where the limitation is accurately specified.

Definition 21.4 (*probabilistic polynomial time verifier*) A *probabilistic polynomial time verifier* p for a problem $A \subseteq \{0, 1\}^*$ is a verifier in the sense of Definition 18.1 with a further (implicit) input sequence of random bits[8] r, such that

$$A = \{d \in \{0, 1\}^* \mid [\![p]\!]^{\mathrm{L}}(d, c, r) = \text{true for some } c \in \{0, 1\}^*\}$$

The *probabilistic verifier* therefore runs *deterministically* with input x, c and implicitly r. It returns true and false, respectively, with a certain probability. A verifier on input (x, c) with implicit random bit input is called $(r(n), q(n))$-restricted if it makes only $q(|x|)$ enquiries about certificate c and uses only $r(|x|)$ random bits.

Definition 21.5 (*PCP[r(n), q(n)]*) The complexity class **PCP**$[r(n), q(n)]$ is the class of problems A that have a $(r(n), q(n))$-restricted probabilistic verifier V (as defined in Definition 21.4) with the following property: if $x \in A$ there is a certificate c such that V returns true with probability 1, and if $x \notin A$, then for each certificate c the probability that V returns true is at most $\frac{1}{2}$.

Form this definition some simple observations follow easily like **PCP**$[0, 0] = $ **P**, **PCP**$[\mathcal{O}(\log_2 n), 0] = $ **P**, **PCP**$[0, n^{\mathcal{O}(1)}] = $ **NP** (see Exercise 5).

The PCP theorem states that every statement $x \in A$ of an **NP**-complete problem A has a probabilistically checkable proof, using "very little randomness" and only reading constant bits of the proof:

Theorem 21.1 (PCP Theorem [5, 6][9]) $NP = PCP[\mathcal{O}(\log_2 n), \mathcal{O}(1)]$.

The PCP Theorem *"...is an amazing result. No matter how large the instance and how long the corresponding proof, it is enough to look at a fixed number of (randomly chosen) bits of the proof in order to determine (with high probability) whether it is valid. ...The relevant property here is that if you break off a tiny corner of a hologram,*

---

[8]One can drop r from the input by running the algorithm probabilistically over a uniformly distributed random bit vector r.

[9]Sanjeev Arora, Uriel Feige, Shafi Goldwasser, Carsten Lund, László Lovász, Rajeev Motwani, Shmuel Safra, Madhu Sudan, and Mario Szegedy won the 2001 Gödel Prize for the PCP theorem and its applications to (hardness of) approximation.

it can still be made to display the image contained in the full hologram, albeit somewhat more fuzzily. Similarly, most small collections of bits from the proofs in question here must in a sense reflect the entire proof, at least insofar as its correctness is concerned" [25]. Note that the PCP theorem does not make any claims about the runtime of the probabilistic verifier other than it is polynomial, so the runtime may actually be worse than that of a traditional non-probabilistic verifier. The focus here is rather on how much of the certificate is actually checked by the verifier. This is not in line with our intuition. Why does a verifier not need to check a large part of the certificate showing that an instance x of an **NP**-complete problem A is actually in A? Imagine x to be a large graph G and a number k and A be TSP. Then the certificate (proof) must contain enough information to show that a tour of length k or less is possible in G. The key lies in the trade-off between probabilistic acceptance and quantity of the certificate inspected.

Probabilistically checkable proofs have applications in cryptography. Even more amazing is the connection of the PCP Theorem to the limitations of approximation algorithms. While for some optimisation problems approximation algorithms with small approximation factors were discovered (mostly in the 1970s), for others intractability results were quickly obtained, like e.g. for TSP (see Theorem 21.3 below). Those *in*approximability results were usually up to **P** = **NP** (or some other very unlikely assumption). But there remained a third group of problems for which the situation was not so obvious. In [17] the connection between approximations and probabilistic provers in the context of the *Clique Problem*[10] was noticed "showing how any sufficiently good approximation algorithm for clique could be used to test whether probabilistically checkable proofs exist, and hence to determine membership in NP-complete languages." [25] The details of this connection are beyond an introductory text and can be found in [17] or summarised in e.g. [25]. From the results of [5] one could conclude that the *Clique Problem* does not admit any approximation and is therefore not in **APX**. Thanks to [6] this could be generalised to a wide variety of problems. For many **NP**-complete problems thus either

1. for every $\alpha > 1$ there is a polynomial time approximation algorithm with approximation factor α or better (so the problem is in **PTAS**) or
2. there is a certain constant $\alpha > 1$, the *approximation threshold*, such that no algorithm can produce a solution with approximation factor α or better in polynomial time (unless **P** = **NP**) so the problem is in **APX**
3. there is no $\alpha > 1$ that fulfills (2), so there is no approximation algorithm at all (unless **P** = **NP**) and thus the problem is not even in **APX**.

The 0-1 Knapsack Problem belongs to the first category, MetricTSP to the second category and TSP to the third category. There are a number of problems that are known to be in **APX** but it is unclear whether they belong to (1).

---

[10]See Exercise 12 in Chap. 20.

21.3 Parallelism

What about using more processors? To add up n numbers sequentially with one processor, we know we need $n-1$ addition operations. But using $n/2$ processors we can add pairs of numbers in parallel, first $n/2$ pairs, then $n/4$ pairs, and finally one pair, so we only need $\log_2 n$ additions per processor which leads to a runtime improvement from $\mathcal{O}(n)$ to $\mathcal{O}(\log_2 n)$. This is excellent speed-up. In a way, one can think of the parallel processors/adders to "take care" of the nondeterministic choices a NTM makes that decides membership of a problem in **NP**. On each processor one can run a different "guess" for the continuation of the nondeterministic program.

That sounds great. Does this parallel execution on several processors then allow us to get a parallel program[11] that solves an **NP**-complete problem in polynomial time? The answer is yes but no. The issue with this is that one does get a speed-up, but to get the expected speed-up one would need exponentially many processors (exponential in the size of the input) and that's not feasible either. The input can be arbitrarily large in size, so where would one get all the required processors from? There is a (maybe large but) fixed amount of processors available. Secondly, all these processors need to communicate somehow to co-operate in the right way to produce the right answer. This communication takes time and this time may not be polynomial either, in particular if exponentially many processors are involved.

With a number of processors *polynomial in the size of the input* one can already achieve significant speed-ups. For instance, matrix multiplication, usually computed in time $\mathcal{O}(n^3)$—or $\mathcal{O}(n^{2.807355}) = \mathcal{O}(n^{\log_2 7})$ using the Strassen algorithm[12]—can be done in time $\mathcal{O}(\log_2 n)$ with $\mathcal{O}(n^3)$ processors [16]. There is a complexity class, **NC**, called Nick's Class after *Nick Pippinger*, which covers those cases. Its definition is based on Boolean circuits,[13] but informally and crudely speaking, **NC** contains all problems that can be decided by a polynomial number of processors in *poly-logarithmic* time, i.e. in time $\mathcal{O}((\log_2 n)^k)$.

With a constant number of processors we can still get a constant guaranteed speed-up. This is still a very useful concept and it is used e.g. in multi-core processors these days.[14]

It is worth pointing out that not all problems are equally amenable to parallelisation. When and how to parallelise computations is a big research topic in itself,

---

[11] Better a program executed in parallel.

[12] The Coppersmith-Winograd algorithm [12] does it in $\mathcal{O}(n^{2.376})$ but is only better for extremely large matrices and thus not used in practice, nor are recent slight improvements by Stothers or Williams.

[13] We don't present details in this introductory book but this will be covered in many textbooks listed in the appendix.

[14] Chip manufacturers have to go down this route as *Moore's law* is reaching its limits, as further miniaturisation of processors leads to transistors of almost atomic size and thus to unwanted and interfering quantum effects. We'll say more about that in Chap. 23.

affecting networking, operating system and algorithm design. Controlled parallelisation is also essential in the Map-Reduce programming model used in key/value
stores for processing large datasets [15].

21.4   Randomization

Another method attempts to get faster algorithms by giving up some "certainty",
either about the runtime or about the optimality of the solution. One distinguishes
actually two different probabilistic approaches:

- Las Vegas
  is always correct but only probably fast
- Monte Carlo
  always fast but only probably correct.

Of course, one attempts to achieve extremely *high* probabilities. We just look at
Monte Carlo approaches.

21.4.1   The Class RP

The class **RP** (for *Randomized Polynomial*) defines a class of problems that are
decided by Monte Carlo algorithms with a "yes"-bias in polynomial time. Let us
define this more precisely and to do so, we need to consider *randomised algorithms*.
A randomised algorithm is one that uses commands that have random outcomes. For
instance a *probabilistic Turing machine* program can be seen to realise a randomised
algorithm. A probabilistic Turing machine is a nondeterministic Turing machine that
chooses its transitions according to a given probability distribution.

Definition 21.6 Let us call a randomised decision algorithm *Monte Carlo* if the
following holds:

- the algorithm returns "yes" only for correct answers
- the error of rejecting a correct answer has a low probability $\frac{1}{2} > p > 0$.

Note that the probability of rejecting correct answers can be made arbitrarily low by
independently running the algorithm n times, as the probability is then p^n which is
a small number (since $p < 1$, see Exercise 6). Therefore, Monte Carlo algorithms
work well in practice.

   Randomised algorithms can be implemented e.g. on a probabilistic version of a
Turing machine [18] (similarly for WHILE's choose command).

Definition 21.7 (*Probabilistic Turing machine*) A *probabilistic* Turing machine M
is a nondeterministic Turing machine where each binary nondeterministic choice

occurs with equal probability $\frac{1}{2}$ and acceptance is defined by the final state (label) in which the program terminates. There are *accepting* states 1 and there are *rejecting* states 0 (and states that are neither accepting or rejecting). The probability that M accepts input x, i.e. halts in a 1 state on input x, is denoted $\Pr[M(x) = 1]$, and the probability that M rejects input x, i.e. halts in a 0 state on input x, is denoted $\Pr[M(x) = 0]$.

By definition, it follows that the probability that a probabilistic Turing machine M accepts x is the sum of the probabilities of all accepting paths of M with input x. The probability of any given path is $\frac{1}{2^n}$ where n is the number of nondeterministic choices (gotos) in the path.

Definition 21.8 (*RP*) The class **RP** is the class of problems (of words over alphabet $\{0, 1\}$ as usual) that are decided by a (probabilistic) Monte Carlo algorithm in the above sense in polynomial time. In terms of probabilistic Turing machines we can define this more precisely as the sets A for which there is a probabilistic Turing machine M with polynomially bounded runtime, such that for all input x it holds that

$$x \in A \Rightarrow \Pr[M(x) = 1] \geq \frac{1}{2}$$
$$x \notin A \Rightarrow \Pr[M(x) = 1] = 0$$

Theorem 21.2 $P \subseteq RP \subseteq NP$.

Proof The first inclusion is trivial as any deterministic decision procedure is automatically a degenerated randomised one that does not make use of probabilistic choices. For the second inclusion we reason as follows: Let $A \in$ **RP**, so A is accepted by a Monte Carlo algorithm that can be implemented by a probabilistic Turing machine M with polynomial time bound according to Definition 21.8. Due to Theorem 18.1 we know already that a problem accepted by a nondeterministic Turing machine in the sense of Definition 18.3 in polynomial time is in **NP**. We must construct such a machine with the right properties. To do that, we modify the given M such that it returns output true if it finishes in a 1-state and something else in all other final states. Of course, M is a nondeterministic Turing machine by definition. It remains to be shown that x is accepted by our modified M in the sense of Definition 18.3 if, and only if, $x \in A$. If x is accepted, then there is a path leading to output true and thus to an accepting state in the original M where the path has obviously a non-zero probability. By assumption that M proves that A is in **RP**, we know that x must be in A. On the other hand, if $x \in A$ we know that for the original M with input x there is a path with non-zero probability leading to a 1-state. The modified machine returns true for this path and thus, viewed as nondeterministic TM, modified M accepts x.

As usual in computational complexity, there are open problems: We don't know whether **P** $=$ **RP** or whether **RP** $=$ **NP**.

The class **RP** can be generalised to class **BPP** (Bounded-error probabilistic polynomial-time):

Definition 21.9 (*BPP*) The class **BPP** is the class of problems (of words over alphabet {0, 1} as usual) that are decided by a probabilistic algorithm that runs in polynomial time and can give incorrect answers on both "Yes" and "No" (so no bias) instances with error probabilities that are at least an ε less than $\frac{1}{3}$ for all instances. With probabilistic Turing machines we can make this more precise (and we should): A problem A is in **BPP** if, and only if, there is a probabilistic Turing machine M with polynomially bounded runtime such that for all input x it holds that for some constant $\varepsilon < \frac{1}{3}$

$$\Pr[M(x) = (x \in A)] \geq \frac{2}{3} + \varepsilon$$

Here $(x \in A)$ is supposed to denote 1 if $x \in A$ and 0 if $x \notin A$. In other words, there is a probabilistic Turing machine that decides whether its input is in A with unbiased error probability bounded by $\frac{1}{3} - \varepsilon$.

What do we know about **BPP**? Well, one can show that **RP** $\subseteq$ **BPP** (see Exercise 7) and so **P** $\subseteq$ **BPP** but it is unknown whether **P** = **BPP** or whether **BPP** $\subseteq$ **NP**. **BPP** is currently considered to be the right notion of what is (probabilistically) "efficiently computable". This is because, as for **RP**, one can repeatedly run the algorithm to reduce the error probability exponentially to get an arbitrarily low error margin.

Finally, it should be pointed out that **RP** and **BPP** can also be defined as classes of problems decided by deterministic Turing machines (or algorithms) if the random choices are made part of the input like we have done in the definition of **NP**.

21.4.2   *Probabilistic Algorithms*

In 1975 Michael Rabin[15] and Gary L. Miller designed a polynomial time Monte Carlo algorithm that tests whether a given number is a prime [33, 40]. Polynomial here means polynomial not just in the number n we test (the input) but in the size of the number n which in this case is the number of digits (thus $\log_{10} n$). At the time no deterministic (non-randomised) polynomial time algorithm existed and this remained the case for almost 30 years until 2002 when Agrawal, Kayal, and Saxen presented their multiple award winning AKS algorithm [1] we already discussed in Sect. 16.4.

Note that there is no polynomial time algorithm yet, probabilistic or not, of the *Factorisation Problem* (Definition 20.15) and so the decision version of the Factorisation Problem is not known to be in **P** nor known to be **NP**-complete.

A specific and well-known Monte Carlo technique for combinatorial optimisation problems is *simulated annealing* developed by Kirkpatrick et al. in 1983 [28].[16]

---

[15]Michael Oser Rabin (born September 1, 1931), is an Israeli computer scientist who won the Turing-Award in 1976.

[16]This paper has an amazing citation count of over 34,000—more than any other publication cited in this book.

It was inspired by "an analogy to the statistical mechanics of annealing in solids" [42]. To find optimal solutions one implements a "random walk" between candidate solutions trying to avoid being caught in a local optimum instead of a global one. So one accepts new solutions with a certain probability during this walk. The better the solution the higher the probability, but with some small probability one accepts solutions that are worse than the best one found so far (this allows one to escape any local optimum to find the global optimum). During the "cooling down" period of the algorithm the probabilities decrease and poorer solutions get accepted less and less. This algorithm is not deterministic, different runs can produce different results. It is easy to use, applies easily to all kinds of optimisation problems, but does not necessarily produce the best results. Therefore, it is usually less efficient than other methods and applied only if other methods fail.

A not too technical introduction to simulated annealing can be found e.g. in [42].

21.5 Solving the Travelling Salesman Problem

We have introduced the *Travelling Salesman Problem* (TSP) in detail in Sect. 17.1 already. As mentioned there, overview of history, methods and results can be found online [45] as well as in popular books [2, 11, 29]. There is a benchmark library called TSPLIB [46] that contains many instances of TSP problems (and actually variations of TSP problems too). It was built and is maintained by the University of Heidelberg. Problems from this library are often used in the literature to test and compare algorithms.

We briefly review how the various approaches mentioned earlier in this chapter fare for TSP. All versions of TSP we encounter will be formulated as a function problem and not as a decision problem.[17]

21.5.1 Exact Solutions

If the graph in question has n vertices (cities) there are $\frac{n-1!}{2}$ possible tours, so the brute-force generate-and-test approach is $\mathcal{O}(n!)$ which is much worse than $\mathcal{O}(2^n)$. With the help of *dynamic programming* techniques, remembering intermediate results, in this case sub-tours, one can however get an algorithm in $\mathcal{O}(n^2 2^n)$ as shown in [21] and independently in [8] where this technique is generalised and applied to other **NP**-complete problems. "This result was published in 1962, and from nowadays point

---

[17]The decision problem versions are mainly for complexity class analysis and not for practical purposes.

of view almost looks trivial. Still, it yields the best time complexity that is known today" [48]. Of course, this refers to *exact solutions only*.

Historically important are the attempts using linear integer programming. Encode the TSP problem as set of linear inequalities on integer variables and solve it with the simplex algorithm encountered already in Sect. 16.6. This has been done a few years after the simplex algorithm was discovered [13] and the optimal tour of 49 cities in the US[18] could be computed this way. The technique used *cutting planes* which were proposed by Ralph Gomory [19] in the 1950s as a method for solving integer programming problems. Cutting planes are linear inequalities, termed cuts, that are added to a linear integer programming problem in cases where the optimal solution does not meet the integer constraint. In this case rounding might not give the best result. The cuts produce a convex polytope consisting of all feasible points. In later attempts this has been combined with a branch-and-bound approach to explore possible relaxations of the problem at hand. An overview of the progress and methods in solving TSP can be found in [32]. "The largest solved instance[19] of the traveling salesman problem consists of a tour through 85,900 cities in a VLSI application that arose in Bell Laboratories in the late 1980s" [37]. It was solved in 2006. The optimal solutions can take a very long time to compute, between several CPU days to several CPU years (!), so these solutions may not be practical for logistics companies.

21.5.2 *Approximative Solutions*

In Sect. 21.2 we already alluded to the fact that if the distances observe the triangle inequalities then a tour can be found in polynomial time that is at most 50 % worse than the shortest one. The corresponding algorithm, *Christofides'* algorithm [9], is therefore a 1.5-approximation algorithm the runtime of which is $\mathcal{O}(|V|^3)$ (where V denotes the set of vertices as usual). There is a simple version that has poorer quality, producing a tour at most 100 % worse than the optimum one, but which runs faster in $\mathcal{O}(|E| + |V| \times \log_2 |V|)$. This 2-approximation algorithm is the topic of Exercise 10.

We also briefly defined **APX**, the set of (function) optimisation problems that allow polynomial time approximation algorithms where the approximation ratio is bounded by a (fixed) constant. We will now define the approximative variant of TSP and present a famous result:

Definition 21.10 Let ApproxTSP(α) denote the problem of solving the (optimising) *standard* Travelling Salesman Problem with approximation factor (or ratio) α.

Theorem 21.3 *There is no polynomial time algorithm that solves ApproxTSP(α) unless $P = NP$.*

---

[18]Actually the capital plus one city from each state of the United States had been chosen, which had only 48 states at the time since Alaska and Hawaii joined the United States only in 1959.

[19]At the time of writing.

This has been shown by Sahni and Gonzalez in 1976 [43][20] by reducing an **NP**-complete problem to ApproxTSP(α), namely the problem of finding a Hamiltonian cycle in a graph. It follows that ApproxTSP(α) $\notin$ **APX** for any α.

21.5.2.1   Metric Version of TSP

In *Euclidean* geometry (as taught at school) the *triangle inequality* is satisfied. The triangle inequality states that, given the three sides of a triangle, it is always shorter to go along one side than along the other remaining two. In other words, from a pedestrian point of view, the triangle inequality justifies cutting corners and walking diagonally across the grass when in a hurry.

Definition 21.11 (*MetricTSP*)

- **instance**: a road map of cities with distances—i.e. a graph $G = (V, E)$—where vertices are cities and edges are roads with weights $w(u, v)$ denoting the length (distance) between vertices (cities) u and v such that

$$w(u, v) + w(v, z) \geq w(u, z) \quad \text{(triangle inequality)}$$

and two vertices a and b denoting start and end of the tour.
- **question**: what is the shortest route from vertex (city) a to b that passes through *all* cities?

In many real-world applications it is actually this *MetricTSP* that requires solving. This excludes, however, situations where, for instance, roads are not very straight and can involve detours, such that the road from u to v can actually be longer than the roads from u to v and v to z taken together. Roads do not follow the beeline, they may have many twists and turns. Note also that if the distance between cities is supposed to be the time it takes to travel between them, the resulting problem is most likely not metric, as on highways one can travel much faster than on small roads.

According to Nicos Christofides [9] we know that MetricTSP $\in$ **APX**. In [39] the (so far) best lower bound for the approximation factor has been given: $\frac{117}{116}$, when distances are symmetric, and $\frac{220}{219}$, if distances are asymmetric.

Even more can be shown:

Theorem 21.4 *The MetricTSP problem (of Definition 21.11) is **APX**-complete under an appropriate notion of reduction[21] that preserves approximations (which is also computable in polynomial time).*

So MetricTSP is one of the "hard" approximation problems.

---

[20]They actually show this result for other **NP**-complete problems as well, not just TSP.

[21]Called PTAS reduction as it preserves the property that a problem has a polynomial time approximation scheme (PTAS).

21.5.2.2   Euclidean TSP and Polynomial Approximation Schemes

MetricTSP can be further restricted by requiring that each city (which in higher dimension we'd better call "location") is identified in a d-dimensional real vector space where distances between these points are fixed by the Euclidean norm on vectors:

Definition 21.12 (*EuclideanTSP*)

- **instance**: a road map of n locations (cities) identified with n points in $\mathbb{R}^d$ (where d is fixed)[22] where the distances between vertices (cities) u and v are computed as follows:

$$w(u, v) = \sqrt{\sum_{i=1}^{d} (u_i - v_i)^2} \quad \text{(Euclidean norm)}$$

- **question**: what is the shortest route from vertex (location/city) u to v that passes through *all* vertices?

EuclideanTSP is also **NP**-hard and with distances rounded to integer it is also **NP**-complete [38]. However, for this problem *arbitrarily good* approximations to the optimal solution can be found in polynomial time as proved by Arora[23] and Mitchell[24]:

Theorem 21.5 *EuclideanTSP is in **PTAS**.*

Proof See [3] and [34]. Let d be the dimension of the underlying vector space. The corresponding randomized algorithm runs in $\mathcal{O}\left(n \times (\log_2 n)^{\mathcal{O}((\sqrt{d}\varepsilon)^{d-1})}\right)$. If d and ε are fixed this is nearly linear in n. The algorithm can be derandomized, increasing the running time by a factor $\mathcal{O}(n^d)$. The algorithm crucially uses the "divide-and-conquer" technique of *dynamic programming*.

In a nutshell, the above theorem showed that for *some* **NP**-hard optimisation problems solutions which are arbitrarily close to optimal can be found in polynomial time. This is quite an exciting theoretical result but in practice often different algorithms are used because they achieve better runtime. Even if they do not produce a guaranteed approximation, the approximation still turns out to be good, usually somewhere between 0.6 % and 5 %. And they also work for more general versions of TSP, not just the Euclidean one.

---

[22] In order to model points on a plane choose $d = 2$ (two-dimensional vector space) and in order to model points in space choose $d = 3$.

[23] Sanjeev Arora (born January 1968) is an Indian American theoretical computer scientist who won the Gödel Prize twice: in 2001 for his work on probabilistically checkable proofs and again in 2010 for the discovery of the PTAS for EuclideanTSP.

[24] Joseph S.B. Mitchell is an American mathematician working in computational geometry and distinguished SUNY professor. He won the Gödel prize for the discovery of the PTAS for EuclideanTSP in 2010.

21.5.2.3 The Lin-Kirnighan Approximation Algorithm

Regarding approximation algorithms for TSP the most famous ("classic") result in this area is due to Lin and Kernighan [31]. "The Lin–Kernighan algorithm (LK) is a variable k-opt algorithm. It decides which k is the most suitable at each iteration step. This makes the algorithm quite complex, and few have been able to make improvements to it. For a more in-depth study of the LK algorithm and possible improvements, see [23]. The time complexity of LK is approximately $\mathcal{O}(n^{2.2})$ [23], making it slower than a simple 2-opt implementation. However the results are much better with LK".

Lin and Kernighan's algorithm gets within 2 % of the Held–Karp lower bound which is a famous lower bound that can be computed quickly (polynomially) [22]. There is a plethora of various optimisation and implementation tricks and often so-called Lin and Kernighan implementations actually diverge from the original algorithm. For details the reader is referred to the literature.

"The World TSP" has been compiled in 2001 using data from the National Imagery and Mapping Agency database of geographic feature names and data from the Geographic Names Information System (GNIS), to locate all populated towns and cities in the world[25] [49]. This provides an extremely large 1,904,711-city instance of TSP. A clever and heavily optimised implementation of Lin-Kernigham by Helsgaun, called LKH-2 [27], found the so-far best tour for this problem[26] on May 24, 2013. This was shown to be at most 0.0474 % greater than the length of an optimal tour [49]. The exact running time of LKH-2 to solve WorldTSP was not published in [27], but for smaller problems with several thousand cities, LKH-2 usually finds solutions within minutes and for extremely large problems, within few hours.

In order to scale the LK-algorithms further, one uses clustering techniques [35] to solve TSP for millions of cities. One gives up precision for speed. "At the 1,000,000 city level, our solution is just 11 % off the Concorde algorithm results, but only takes 16 % as long to run." [35].

21.6 When Bad Complexity is Good News

The fact that no polynomial time algorithms have been discovered yet that solve important problems like TSP, Colouring, Knapsack etc., is bad news. At least, if one has to find optimal solutions for larger input in a decent amount of time and the above approximation techniques do not help. But it turns out that this also has some positive ramifications. More precisely, let us consider the *Factorisation Problem* from Definition 20.15. This problem is in **NP** but no polynomial time algorithm is known. And this fact actually is *crucial* for *public key cryptography*.

---

[25]Including research bases in Antarctica.
[26]Length 7,515,772,212.

This sort of cryptosystem uses a public key for encryption and a private key for decryption, it's thus called an *asymmetric* cryptosystem. Ron Rivest, Adi Shamir and Leonard Adleman[27] developed the most famous version in 1977, the "RSA" algorithm named after their initials [41], based on ideas of Whit Diffie,[28] Marty Hellman[29] and Ralph Merkle[30] first published in 1976 [14]. Many protocols like SSH (secure shell), PGP (pretty good privacy), and SSL/TLS (secure sockets layer/ transport layer security) rely on RSA. James Ellis, Clifford Cocks and Malcolm Williamson, English mathematicians working for GCHQ, apparently found such a public cryptosystem independently [30] in the 1970s, but this could not be publicly announced until it was declassified in 1997.[31]

However, note that *Factorisation* is *not known to be* **NP**-*complete* either. And even if it turned out to be in **P**, the real threat to public key cryptography is finding a polynomial algorithm for Factorisation with *good runtime behaviour*, and this is very different from just showing that it is in **P**. So public encryption is safe until such an algorithm has been found.

RSA in a Nutshell

For readers with some interest in basic algebra the following section describes how RSA actually works. This section may be skipped, it requires some basic mathematical knowledge. The most complicated result used (Fermat's Little Theorem) will not be proved.

The system RSA is asymmetric in the sense that encryption and decryption use a different key. The sender uses the receiver's public key to encrypt, the receiver then uses their own private key to decrypt. Private and public key are constructed with the help of two large prime numbers with certain properties. One chooses two primes p and q and computes the products $n = p \times q$ and $\phi = (p - 1) \times (q - 1)$. Next, one destroys the p and q as they are not needed any longer and pose a threat to security of the code. Typically, n has 1024 binary digits. The user now chooses numbers d and e such that

$$e \times d = 1 \ (\mathrm{mod}\ \phi) \tag{21.1}$$

which is important to ensure that an inverse to encryption exists. As usual, $n \ (\mathrm{mod}\ m)$ denotes the remainder of dividing integer n by integer m. Next, also ϕ is safely discarded, only n, e and d will be used afterwards. The public key resulting from

---

[27]They won the Turing Award in 2002.

[28]Bailey Whitfield 'Whit' Diffie (born June 5, 1944) is an American cryptographer who also worked for Sun Microsystems. He is one of the inventors of public-key cryptography for which he has won several prizes.

[29]Martin Edward Hellman (born October 2, 1945) is an American cryptographer, and one of the inventors of public key cryptography for which he won numerous awards.

[30]Ralph C. Merkle (born February 2, 1952) is a computer scientist and was a PhD student of Martin Hellman. He is one of the inventors of public key cryptography, for which he was awarded numerous prizes.

[31]The Institute of Electrical and Electronics Engineers (IEEE) has granted its 100th Milestone Award to the three scientists and a plaque has been also revealed at Cheltenham in 2010.

this is the pair (e, n) where n is called the modulus and e the exponent. The private key is d.

How do we encrypt our message to the user with public key (e, n)? First, the ASCII message is represented as an integer with base 256 as each character is a number between 0 and 255. The resulting number M must be smaller than n, otherwise the message has to be split up. Furthermore the number M must not be a divisor of n (if this should happen then one must use some padding etc.). Now we can encrypt the message M to get the *cyphertext* C as follows:

$$C := M^e \ (\text{mod } n)$$

or in other words, the encoded message C is the remainder of dividing M^e by n.

How do we decrypt the cyphertext C then? The receiver uses their private key d and computes $C^d \ (\text{mod } n)$ which equals $(M^e)^d \ (\text{mod } n)$. The claim now is that the result reveals the original message M. Why does this hold? This relies on some algebra (number theory):

Lemma 21.1 *For primes p and q it holds that*

$$a^{(p-1)\times(q-1)} = 1 \ (\text{mod } p \times q)$$

if a has no common divisors with $p \times q$.

Proof This can be shown with the help of a fundamental result in algebra. *Fermat's Little Theorem* that for integers x and prime p states

$$x^{p-1} = 1 \ (\text{mod } p) \qquad \text{if } p \text{ does not divide } x \qquad (21.2)$$

We now apply this theorem twice (with the object to express $a^{(p-1)\times(q-1)}$ in two different ways) and obtain

$$(a^{(p-1)})^{(q-1)} = 1 \ (\text{mod } q) \qquad \text{and} \qquad (a^{(q-1)})^{(p-1)} = 1 \ (\text{mod } p)$$

Since $(a^{(p-1)})^{(q-1)} = a^{(p-1)\times(q-1)} = a^{(q-1)\times(p-1)} = (a^{(q-1)})^{(p-1)}$ it follows that there are k_1 and k_2 such that $k_1 \times p = k_2 \times q$. Since p and q are primes this is only possible if $k_1 = b \times q$ for some natural number b. In this case, however, we know that $a^{(p-1)\times(q-1)} - 1 = k_1 \times p = b \times q \times p = b \times p \times q$ and thus $a^{(p-1)\times(q-1)} = 1 \ (\text{mod } p \times q)$.

We can now show that the decrypted cyphertext is the original message again and thus the correctness of RSA:

$$
\begin{aligned}
(M^e)^d \ (\mathrm{mod}\ n) &= M^{e\times d} \ (\mathrm{mod}\ n)\\
&= M^{1+k\times(p-1)\times(q-1)} \ (\mathrm{mod}\ n) \qquad \text{Eq. 21.1}\\
&= M \times (M^{(p-1)\times(q-1)})^k \ (\mathrm{mod}\ n)\\
&= M \times 1^k \ (\mathrm{mod}\ n) \qquad\qquad \text{Lemma 21.1}\\
&= M \ (\mathrm{mod}\ n)
\end{aligned}
$$

For details please consult a cryptography textbook, for instance the classic [44].

What Next?

We have seen how one can try to attack **NP**-complete problems in practice to get decent results in a reasonable amount of time. But maybe harnessing natural phenomena (in the small) can help? We will have a look at emerging models of computation: molecular and quantum computing.

Exercises

1. A set $S \subseteq V$ of vertices in a graph $G = (V, E)$ is called *independent* if no two of its nodes are adjacent, i.e. no two nodes in S are connected by an edge $e \in E$. Consider the *Independent Set Problem*:

   - **instance**: a graph $G = (V, E)$ and a number k
   - **question**: is there an independent set in G of size at least k?

   The optimisation version of the problem would require to find an independent set of maximum size in G.

   a. Show that the Independent Set Problem is in **NP**.
   b. Show that Independent Set is **NP**-hard via a reduction:
      3SAT $\leq_P$ IndependentSet.
      Hint: Use a similar technique as for the reduction SAT $\leq_P$ Clique in Chap. 20, Exercise 12.

2. Consider the dynamic programming algorithm for 0-1 Knapsack given in the proof of Proposition 17.1 in Chap. 17.

   a. First explain why we can assume that all items have a weight that is less or equal the capacity W.
   b. Explain why the algorithm is correct.
   c. Use the algorithm to decide the 0-1 Knapsack problem for the following instance

   $$n = 5,\ W = 12,\ K = 34,\ \mathbf{p} = (2, 6, 18, 20, 26),\ \mathbf{v} = (1, 2, 5, 6, 8)$$

   d. How does one have to change the algorithm to compute also the solution $\mathbf{x}$?

3. Show that solutions to the optimisation version of the 0-1 Knapsack problem (presented as decision problem in Definition 17.4 which means dropping K) can be efficiently approximated:

  a. Consider the following "greedy" algorithm that produces a solution:

   i. sort all the items by the ratio of their value to their size such that

$$\frac{p_1}{w_1} \geq \frac{p_2}{w_2} \geq \ldots \geq \frac{p_n}{w_n}.$$

   ii. greedily pack items in the given order as long as the capacity W of the knapsack is not exceeded.

   Show that this algorithm can produce arbitrarily bad results (packings) using the following example: let $n = 2$, $w_1 = 1$, $w_2 = W$, $p_1 = 2$ and $p_2 = W$.

  b. One can modify the above algorithm to provide a 2-approximation algorithm. Run the above algorithm and then take its result or the first item (according to the ordering) that could not be included, whatever is more valuable. Show that this modified algorithm produces a result that is at least $\frac{1}{2}$ of the optimal solution.

  c. Use the modified algorithm to compute the instance given in Exercise 2c.

  d. Conclude that the 0-1 Knapsack problem is in **APX**.

4. In Exercise 3 we have shown that 0-1 Knapsack is in **APX**. One can do better and show it is in **PTAS**, but one can do better still: There is actually a stronger class **FPTAS**:

Definition 21.13 A *fully polynomial time approximation scheme* or *FPTAS* is a polynomial time approximation scheme which requires the algorithm in question to be polynomial in both the problem size and $\frac{1}{\varepsilon}$ where $1 > \varepsilon > 0$. Let **FPTAS** be the class of optimisation problems with a fully polynomial time approximation scheme.

a. Show that **FPTAS** $\subseteq$ **PTAS**.

b. We have already seen a pseudo-polynomial algorithm for 0-1 Knapsack in the proof of Proposition 17.1 based on dynamic programming, which is $\mathcal{O}(n^2 \times p_{max})$. We now need to get rid of factor p_{max} by scaling the profits down. This will require rounding and therefore loss of accuracy. Consider the FPTAS in Fig. 21.1, where *trunc* does rounding down to integer values. Observe how the maximum profit value p'_{max} is now $trunc(\frac{n}{\varepsilon})$ which depends solely on n and ε.

  i. Show that the above algorithm is a $(1 + \varepsilon)$-approximation algorithm for 0-1 Knapsack.

  ii. Show that the runtime of the above algorithm is $\mathcal{O}(n^3 \times \frac{1}{\varepsilon})$.

5. Explain informally the following characterisations of simple instances of the class **PCP**$[r(n), q(n)]$

  a. **PCP**$[0, 0] = $ **P**.

```
k  := p_max × ε/n
for i := 1 to n do
      p'_i := trunc(p_i/k)
S' := result of dynamic programming algorithm from Prop. 17.1
             applied to new p'
return S'
```

Fig. 21.1 FPTAS for 0-1 Knapsack

    b. **PCP**$[\mathscr{O}(\log_2 n), 0] = $ **P**
    c. **PCP**$[0, n^{\mathscr{O}(1)}] = $ **NP**
    d. **PCP**$[n^{\mathscr{O}(1)}, 0] = $ **BPP**

6. Show that in the definition of **RP** it does not matter whether one uses $\frac{1}{2}$ or some other p such that $1 > p > 0$, the class of problems remains the same.
   Hint: Repetition!
7. Show that **RP** $\subseteq$ **BPP**. This is not immediate, one has to obtain the required ε first.
   Hint: Repetition!
8. Show that **RP** and **BPP** are closed under intersection and union.
9. No **BPP**-complete problem is known (using deterministic polynomial time reduction). "One reason for this difficulty is that the defining property of BPTIME machines is semantic, namely, that for every string they either accept with probability at least 2/3 or reject with probability at least 1/3. Given the description of a Turing machine M, testing whether it has this property is undecidable" [7].

   a. Explain why the property of being a probabilistic Turing machine is undecidable.
   b. Explain why, on the other hand, the property of being a nondeterministic Turing machine is syntactic und thus decidable.

10. Develop another algorithm for *MetricTSP* (see Definition 21.11) that is guaranteed to produce a tour that is at most twice as bad as the optimal solution, and that runs in polynomial time. Proceed as follows:

    a. First, we compute a *minimum spanning tree* for graph G. We know from Exercise 13 of Chap. 16 that this can be computed in polynomial time (and more details can be found there).
    b. Show that from the minimum spanning tree T for G one can compute a tour that visits every city (i.e. vertex of G) exactly once.
    c. For the above tour show that it is by construction at most twice the weight of the spanning tree.
       Hint: use triangle inequalities!

d. Show that the weight of the minimum spanning tree T for G is not larger than the length of an optimal tour for *MetricTSP*.

e. From the above conclude that the computed tour's length is at most twice the length of an optimal tour.

References

1. Agrawal, M., Kayal, N., Saxena, N.: PRIMES is in P. Ann. Math. **160**(2), 781–793 (2004)
2. Applegate, D.L., Bixby, R.E., Chvátal, V., Cook, W.J.: The Traveling Salesman Problem: A Computational Study. Princeton University Press, Princeton (2007)
3. Arora, S.: Polynomial time approximation schemes for Euclidean traveling salesman and other geometric problems. J. ACM **45**(5), 753–782 (1998)
4. Arora, S.: The approximability of NP-hard problems. In Proceedings of Symposium on Theory of computing, pp. 337–348. ACM (1998)
5. Arora, S., Safra, S.: Probabilistic checking of proofs: a new characterization of NP. J. ACM **45**(1), 70–122 (1998)
6. Arora, S., Lund, C., Motwani, R., Sudan, M., Szegedy, M.: Proof verification and hardness of approximation problems. J. ACM **45**(3), 501–555 (1998)
7. Arora, S., Barak, B.: Computational Complexity: A Modern Approach. Cambridge University Press, Cambridge (2009)
8. Bellman, R.: Dynamic programming treatment of the travelling salesman problem. J. Assoc. Comput. **9**, 61–63 (1962)
9. Christofides, N.: Worst-case analysis of a new heuristic for the travelling salesman problem. Report 388, Graduate School of Industrial Administration, Carnegie Mellon University (1976)
10. Concorde Home Page. Available via DIALOG. http://www.math.uwaterloo.ca/tsp/concorde (2015). 19 June 2015
11. Cook, W.J.: In Pursuit of the Traveling Salesman: Mathematics at the Limits of Computation. Princeton University Press, Princeton (2012)
12. Coppersmith, D., Winograd, S.: Matrix multiplication via arithmetic progressions. J. Symb. Comput. **9**, 251–280 (1990)
13. Dantzig, G., Fulkerson, R., Johnson, S.: Solution of a large-scale traveling-salesman problem. J. Op. Res. Soc. Am. **2**(4), 393–410 (1954)
14. Diffie, W., Hellman, M.E.: New directions in cryptography. IEEE Trans. Inf. Theory **22**(6), 644–654 (1976)
15. Dean, J., Ghemawat, S.: MapReduce: simplified data processing on large clusters. Commun. ACM **51**(1), 107–113 (2008)
16. Dekel, E., Nassimi, D., Sahni, S.: Parallel matrix and graph algorithms. SIAM J. Comput. **10**(4), 657675 (1979)
17. Feige, U., Goldwasser, S., Lovász, L., Safra, S., Szegedy, M.: Interactive proofs and the hardness of approximating cliques. J. ACM **43**(2), 268–292 (1996) (Preliminary version in Proc. of FOCS91)
18. Gill, J.: Computational complexity of probabilistic Turing machines. SIAM J. Comput. **6**(4), 675–695 (1977)
19. Gomory, R.E.: Outline of an algorithm for integer solutions to linear programs. Bull. Amer. Math. Soc. **64**(5), 275–278 (1958)
20. Harel, D.: Computers LTD: What they Really Can't Do. Oxford University Press, Oxford (2000)

21. Held, M., Karp, R.M.: A dynamic programming approach to sequencing problems. J. SIAM **10**, 196–210 (1962)

22. Held, M., Karp, R.M.: The traveling salesman problem and minimum spanning trees, part II. Math. Program. **1**, 6–25 (1971)

23. Helsgaun, K.: An effective implementation of the Lin-Kernighan traveling salesman heuristic. Eur. J. Op. Res. **126**, 106–130 (2000)

24. Johnson, D.S.: Approximation algorithms for combinatorial problems. J. Comput. Syst. Sci. **9**, 256–278 (1974)

25. Johnson, D.S.: The NP-completeness column: an ongoing guide—the tale of the second prover. J. Algorithms **13**, 502–524 (1992)

26. Johnson, D.S.: A brief history of NP-completeness 1954–2012. In: Götschel, M. (ed.) Optimization Stories, Book Series, Vol. 6, pp. 359–376, Documenta Mathematica (2012)

27. Helsgaun, K.: LKH-2 Website. Available via DIALOG. http://www.ruc.dk/keld/research/LKH/. Accessed 20 June 2015

28. Kirkpatrick, S., Gelatt, C.D., Vecchi, M.P.: Optimization by simulated annealing. Science **220**(4598), 671–680 (1983)

29. Lawler, E.L., Lenstra, J.K., Rinnooy Kan, A.H.G., Shmoys, D.B. (eds.): The Traveling Salesman Problem: A Guided Tour of Combinatorial Optimization. Wiley, New York (1985)

30. Levy, S.: The Open Secret. Wired Issue 7.04. Available via DIALOG. http://archive.wired.com/wired/archive/7.04/crypto.html (1999). Accessed 19 June 2015

31. Lin, S., Kernighan, B.W.: An effective heuristic algorithm for the traveling-salesman problem. Op. Res. **21**(2), 498–516 (1973)

32. Milestones in the Solutions of TSP Instances. Available via DIALOG. http://www.math.uwaterloo.ca/tsp/history/milestone.html (2015). Accessed 20 June 2015

33. Miller, G.L.: Riemann's hypothesis and tests for primality. In: Proceedings of Seventh Annual ACM Symposium on Theory of Computing, pp. 234–239. ACM (1975)

34. Mitchell, J.: Guillotine subdivisions approximate polygonal subdivisions: a simple polynomial-time approximation scheme for geometric TSP, k-MST, and related problemsd. SIAM J. Comput. **28**(4), 1298–1309 (1999)

35. Mulder, S.A., Wunsch II, D.C.: Million city traveling salesman problem solution by divide and conquer clustering with adaptive resonance neural networks. Neural Netw. **16**(5–6), 827–832 (2003)

36. Nash, J.C: The (Dantzig) simplex method for linear programming. Comput. Sci. Eng. **2**(1), 29–31 (2000)

37. Optimal 85,900-City Tour. Available via DIALOG. http://www.math.uwaterloo.ca/tsp/pla85900/index.html (2015). Accessed 20 June 2015

38. Papadimitriou, C.H.: The Euclidean traveling salesman problem is NP-complete. Theor. Comput. Sci. **4**(3), 237–244 (1977)

39. Papadimitriou, C.H., Vempala, S.: On the approximability of the traveling salesman problem. Combinatorica **26**(1), 101–120 (2006)

40. Rabin, M.O.: Probabilistic algorithm for testing primality. J. Number Theory **12**(1), 128–138 (1980)

41. Rivest, R.L., Shamir, A., Adleman, L.: A method for obtaining digital signatures and public-key cryptosystems. Commun. ACM **21**(2), 120–126 (1978)

42. Rutenbar, R.: Simulated annealing algorithms: an overview. IEEE Circuits Devices Mag. **5**(1), 19–26 (1989)

43. Sahni, S., Gonzalez, T.: P-complete approximation problems. J. ACM **23**(3), 555–565 (1976)

44. Scheier, B.: Applied Cryptography. Wiles, New York (1996)

45. The Traveling Salesman Problem. Available via DIALOG. http://www.math.uwaterloo.ca/tsp/index.html (2015). Accessed 19 June 2015

46. TSPLIB. Available via DIALOG. http://www.iwr.uni-heidelberg.de/groups/comopt/software/TSPLIB95/ (2015). Accessed 19 June 2015

47. Williamson, D.P., Shmoys, D.B.: The Design of Approximation Algorithms. Cambridge University Press, Cambridge (2011)

48. Woeginger, G.J.: Exact algorithms for NP-hard problems: A survey. In: Juenger, M., Reinelt, G., Rinaldi, G. (eds.) Combinatorial Optimization—Eureka! You shrink! LNCS, vol. 2570, pp. 185–207. Springer, Berlin (2003)

49. World TSP. Available via DIALOG. http://www.math.uwaterloo.ca/tsp/world/ (2015). Accessed 20 June 2015

50. Vazirani, V.V.: Approximation Algorithms. Springer, Berlin (2001)

Chapter 22
Molecular Computing

What is molecular computing and can it help finding feasible solutions to NP-complete problems? What other benefits does it have?

DNA (Deoxyribonucleic acid) contains the genetic code that is responsible for the development of all known living organisms. Every living cell[1] contains a nucleus with DNA. The word "code" already indicates that DNA might be appropriate to replace the binary codes known from the silicon world of computing.[2]

Until recently, DNA was thought to be the *same* for every cell.[3] Strands of DNA are long polymers of millions of linked nucleotides which then appear in its famous double helix form as outlined in Fig. 22.1. This helix has a sugar and phosphate backbone and the links between the two backbone strands are complementary C–G (Cytosine–Guanine) and A–T (Adenine–Thymine) base pairs.

DNA is—necessarily—very small, yet evolution has come up with sophisticated and *almost* error-free biochemical replication and manipulation processes which makes it amenable to computations. *"DNA replication is a truly amazing biological phenomenon. The polymerase enzymes …do make mistakes at a rate of about 1 per every 100,000 nucleotides. That might not seem like much, until you consider how much DNA a cell has. In humans, with our 6 billion base pairs in each diploid cell, that would amount to about 120,000 mistakes every time a cell divides! Fortunately, cells have evolved highly sophisticated means of fixing most, but not all, of those mistakes"* [23].

---

[1] A human cell contains almost two meters of DNA [19] strand rolled up into a small ball, i.e. nucleus, of $5\,\mu m$ [13] structured into 23 pairs of chromosomes.

[2] The complete human genome needs about 3.08 GB of storage space on a conventional hard drive [22].

[3] Latest research seems to suggest that there is rather a "mosaic" of DNA in the cells [21].

© Springer International Publishing Switzerland 2016
B. Reus, *Limits of Computation*, Undergraduate Topics in Computer Science,
DOI 10.1007/978-3-319-27889-6_22

Fig. 22.1 Schematic
illustration of DNA double
helix with base pairs

Despite the correction mechanisms errors do happen, causing mutations[4] which
in moderation are well known to be essential for evolution.

In this chapter we will first look at the historic beginnings and the attempts to solve
NP-complete problems efficiently (Sect. 22.1), discuss its potential (Sect. 22.2) and
the major challenges (Sect. 22.3). The main topic of this chapter is the presentation of one particular abstract model of molecular computing (Sect. 22.4) which will
introduce *Chemical Reaction Networks* and compare them to other notions of computations. We also briefly discuss how they can be implemented with the help of
DNA strands.

22.1 The Beginnings of DNA Computing

In 1994, *Leonard Adleman*[5] started the field of *DNA computing* by attempting to find
a solution for the Travelling Salesman Problem[6] with 7 cities. He encoded the graph
with DNA and the computation via enzymes. Each city had its own "CTGA" code,
and edges were encoded as bridges consisting of the end of the (matching) code of
one city attached to the start of the (matching) code of the other city. Bits of "city"
DNA can only stick together if there is a matching "bridge" DNA piece (encoding
an edge in the graph) which bonds with both the complementary city pieces to form
a double strand. Putting many copies of such prepared DNA in a test tube, allowing
the natural bonding tendencies of the DNA building blocks to occur, vast number
of strands build up that therefore encode paths through cities. In a separate step one
must then chemically isolate those strands that correspond to tours.[7]

So can we obtain polynomial time solutions for **NP**-complete problems with the
help of DNA computing? The situation is analogous to parallel computing which
required an unrealistically vast amount of processors. Similarly, using DNA requires

---

[4]*"Mutation rates vary substantially …even among different parts of the genome in a single organism. Scientists have reported mutation rates as low as 1 mistake per 100 million …to 1 billion
…nucleotides, mostly in bacteria, and as high as 1 mistake per 100 …to 1,000 …nucleotides, the
latter in a group of error-prone polymerase genes in humans"* [23].

[5]Yes, the one from the RSA algorithm!.

[6]More precisely the *Directed Hamiltonian Path Problem* version: "does a tour exist that visits each
city exactly once?".

[7]Using and combining several techniques: polymerase chain reaction, gel electrophoresis, and
affinity purification.

at least "bathtubs of DNA" [1] to encode the exponentially many possible candidate solutions. The "molecular soup" allows a huge number of candidate solutions to form quickly, but any "generate-and-test" kind of solution still won't do the job due to the exponential growth of required DNA material. *"In fact, it seems impossible for graphs with more than 30 vertices to be handled by this approach"* [20].

However, there is another way to use DNA inspired by biology. Biological systems are molecularly programmed. One can try to copy nature's ideas in order to *write* molecular programs. This has become easier nowadays, as DNA molecules are relatively easy to produce and encode.

22.2 DNA Computing Potential

DNA computing has enormous potential because:

- it uses one billion times less energy than silicon computers (it is often self-powered)
- it uses one trillion times less space than silicon computers (thus allowing the construction of nano-computers)
- it has the ability to self-assemble
- it interfaces with living organisms and thus can be used "in vivo" to "run programs" inside an organism to deliver for instance *smart drugs* or kill cancerous cells (nanomedicine)
- it is easily available and cheap.

DNA storage is already possible in the lab. In January 2013, a team from the European Bioinformatics Institute (EBI) at Hinxton, near Cambridge, *"as a proof of concept for practical DNA-based storage, ...selected and encoded a range of common computer file formats to emphasize the ability to store arbitrary digital information"* [15].[8]

Church, Kosuri et al. had already achieved a similar objective in 2012 [10]. According to them, apparently one gram of DNA can store 700 terabyte of data, which is equivalent to 14,000 Blu-ray discs. DNA storage offers an excellent way of archiving, as DNA requires no energy and, adequately stored, lasts for thousands of years. It may take some time to retrieve the information again though, as this requires DNA sequencing. Sequencing, however, has made great progress. The cost reduction of DNA sequencing has long been following Moore's Law, but since 2008 has significantly outperformed it.[9] Therefore, the high expectations of DNA computing and data storage are not unfounded. The interfacing with living organisms and the nano-scale of this paradigm allows for all kinds of fascinating applications which are currently still science fiction.

---

[8]This included all 154 of Shakespeares sonnets (ASCII text) and a 26-s excerpt from Martin Luther King's 1963 I have a dream speech (MP3 format), and with some other data used a total of 757,051 bytes.

[9]According to the National Human Genome Research Institute.

22.3   DNA Computing Challenges

Inherent limitations of DNA computing for solving hard optimisation problems are:

- scaling: a large number of DNA molecules is needed for large problem instances;
- accuracy: depending on the complexity of DNA copying mechanisms used, errors in pairings of DNA strands themselves can occur (this is normal and happens in organisms, leading, e.g., to mutations). Therefore, one usually gets some wrong results along with the correct solutions unless one invests in error correction. For a simple experiment (encoding search), [12] report an error rate of 1 in 2824 (if no error correction is applied).
- simple operations may take hours to days (depending on chemical reaction needed);
- most practical DNA computers are not universally programmable. Charles Bennett has shown that in principle a chemical (reversible!) Turing machine can be constructed in a thought experiment [3] (see also [4]), but nobody has actually built such a universal machine yet. Moreover, and one cannot stress this enough, the interesting applications do not require such a universal DNA computer anyway.

"The problems inherent in nucleic-acid-based computation make it unlikely that time-consuming or complex algorithms will ever be conveniently addressed. Instead, nucleic acid computation may find extremely important reflexive applications, in which they will serve as integrated circuits that measure biological signals (e.g. insulin levels), decide between a limited set of desired outcomes (e.g. too little or too much insulin) and transduce biological function (e.g. release or retain insulin)" [12]. This line of research will be discussed in the following Sects. 22.4.5.1 and 22.4.5.2. First, we define a formal computational model that can serve as an abstract model of Molecular computing.

22.4   Abstract Models of Molecular Computation

There exist various approaches to define models of chemical computing. The idea is to abstract away from concrete chemistry similarly to the way the computation models presented in Chap. 11 abstract away from the concrete physical implementation of machines.

In this chapter we just present one popular approach, although there are others like, e.g., the "blob model" of [16] which is based on "bond sites" and adjacency. The fact that only blobs that are close to each other can interact, i.e. locality, is a concept already seen in cellular automata (see Sect. 11.7). The approach we focus on, are *Chemical Reaction Networks* (CRN). Here one uses *stochastic chemical kinetics*[10] to model computation *"in which information is stored and processed in the integer counts of molecules in a well-mixed solution"* [27]. The model is particularly

---

[10]In chemistry the term "kinetics" refers to the study of reaction rates.

appropriate for small molecular counts. But *"small molecular counts are prevalent in biology: for example over 80 % of the genes in the* E. coli *chromosome are expressed at fewer than a hundred copies per cell, with some key control factors present in quantities under a dozen"* [27].

Normally the model is used to *describe* chemical reactions or other processes in population protocols in biology, sensor networks and various areas in mathematics. Here it is rather used to *prescribe* reactions to happen as intended (programs). Of course, it is expected that these can then be implemented in actual chemistry. This will be briefly discussed in Sect. 22.4.5.

The programming of CRNs is quite different from programming RAMs or TMs or writing WHILE-programs. The reasons is that one has no control over the order of reactions executed. It would help the programmer to be able to prioritise the chemical reaction rules. However, these rules will be executed randomly in a well-mixed solution that contains the necessary chemicals.[11] On the other hand, this also provides us with some guarantees, namely that nobody and nothing else can control the order of execution either, implementing some basic notion of fairness.

It is worth pointing out that some implicit assumptions are made when employing the model for the purposes of computation which are e.g. explained in [27].

22.4.1 Chemical Reaction Networks (CRN)

Next we briefly outline the definition of this notion of computation following [11, 14]. We will make use of vectors that are actually like the sets of key-value pairs we introduced for stores in Sect. 4.1. Let S be a finite set of size $|S| = k$, then we write $\mathbb{N}^S$ to denote the type of all vectors that contain k natural numbers, indexed by the elements in S. Vectors we usually write boldface. So if $\mathbf{x} \in \mathbb{N}^S$ and $s \in S$, then $\mathbf{x}(s) \in \mathbb{N}$ contains the natural number associated with s. As the vectors used contain natural numbers, we also write $\#_{\mathbf{x}}s$ for $\mathbf{x}(s)$.

Definition 22.1 (*Chemical Reaction Network* (CRN)) A (finite) *chemical reaction network* (CRN) is a pair $\mathscr{C} = (\Lambda, R)$, where Λ is a finite set of chemical species, and R is a finite set of reactions over Λ. A *reaction* over Λ is a triple $\beta = (\mathbf{r}, \mathbf{p}, k) \in \mathbb{N}^\Lambda \times \mathbb{N}^\Lambda \times \mathbb{R}^+$, specifying quantitative data, also called the *stoichiometry* of the *reactants* and *products*, respectively, and the *rate constant* k.

A *configuration* of a CRN $\mathscr{C} = (\Lambda, R)$ is a vector $\mathbf{c} \in \mathbb{N}^\Lambda$.

One usually distinguishes types of reactions as follows:

Definition 22.2 (*Type of reactions*) A reaction β of a CRN (Λ, R) is called *unimolecular* if it has one reactant (of one species) only. It is called *bimolecular* if it has two reactants. Commonly, higher numbers of reactants are not needed.

---

[11] However, there will be reaction rates associated to reactions that depend on the concentrations of the molecules in use and thus vary during computation. They can be used to implicitly prioritise reactions by increasing or decreasing the probability to fire.

A reaction is called *catalytic* if any species appears as reactant *and* as product (and thus is not consumed by the reaction). A catalytic reaction is called *autocatalytic* if the catalyst species is self-reproducing, i.e. the produced quantity is higher than the consumed quantity.

One can then execute a CRN as follows:

Definition 22.3 (*Deterministic semantics of CRN reactions*) Like in many of the other notions of computation discussed earlier, there is a notion of state that is modified during execution. The *state of a chemical reaction network* is defined as configuration, specifying the molecular counts of each species currently present. Given a configuration c of a CRN $\mathscr{C} = (\Lambda, R)$ and reaction $\beta = (\mathbf{r}, \mathbf{p}, k) \in R$, we say that β is *applicable* to c if $\mathbf{r} \leq c$ (or equivalently if $c - \mathbf{r} > 0$).[12] In other words, configuration c contains enough of each of the reactants for the reaction to occur. If β is applicable to c, we write $\beta(c)$ to denote the configuration $c - \mathbf{r} + \mathbf{p}$. This means that $\beta(c)$ is the result reaction β "happening to" c. If $c_2 = \beta(c_1)$ for some reaction $\beta \in R$, we have observed a computation step in a CRN and write $c_1 \rightarrow_{\mathscr{C}} c_2$ or simply $c_1 \rightarrow c_2$ when the CRN in question is clear from the context. An execution (sequence) of $\mathscr{C}$ is a finite or infinite sequence of one or more configurations $(c_0, c_1, c_2, \ldots)$ such that $c_i \rightarrow_{\mathscr{C}} c_{i+1}$ for all i. A finite execution sequence from c to c' is denoted $c \rightarrow_{\mathscr{C}}^{*} c'$, and in this case c' is called *reachable* from c.

If $\Lambda = \{X_1, X_2, \ldots, X_m\}$, a rule $\beta = (\mathbf{r}, \mathbf{p}, k)$ is often denoted in a slightly more readable form:

$$r_1 X_1 + r_2 X_2 + \cdots + r_m X_m \xrightarrow{k} p_1 X_1 + p_2 X_2 + \cdots + p_m X_m$$

where $r_i = \#_{\mathbf{r}} X_i$ and $p_i = \#_{\mathbf{p}} X_i$

Example 22.1 The following is a deterministic CRN that computes whether there are more molecules of species A than B in the current state. The CRN should end[13] with one molecule Y if this is the case, otherwise with one molecule N. Let $\Lambda = \{A, B, Y, N\}$ and the rule set R:

$$A + N \xrightarrow{1} Y \qquad B + Y \xrightarrow{1} N$$

Both reactions are non-catalytic and can be expressed in the form of Definition 22.1 as follows:

$$\beta_1 = (\mathbf{r}_1, \mathbf{p}_1, 1) \qquad\qquad \beta_2 = (\mathbf{r}_2, \mathbf{p}_2, 1)$$
$$\mathbf{r}_1 = \{A : 1, B : 0, Y : 0, N : 1\} \qquad \mathbf{r}_2 = \{A : 0, B : 1, Y : 1, N : 0\}$$
$$\mathbf{p}_1 = \{A : 0, B : 0, Y : 1, N : 0\} \qquad \mathbf{p}_2 = \{A : 0, B : 0, Y : 0, N : 1\}$$

---

[12] The minus symbol here denotes (pointwise) vector subtraction. Similarly, we use $+$ for vector addition.

[13] This is a very particular CRN, and we will have to discuss below how to read off results or outputs from a CRN in a more general setting.

Now let the initial configuration be the vector $\{A : a, B : b, Y : 1, N : 0\}$. In this initial state, the second reaction β_2 can fire using up the one Y molecule (and one B) to produce one N. Next, the first reaction β_1 can fire, consuming the N molecule and also one A molecule, thus re-establishing one Y molecule. Then this process can repeat itself. In this case we actually have a deterministic order of execution[14] and we can continue until no B molecule is left. If any more A molecules are available after that, the first reaction can fire one last time to produce a final Y, otherwise no more reactions are possible and one ends up with one N.

We observe that in the above example the count of species Y and N continuously changes with every reaction until no further reaction can occur. In this case the CRN halts, so to speak. Only then can we read off the result. We will see below that there is also another way of "producing a result" with a CRN.

   Often, however, one does not have a deterministic CRN but rather a high volume of molecules and many (overlapping) reactions. In this case many reactions will take place concurrently in a random fashion. The likelihood of a reaction to occur will be proportional to the number of its reactants available. For such applications one needs a model of *stochastic chemical kinetics* which takes all this into account. One has to assign probabilities to execution sequences and to consider the time of reactions which is important for computational complexity considerations. This involves Markov's processes and some slightly more advanced probability theory, so it is beyond the scope of this book. The interested reader is referred to, e.g., [11] for details and more references.

22.4.2 *CRNs as Effective Procedures*

How powerful a means of computation are CRNs? A first important observation is that one can distinguish different kinds of CRN according to [27]:

- discrete versus continuous: are the molecule counts discrete (natural numbers) or real numbers? We only consider discrete models;
- uniform versus non-uniform: is a single CRN supposed to handle all inputs, or just a finite number (size) of inputs. We only consider uniform models which corresponds to the fact that also the other notions of computation considered in Chap. 11 use unbounded storage (unbounded tapes, unbounded content in registers and variables);
- deterministic versus probabilistic: is the correct output *guaranteed* or just, probabilistically speaking, *likely*?
- halting versus stabilizing: is the CRN "aware" it has finished, by entering an irreversible final state, or is a *stable* state eventually reached from which reactions can continue but cannot further change the result.

---

[14]This is a *special* case for CRNs.

When we use CRNs for probabilistic computation, we use the term stochastic CRN, or shorter SCRN, as stochastic semantics will be required in this case. But in the following we will focus on deterministic computation (as it is easier and does not require much stochastic and probability theory).

We thus define more precisely what it means for a problem to be (deterministically) decided or a function to be (deterministically) computed by a CRN. As for our other notions of computation, we need to use the semantics at hand, but also decide how input and output are dealt with, and how the counts of any auxiliary species are initialised.

Definition 22.4 (*Deterministic decidability in CRN* [8]) Let $\mathscr{C} = (\Lambda, R)$ be a CRN and let $Y, N \in \Lambda$ be distinguished species (used for output purposes). A decision procedure in CRN, called *chemical reaction decider*, is a tuple $p = (\mathscr{C}, I, \mathbf{s})$ where $I \subseteq \Lambda$ is the set of *input species*[15] and $\mathbf{s} \in \mathbb{N}^{\Lambda \setminus I}$ is the vector of initial counts for all non-input species. If $\mathbf{i} \in \mathbb{N}^I$ denotes the vector of counts of input species, then the *initial configuration* $\mathbf{c}_0^{\mathbf{s}}(\mathbf{i}) \in \mathbb{N}^\Lambda$ is defined as follows:

$$\mathbf{c}_0^{\mathbf{s}}(\mathbf{i})(X) := \begin{cases} \#_{\mathbf{i}} X & \text{if } X \in I \\ \#_{\mathbf{s}} X & \text{if } X \in \Lambda \setminus I \end{cases}$$

The initial configuration maps the input species to the input counts and the other species to the counts prescribed in vector $\mathbf{s}$. We further define an output function out $: \mathbb{N}^\Lambda \to \{0, 1\}$ as follows:

$$\text{out}(\mathbf{c}) := \begin{cases} 1 & \text{if } \#_{\mathbf{c}} Y > 0 \text{ and } \#_{\mathbf{c}} N = 0 \\ 0 & \text{if } \#_{\mathbf{c}} Y = 0 \text{ and } \#_{\mathbf{c}} N > 0 \\ \text{undefined} & \text{otherwise} \end{cases}$$

A configuration $\mathbf{c}$ is called *output stable* if out$(\mathbf{c})$ is defined and for all configurations $\mathbf{c}'$ such that $\mathbf{c} \to_{\mathscr{C}}^* \mathbf{c}'$ it holds that out$(\mathbf{c}') = $ out$(\mathbf{c})$. So from an output stable configuration one can reach only configurations that do not change the result of the computation. Finally, a set $A \subseteq \mathbb{N}^k$ is called *(deterministically) stably CRN-decidable*, if there is a *chemical reaction decider* $p = (\mathscr{C}, I, \mathbf{s})$ such that

$$\langle n_1, n_2 \ldots n_k \rangle \in A \text{ if, and only if, } \begin{pmatrix} \mathbf{c}_0^{\mathbf{s}}\{X_1 : n_1, \ldots, X_k : n_k\} \to_{\mathscr{C}}^* \mathbf{c} \text{ implies} \\ \exists \text{ output stable } \mathbf{c}'. \mathbf{c} \to_{\mathscr{C}}^* \mathbf{c}' \text{ s.t. out}(\mathbf{c}') = 1 \end{pmatrix}$$

In other words, tuple $\langle n_1, n_2 \ldots n_k \rangle$ is in set A iff running CRN $\mathscr{C}$ in the initial configuration (setting input species counts to $n_1, n_2 \ldots n_k$ with other counts as prescribed in $\mathbf{s}$) always reaches an output stable configuration with output 1. Note that this definition does not require the execution to terminate, but it prescribes that it stabilises w.r.t. the result. It is guaranteed that the expected result is produced eventually and it never changes thereafter, even if more rules are executed.

---

[15] We allow more than one input here as the encoding of tuples in CRNs is not as easily programmable as for our other notions of computations.

The data type in use here are the natural numbers, $\mathbb{N}$, but as for other notions of computation these can encode any datatype, by looking, e.g., at their binary representation.

Similarly, we define function computability:

Definition 22.5 (*Deterministic function computability in* CRN [8]) Let $\mathscr{C} = (\Lambda, R)$ be a CRN. A *chemical reaction computer* is a tuple $p = (\mathscr{C}, I, O, \mathbf{s})$ where $I \subseteq \Lambda$ is the set of *input species*, $O \subseteq \Lambda$ is the set of *output species*,[16] and $\mathbf{s} \in \mathbb{N}^{\Lambda \setminus I}$ is the vector of initial counts for all non-input species. A configuration $\mathbf{c}$ is *output count stable* if, for every $\mathbf{c}'$ such that $\mathbf{c} \rightarrow^*_{\mathscr{C}} \mathbf{c}'$ it holds that $\mathbf{c}(Y) = \mathbf{c}'(Y)$ for all $Y \in O$. In other words, any corresponding reaction sequence from an output count stable configuration will have no effect on the result as the result is read off from the output variables.

For the following assume that $O = \{Y_1, Y_2, \ldots, Y_l\}$. We say that p as above *deterministically stably computes* a function $f : \mathbb{N}^k \to \mathbb{N}^l$ if, and only if, for all $\mathbf{c}$ such that $\mathbf{c}_0^s\{X_1 : n_1, \ldots X_k : n_k\} \rightarrow^*_{\mathscr{C}} \mathbf{c}$, there exists an output count stable configuration $\mathbf{c}'$ such that $\mathbf{c} \rightarrow^*_{\mathscr{C}} \mathbf{c}'$ and $f(n_1, n_2, \ldots, n_k) = (\mathbf{c}'(Y_1), \mathbf{c}'(Y_2), \ldots, \mathbf{c}'(Y_l))$. This implies that no incorrect output count stable configuration is reachable from the initial configuration.

The corresponding notion of computation is called deterministic because the correct result is guaranteed. The probability for error is zero. It is called stabilising since the correct result is eventually reached and then maintained whatever happens afterwards. In particular, the output may oscillate during execution before it eventually stabilizes. The network does not "know" when this has happened [27]. By contrast, one can write chemical reaction deciders and computers that stop once a (valid) result has been produced. In the deterministic setting this would be, however, quite restrictive as will be discussed below.

Example 22.2 Consider a chemical reaction decider that deterministically stably decides the predicate $even(x)$. We need three species, X representing the input parameter, Y for the output "yes" (if $x = y$) and N for output "no". We define:

$$\Lambda = \{X, Y, N\} \qquad\qquad \mathbf{s} = \{Y : 1, N : 0\}$$
$$R = \{ X + Y \xrightarrow{1} N \quad, \quad X + N \xrightarrow{1} Y \}$$

Assume we have n molecules of species X. Hence, the initial configuration is $\mathbf{c}_0^s\{X : n\} = \{X : n, Y : 1, N : 0\}$. Initially it holds that $out(\mathbf{c}_0^s\{X : n\}) = 1$ which would only be the correct result if n was even. The first rule can fire to reach configuration $\{X : n-1, Y : 0, N : 1\}$, then only the second rule can fire, and so on. We get a deterministic sequence very similar to the one in Example 22.1. After all molecules

---

[16]We allow more than one input and output here for the same reasons as for chemical reaction deciders.

of the input species are used up, the correct output has been produced. The CRN does not "know" internally *when* this happened and the right result has been produced, but we are guaranteed that the right result is produced eventually.

In the exercises of this chapter we will encounter some other examples.

22.4.3   Are CRNs Equivalent to Other Notions of Computation?

In Chap. 11 we have shown that all our presented notions of computation are of equivalent power. An obvious question is whether this holds also for CRNs. It is important here that CRNs perform probabilistic computation and probabilistically they are Turing-complete as [27] write: "*Well-mixed finite stochastic chemical reaction networks with a fixed number of species can perform Turing-universal computation with an arbitrarily low error probability. This result illuminates the computational power of stochastic chemical kinetics: error-free Turing universal computation is provably impossible, but once any non-zero probability of error is allowed, no matter how small, stochastic chemical reaction networks become Turing universal.*"

Theorem 22.1 (Turing-universal computation [27]) *For any TM, there is a SCRN such that for any non-zero error probability ε, and any bound s_0 on the size of the input, there is an initial amount of a chosen species A (called* accuracy *species) that allows simulation of the TM on inputs of size at most s_0 with cumulative error probability at most ε over an unbounded number of steps and allowing unbounded space usage. Moreover, in the model where the volume grows dynamically in proportion with the total molecular count, t steps of the TM complete in expected time (conditional on the computation being correct) of $\mathcal{O}\left(\left(\frac{1}{\varepsilon} + s_0 + t \cdot s\right)^5 \cdot \frac{t \cdot s}{k}\right)$ where s is the space used by the TM, and k is the rate constant (of all reactions).*[17]

If s_0 is roughly t, then this gives polynomial runtime of degree 12. According to this theorem, one can compile from TM to CRN, however up to an arbitrarily small error. It is worth pointing out that the error is independent of the number of computation steps. Of course, one can also go the other way. Any CRN can be simulated by a classic computer. The Church-Turing thesis therefore also encompasses CRNs if (arbitrary small) errors are permitted. Interestingly, with CRNs one cannot achieve Turing-universal computation without error. The reason is that the CRN can "never

---

[17]Here all reactions are assumed to have the same rate constant.

be sure" whether it has dealt with all molecules of a certain species. In other words, with a CRN one cannot test *deterministically* whether there are *no* molecules of a certain kind.

Theorem 22.1 (and similar results in the literature) are conceptual proof of the power of CRNs, but this does not mean that they are practical. It is difficult in practice to control small counts of a large amount of species as required for CRNs that perform the action of a universal TM or RAM for instance. For practical purposes, the robustness of CRN implementations is important. This means that computations need to perform correctly even if molecular counts are "perturbed by small amounts" [17].

22.4.4 Time Complexity for CRNs

Complexity theory for (stochastic) CRN is much more complicated than computability and results appear to very much depend on the shape of concrete reactions. There are two sources for slow execution speed: slow reactions and too many reactions. Therefore, one is looking for (average) execution times in $\mathcal{O}(\log_2 n)$ where n is the number of input molecules. Linear (average) execution times are already considered slow in this context due to the potentially high volume of molecules involved and the high number of possible combinations of reactions.

Some results for implementing a particular kind of functions or predicates are already known, namely *semilinear* ones. Semilinear predicates can be described as union of linear sets $\mathbb{N}^d$. For instance, the predicate "$A \geq B$" in state x (where A is the first and B the second species) is semilinear as it can be expressed as $x \in \{n_1 \times (1, 0) + n_2 \times (1, 1) \mid n_1, n_2 \in \mathbb{N}\}$. In [2] it has been shown that every stably computable predicate is semilinear. But one can show more.

Theorem 22.2 ([8]) *Every semilinear predicate (or function with semilinear graph) can be stably deterministically computed by a CRN that stabilizes in expected time* $\mathcal{O}\left((\log_2 n)^c\right)$ *for a constant c.*

22.4.5 Implementing CRNs

A program can be written abstractly as a CRN, its semantics (behaviour) is then described by either a Markov process or differential equations. The modelling can thus be supported by a number of tools. A major challenge is still the implementation. One needs to find ways to run programs written in such molecular languages, build an interpreter so to speak. There are various ways to build such implementations.

22.4.5.1   Synthetic Biology

As recent as 2013, a biological transistor was created at Stanford University, called the "transcriptor" which constitutes another step into the direction of a biological computer. *"Where transistors control the flow of electricity, transcriptors control the flow of RNA polymerase*[18] *as it travels along a strand of DNA. The transcriptors do this by using special combinations of enzymes (integrases) that control the RNAs movement along the strand of DNA"* [5]. With the help of transcriptors one can implement boolean gates as with traditional silicon-based hardware. For the purpose of implementing CRNs, however, other techniques and approaches might be more appropriate, depending on which abstract biological machine one is using: gene, protein, or membrane machines [7]. Yet, large parts of the inner workings of those biological machines are still not fully understood, and design methods are therefore lacking. There exists another approach that does *not rely on "alien technology"* [26][19]: an approach based on first principles of chemistry, *Synthetic Chemistry*.

22.4.5.2   Synthetic Chemistry

This approach uses *chemistry as "assembly language"* [7]. DNA computing attempts to *"precisely control the organization and dynamics of matter and information at the molecular level"* [7]. It is crucial here (in contrast to "Synthetic Biology" Sect. 22.4.5.1) that DNA is used as engineering material. The fact that its origin is biological can be seen as "accident" [7].

If we consider CRNs we observe that we can implement a reaction $A + B \rightarrow C$ with two special gates, called *join* and *fork* gate, respectively. *"The join gate consumes (and thus "joins") the two signals A and B and the fork gate releases the signal C, which is initially bound to the fork gate ...and thus inactive. (The name "fork gate" derives from the fact that multiple signal strands can be released"* [9]. These signal processing gates are also known from so-called *Petri nets*[20] used to model parallel and distributed computation (see [24]).

We therefore focus on one particular technique for the implementation of those join and fork gates: *strand displacement.*

DNA strand displacement. The idea is to use single-stranded "bits" of DNA[21] strings as signals to encode input and output and rely on the "predictability of

---

[18]This is an enzyme that produces transcript RNA.

[19]Here "alien" alludes to any technology one does not understand.

[20]Carl Adam Petri (12 July 1926—2 July 2010) was an award-winning German mathematician and computer scientist who invented Petri nets exactly for the purpose of describing chemical reactions.

[21]DNA occurs most of the time as double stranded (double helix) "ladder".

Watson[22,23]–Crick[24,25] base pairing"[26] [25] as "engine". The input strands react (in a predictable way) with double-stranded DNA in which parts are displaced, releasing some other single-stranded piece of DNA as output. This means computations (gates) can be compositional and no enzymes are required to drive the computation.

The double-stranded DNA acts as gates. Subsequences of DNA are called (recognition) domains. The subsequences must be designed so that they cannot accidentally "hybridize", i.e. bind together according to the Crick–Watson base pairing. They should also be reasonably short. The displacement is mediated by so-called *toeholds*, which are initial bits of the subsequences that are used to initiate the binding process. The double-stranded gates consist of hybridized single strands with opposite orientations. Single strands can only be synthesized in vivo in a certain direction: 5'-to-3'.[27] The numbers 5' and 3' refer to the number of the carbon atom in the sugar-ring of the deoxyribose at one of the two ends of the molecule. In double-stranded DNA the 5' end of one strand pairs with the 3' end of the opposite (and complementary) strand.

Displacement of a subsequence in a double strand can occur if a complementary single DNA strand first binds at the toehold followed by so-called *branch migration*. Migration means that a specific subsequence (also called *domain*) of the single strand replaces the original one. Seelig et al. have built logic circuits with this technique [25] interpreting high quantities of input strands as "1" and low quantities as "0". Reading off the result is slightly more complicated than on a desktop computer. Output strands are modified with a dye so fluorescence can be observed and measured. Debugging is also difficult, one uses *gel electrophoresis*[28] to see how many molecules (of which size, i.e. length in terms of number of base pairs) of each reactant are present. The energy for the computation comes from the destruction of the gates themselves. Toehold binding and branch migration are reversible until branches are physically detached. Reversibility is *essential* in order to undo *wrong* matches. After detachment, due to the orientation of toeholds, the process cannot be reversed (in this case the corresponding gate is consumed and cannot be used any longer). In order

---

[22]James Dewey Watson (born April 6, 1928) is an American molecular biologist who co-discovered the structure of DNA and was awarded the 1962 Nobel Prize for Physiology or Medicine with Crick and Wilkins (see footnotes 23–25) among numerous other prizes. He is the first Nobel laureate to have sold his Nobel Prize medal at auction to raise funds in 2014.

[23]Maurice Hugh Frederick Wilkins (15 December 1916—5 October 2004) was a New Zealand-born English molecular biologist, who was co-awarded the 1962 Nobel Prize for Physiology or Medicine among numerous other prizes (See footnote 25).

[24]Francis Harry Compton Crick (8 June 1916—28 July 2004) was a British molecular biologists who co-discovered the structure of DNA and was co-awarded the 1962 Nobel Prize for Physiology or Medicine (see footnote 25).

[25]The discovery of the double helix involved many more researchers who were not mentioned by the Nobel Committee. The most famous ones, *Rosalind Franklin* and *Oswald Avery*, were already deceased in 1962 and apparently, according to the Nobel Prize rules, could not be mentioned.

[26]Watson–Crick base pairing refers to the complementary C–G (Cytosine-Guanine) and A–T (Adenine–Thymine) base pairs, see also Fig. 22.1.

[27]Pronounced "five-prime to three-prime".

[28]Gel electrophoresis is a laboratory technique for visualising DNA or proteins. DNA fragments are separated according to their molecular size as they migrate through a gel.

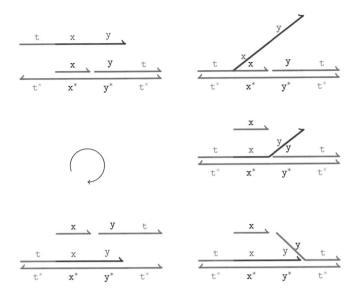

Fig. 22.2   Toehold-mediated strand displacement in 4 steps according to [18]

for toehold matchings to be reversible, toeholds must not be very long, only a few nucleotides.

Figure 22.2 (adapted from [18]) shows graphically how strand displacement works. The two-stranded gate uses domains x and y as well as toehold t (coloured red). The Watson–Crick duals of x and y are called x* and y*, respectively. The orientation of the strands is highlighted by the ⇀ arrowheads. The picture shows the various phases of strand displacement, starting from the top left corner following in clockwise direction. The extra single DNA strand that is displacing the upper strand of the given gate is coloured blue. The second step shows the branch migration. At some point strand x is replaced by the newly aligned blue version. Once disconnected, the isolated x has no toehold to align to and cannot move back into its old position. All other steps are reversible. At the end the result is the single strand produced that contains y and the toehold t.

Luca Cardelli proposed *two* domain strand displacement [6] to build transducers that can be easily extended to fork and join gates in order to implement Chemical Reaction Networks. The restricted and special form of allowed strands leads to particularly simple gates and reactions where all the gates are always double strand helixes with nothing "dangling off", i.e. protruding. In Fig. 22.2 the initial and final gate (as seen on the left hand side of the figure) are of this shape but the intermediate ones are not. A two-domain strand gate from [6], which forms half of a signal transducer, is depicted in Fig. 22.3 together with an input signal (blue) *tx*. The top strand

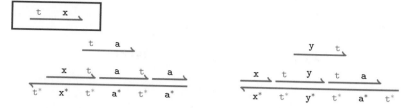

Fig. 22.3 Two domain gate according to [6]

of each two-stranded gate has two so-called "nicks" where parts can naturally break off.[29]

There are still many problems to consider when designing gates: they must not interfere with each other or the environment, and the waste material (garbage) of the reactions must not slow down reactions. In any case, whatever concrete technique and methods molecular computers will be realised with, they have much potential, in particular for the medical sciences.

> I believe things like DNA computing will eventually lead the way to a molecular revolution, which ultimately will have a very dramatic effect on the world.
>
> Leonard Adleman 1995

Of course, DNA computing cannot solve more problems than any other computing paradigm. It can always be simulated on a conventional computer. It should be emphasised once more that it is neither intended to replace silicon-based computers nor to solve **NP**-complete problems in polynomial time.

The *Economist* wrote in an article entitled *Computing with Soup* [28]:

> ... the border between computing and biology is vanishing fast, and the process of hijacking the information-processing potential of DNA to build logic circuits has only just begun.

There is of course much interest in industry. For example, *Microsoft Research Cambridge* has a *Biological Computation* group looking into molecular DNA versions of logic gates and circuits (sometimes called "liquid logic"). IBM Research has a Computational Biology Centre.

What Next?

We have seen the possibilities of molecular computing and some concrete examples in terms of CRNs. We also have seen that there is not much hope that molecular computing can help us solve **NP**-complete problems efficiently. However, there are great opportunities for molecular computing to lead to miniature computing and storage devices and to embed computing in organic tissue, thus changing the world of medicine forever. Molecules are certainly tiny, even smaller are atoms, and ever so slightly more are the subatomic particles like electrons. In the next chapter we will therefore have a look at the subatomic level: quantum computing. Does that allow us

---

[29]In two domain strand displacement the nicks are always on the top strand.

to solve **NP**-complete problems in polynomial time? What can quantum computers potentially achieve?

Exercises

1. Explain the difference between Synthetic Biology and Synthetic Chemistry.
2. Why is DNA a good storage medium? Why is it a good choice as basis for computation?
3. For the CRN in Example 22.2 that deterministically stably decides the predicate $even(x)$ write down a few configuration sequences starting with the initial configuration with input vector: $\{X : n\}$ for $n = 0, 1, 2, 3,$ and 4.
4. Define a CRN that deterministically stably decides the equality predicate $x = y$. Use four species, two representing input parameters, Y for the output "yes" (if $x = y$) and N for output "no". This requires several reaction rules.
5. Define a CRN that deterministically stably computes the function $f(x) = 2x$. Use two species, one representing the input parameter and one for the output parameter. This requires only one reaction rule.
6. Define a CRN that deterministically stably computes the function $f(x) = x \div 2$ (integer division). Use two species, one representing the input parameter and one for the output parameter. This requires only one reaction rule.
7. Define a CRN that deterministically stably computes the function $f(x, y) = min(x, y)$. Use three species, two representing input parameters and one for the output parameter. This requires only one reaction rule.
8. Define a CRN that deterministically stably computes the function $f(x, y) = x + y$. Use three species, two representing input parameters and one for the output parameter. This requires two reaction rules.
9. Define a CRN that deterministically stably computes the function $f(x, y) = max(x, y)$. Use three species, two representing input parameters, and one for the output parameter and two auxiliary species. They are used to remove superfluous molecules from the output. This requires several reaction rules.
10. Show that the function $f(x) = x \div 2$ from Exercise 6 has a graph that is semilinear.
11. Show that the function $f(x_1, x_2) = \begin{cases} x_2 & \text{if } x_1 > x_2 \\ 0 & \text{otherwise} \end{cases}$ has a graph that is semilinear.
12. Give an example of a predicate that is *not* semilinear.
13. Show (by giving a counterexample) that the composition of two deterministic CRNs does not necessarily produce a deterministic CRN. Here the species names of the two CRNs can be assumed to be disjoint but, of course, the output species of the first must be identical to the input species of the second CRN.
    Hint: Since the reaction rules of both CRNs are present to act concurrently, consider a situation where the second CRN rules use the input species before the first one has actually produced the right result. This is similar to "race conditions" in parallel computations.

14. By contrast to Exercise 13, the composition of two deterministic CRNs does yield a deterministic CRN if the first CRN *never consumes any of its output species* and thus produces its output in a monotonic way. Give an example and informally explain why the resulting CRN is deterministic.
15. Explain why it is impossible to perform Turing machine computation in a CRN with probability 1.
16. Sketch how a RAM program could be compiled into a CRN. To do that consider first how to represent registers (and their values) and instructions (and their labels) in a CRN.

References

1. Amos, M.: DNA computing. In: Encyclopædia Britannica, Inc., Available via DIALOG. http://www.britannica.com/EBchecked/topic/941575/DNA-computing (2015). Accessed 22 June 2015
2. Angluin, D., Aspnes, J., Eisenstat, D.: Stably computable predicates are semilinear. In: Proceedings of the Twenty-Fifth Annual ACM Symposium on Principles of Distributed Computing, pp. 292–299. ACM, New York (2006)
3. Bennett, C.H.: On constructing a molecular computer. IBM J. Res. Dev. **17**, 525–532 (1973)
4. Bennett, C.H., Landauer, R.: The fundamental physical limits of computation. Sci. Am. **253**(1), 48–56 (1985)
5. Bonnet, J., Yin, P., Ortiz, M.E., Subsoontorn, P., Endy, D.: Amplifying genetic logic gates. Science **340**(6132), 599–603 (2013)
6. Cardelli, L.: Two-domain DNA strand displacement. In: Cooper, S.B., Kashefi, E., Panangaden, P. (eds.) Developments in Computational Models (DCM 2010). EPTCS 26, pp. 47–61 (2010)
7. Cardelli, L.: Molecular programming. Invited talk at ECOOP 2014. Available via DIALOG. http://lucacardelli.name/Talks/2014-07-30%20Molecular%20Programming%20(ECOOP).pdf (2015). Accessed 22 June 2015
8. Chen, H.-L., Doty, D., Soloveichik, D.: Deterministic function computation with chemical reaction networks. Nat. Comput. **13**(4), 517–534 (2013)
9. Chen, Y.-J., Dalchau, N., Srinivas, N., Phillips, A., Cardelli, L., Soloveichik, D., Seelig, G.: Programmable chemical controllers made from DNA. Nat. Nanotechnol. **8**(10), 755762 (2013)
10. Church, G.M., Gao, Y., Kosuri, S.: Next-generation digital information storage in DNA. Science **337**(6102), 1628 (2012)
11. Cook, M., Soloveichik, D., Winfree, E., Bruck, S.: Programmability of chemical reaction networks. In: Condon, A., Harel, D., Kok, J.N., Salomaa, A., Winfree, E. (eds.) Algorithmic Bioprocesses, pp. 543–584. Springer, Heidelberg (2009)
12. Cox, J.C., Cohen, D.S., Ellington, A.D.: The complexities of DNA computation. Trends Biotechnol. **171**, 151–154 (1999)
13. Diameter of erythrocytes: B10Numbers, the database of useful biological numbers. Available via DIALOG. http://bionumbers.hms.harvard.edu/bionumber.aspx?&id=100509 (2015). Accessed 23 June 2015
14. Doty, D.: Timing in chemical reaction networks. In: Proceedings of the Twenty-Fifth Annual ACM-SIAM Symposium on Discrete Algorithms. SIAM, pp. 772–784 (2014)
15. Goldman, N., Bertone, P., Chen, S., Dessimoz, C., LeProust, E.M., Sipos, B., Birney, E.: Towards practical, high-capacity, low-maintenance information storage in synthesized DNA. Nature **494**, 77–80 (2013)
16. Hartmann, L., Jones, N.D., Simonsen, J.G., Vrist, S.B.: Programming in biomolecular computation: programs, self-interpretation and visualisation. Sci. Ann. Comput. Sci. **21**(1), 73–106 (2011)

17. Klinge, T.H., Lathrop, J.I., Lutz J.H.: Robust biomolecular finite automata. Preprint. Available via DIALOG. http://arxiv.org/abs/1505.03931 (2015). Accessed 6 Aug 2015
18. Lakin, M.R., Parker, D., Cardelli, L., Kwiatkowska, M., Phillips, A.: Design and analysis of DNA strand displacement devices using probabilistic model checking. J. R. Soc. Interface **9**(72), 1470–1485 (2012)
19. Length of DNA in nucleus: B10Numbers, the database of useful biological numbers. Available via DIALOG. http://bionumbers.hms.harvard.edu/bionumber.aspx?&id=104208&ver= 9&trm=length%20DNA (2015). Accessed 23 June 2015
20. Linial, M., Linial, N.: On the potential of molecular computing. Science-AAAS-Wkly. Pap. Ed. **268**(5210), 481 (1995)
21. Lupski, J.R.: Genome mosaicism—one human, multiple genomes. Science **341**(6144), 358–359 (2013)
22. Overall genome size: B10Numbers, the database of useful biological numbers. Available via DIALOG. http://bionumbers.hms.harvard.edu/bionumber.aspx?&id=101484&ver= 19&trm=Human%20Homo%20sapiens%20dna (2015). Accessed 23 June 2015
23. Pray, L.A.: DNA replication and causes of mutation. Nat. Educ. **1**(1), 214 (2008)
24. Peterson, J.L.: Petri Net Theory and the Modeling of Systems. Prentice-Hall Inc., Englewood Cliffs (1981)
25. Seelig, G., Soloveichik, D., Zhang, D.Y., Winfree, E.: Enzyme-free nucleic acid logic circuits. Science **314**(5805), 1585–1588 (2006)
26. Soloveichik, D.: The programming language of chemical kinetics, and how to discipline your DNA molecules with strand displacement cascades. Talk at DNA 17. Available via DIALOG. http://dna17.caltech.edu/Soloveichik_DNA_17_Tutorial_(2011-09-20).pdf (2011). Accessed 20 Jan 2016
27. Soloveichik, D., Cook, M., Winfree, E., Bruck, S.: Computation with finite stochastic chemical reaction networks. Nat. Comput. **7**(4), 615–633 (2008)
28. The Economist: Computing with soup. Technology Quarterly: Q1 2012, 03/03/2012. Available via DIALOG. http://www.economist.com/node/21548488 (2014). Accessed 22 June 2015

Chapter 23
Quantum Computing

*What is Quantum Computing and can it help solving
NP-complete problems? What advantages does it have?
What limitations does it have?*

Can one harness the powers of particle physics and use them for computation? Particle physics obviously happens in the very small just like molecular computing discussed in the previous chapter. In this chapter, we go further, to the level of subatomic particles.

It seems that the idea of a *quantum computer* was first suggested by Richard Feynman[1] [8, 9] at the beginning of the 1980s. Before that, quantum effects and quantum information theory were already established. The idea of a quantum computer is to use quantum effects of superposition to execute massively parallel computation. Feynman wanted to simulate quantum physics and quantum chemistry.

In conventional computing, a bit can only be in one state at a time, either 0 or 1 and computations on bits occur sequentially. In quantum computing, thanks to quantum effects, a so-called *qubit* (quantum bit) can be in (a mixture of) states 0 *and* 1 *at the same time. Schrödinger*[2] devised a famous thought experiment, called *Schrödinger's cat*,[3] to highlight what he considered the absurdity of superposition. A cat is placed in a box with a flask of poison and a radioactive source the decay of which controls

---

[1]Richard Phillips Feynman (May 11, 1918–February 15, 1988) was an American theoretical physicist known for his work in quantum mechanics, quantum electrodynamics, particle physics and other areas. For his contributions to the development of quantum electrodynamics, Feynman, jointly with Julian Schwinger and Sin-Itiro Tomonaga, received the Nobel Prize in Physics in 1965. During his lifetime, Feynman became one of the best-known scientists in the world, helped by the fact that he was a supporting member of the Manhattan project responsible for the development of the atomic bomb, and a member of the Rogers Commission that investigated the 1986 Space Shuttle Challenger disaster. His fame also relies on the significant number of (popular) physics books he published. He is considered one of the greatest physicists of all time.

[2]Erwin Rudolf Josef Alexander Schrödinger (12 August 1887–4 January 1961) was a famous Austrian physicist who developed a number of fundamental results in the field of quantum theory. He received the Nobel Prize for Physics in 1933.

[3]Mentioned also in the episode "The Tangerine Factor" of the TV sitcom "The Big Bang Theory".

© Springer International Publishing Switzerland 2016
B. Reus, *Limits of Computation*, Undergraduate Topics in Computer Science,
DOI 10.1007/978-3-319-27889-6_23

the release of the poison. The cat can be considered dead *and* alive at the same time until one opens the box to see whether the radioactive material has decayed, released the poison and killed the cat.

A more constructive way of looking at superposition is Young's well-known "double-slit experiment", where light (photons) is shone through two parallel slits onto a screen. The interference pattern that can be observed on the screen is rather surprising: there are not just two stripes visible, but many stripes in a wave-like pattern. This is true even if one shoots photons at the screen one at a time, proving that the particles must be passing through *both* slits such that a photon can interfere with itself. Harnessing such interference of qubits and the underlying parallelism in a controlled manner is the idea of quantum computing. There are a number of problems though: as soon as one measures (observes) a result, the state space collapses into one. Even worse, any interaction of the qubits with the environment can cause this collapse. This is called *decoherence* and is one of the several reasons why it is so difficult to build quantum computers (and to predict what is happening at quantum scale).

Before we discuss the underlying mathematics and the role of probabilities, (Sect. 23.2), quantum computability and complexity (Sect. 23.3), quantum algorithms (Sect. 23.4), address the difficulties of building quantum computers (Sect. 23.5), and look into fascinating philosophical and physical discussions that arise from quantum computation (Sect. 23.7), we begin with the question how *going small* affects the traditional way of silicon-based RAM-style computing (Sect. 23.1).

23.1 Molecular Electronics

Molecular electronics is a branch of nanotechnology that aims at constructing electronic components with the help of single molecules (or even atoms). We know that current semiconductor technology cannot shrink much further. Latest microchips shipped from 2015 use 14 nm technology [19], meaning that transistors are 14 nm wide.[4] How big is 14 nm? It is the width of a human hair—divided 5,000 times. In 1971, the Intel® 4004 held 2,300 transistors each of a size of ca. 10,000 nm. The latest 14 nm-technology chips contain several billion transistors.[5] These chips "run 4,000 times faster and use 5,000 times less energy than the 4004 microchip" [25].

Moore's law is the observation that the number of transistors in an integrated circuit doubles approximately every two years. This observation is named after

---

[4]This already uses *multiple gate transistors* that use a three-dimensional design to save space. Planar designs tend to leak current if they are too small. To reduce leaking the so-called *FinFET* architecture is used where the conducting channel is wrapped in a silicon-fin that gives the three-dimensional shape and the name. FET stands for field-effect transistor.

[5]At the time of writing the Intel® 18-core Xeon® E5-2699 v3 ("Haswell-EP") CPU seems to be among the top CPUs regarding transistor count with 5.56 billion (clock speed of 2.3 GHz) [22]. This is still 22 nm technology though, the latest 2015 releases use 14 nm technology but appear to have fewer transistors due to fewer cores.

Gordon E. Moore, co-founder of the Intel Corporation and Fairchild Semiconductor.[6] Interestingly, this trend has continued over the last 50 years. A consequence of this rapid progress is, however, that transistors will reach atomic size (about 1.5 nm) soon, when quantum effects will kick in and require new designs or even technologies if Moore's Law is to be upheld in the medium term. Therefore, there is a huge amount of research in the area of molecular electronics. Although this technology is used to build (complex versions of) our well known register machines (RAM), thus providing no new notion of computation, there have been a number of success stories that are worth pointing out:

- In 2009, Mark Reed and his team at Yale University succeeded in constructing a transistor made from a single benzene molecule. It is smaller than 1 nm [24]. However, the technology to connect several those transistors in an integrated circuit does not yet exist.
- The disadvantages of very small silicon transistors (leakage of current) may be overcome by carbon-based chips. In 2013, researchers at Stanford University [23] managed to produce a chip with carbon nano-tubes.[7]
- Also in 2013, a new chip design using layers of gold, molecular components and graphene has been built by a Danish-Chinese team [15] . The transistors are triggered by light (not electricity) using dihydroazulene/vinylheptafulvene molecules.

23.2 The Mathematics of Quantum Mechanics

Quantum computing can be elegantly modelled mathematically. This book is for computer scientists with limited mathematical background, so we will not go into too much detail, but give just a short overview.[8] We follow the presentation in [1, Chap. 10] where the question "What is quantum mechanics about?" is answered as follows: "...*it's about information and probabilities and observables, and how they relate to each other*".

A classical bit can be in two states 0 and 1. In quantum mechanics we add *probabilities*. Therefore, a bit is in state 0 with probability α and in state 1 with probability β. Classically, α and β would be real numbers, but in any case they have to add up to 1. Viewing (α, β) as a vector in $\mathbb{R}^2$ this means its 1-norm (the sum of absolute values of the vector) must be 1. In the quantum world, one uses, however, the 2-norm,[9]

---

[6]In 1965 Moore predicted the number of transistors on a chip to double every year, which was then actually corrected to 'every two years' in 1975. Apparently, David House, an Intel executive at the time, then concluded "that the changes would cause computer performance to double every 18 months" [20]. This is the reason why sometimes 18 months is attributed wrongly to Moore's statement.

[7]Called "Cedric". Note however that currently the transistors are still of "monster size": 8000 nm.

[8]This section can be skipped by readers not interested but is intended to motivate the study of linear algebra and complex numbers.

[9]Sometimes also referred to as *Euclidean norm*.

which means now that $\alpha^2 + \beta^2 = 1$ and thus the probability of observing 0 is α^2 and that of observing 1 is β^2. The numbers α and β are called probability *amplitudes*. A *qubit* is just such a probability vector with 2-norm. Operations on qubits must therefore preserve the property of being a unit vector in the 2-norm. If we consider only *linear* operations, then those are the ones whose inverse is the transpose. For amplitudes α and β one actually does not use real numbers but complex numbers (so the probabilities of the *observations* 0 and 1 are actually $|\alpha|^2$ and $|\beta|^2$, respectively). One reason appears to be continuity of the linear operations which requires another dimension[10] or, alternatively, the use of complex numbers. Another reason is that the complex numbers can describe wave functions[11] which is essential for modelling quantum mechanics as small particles express wave-like behaviour.

Accordingly, the states of a qubit are unit vectors in the two-dimensional space of complex numbers. These states can be represented geometrically as the surface of a ball with radius 1 in the three-dimensional space of real numbers, the so-called *Bloch-spere*[12] depicted in Fig. 23.1 which shows a red unit vector whose position is determined by the two angles ϕ and θ. The base vectors representing 0 and 1, respectively, are the north and south pole of this sphere (highlighted green in Fig. 23.1). The image should convey the fact that a qubit represents more information than an ordinary bit, which has only two possible states. Although, when we measure the qubit, the result is again either 0 or 1.

Because one qubit can have two states simultaneously, n qubits can be in 2^n states simultaneously, highlighting the potential in parallelisation. Associating quantum wave functions with quantum entities allows one to describe the probability of their position, and the interaction of such waves allows one to express their superposition.

One might wonder why quantum mechanics requires the 2-norm for vectors, i.e. $|\alpha|^2 + |\beta|^p = 1$ and not e.g. $|\alpha|^p + |\beta|^p = 1$ for any $p > 2$, i.e. the p-norm for a p larger than 2. Only for $p = 1$ or $p = 2$ one does get non-trivial linear transformations that preserve the p-norm [1]. So it seems that Nature prefers[13,14,15] case $p = 1$ and $p = 2$.

---

[10] Aaronson [1] uses the example of mirror flipping a two-dimensional shape. The movement must pass through the third dimension if done continuously.

[11] A wave function is a periodic function in time and space.

[12] Felix Bloch (23 October 1905–10 September 1983) was a Swiss born American physicist who won the Nobel Prize with Edward Mills Purcell in 1952. From 1954 to 1955 Bloch was the first Director-General of CERN (European Organization for Nuclear Research).

[13] In [1], this is compared with the equation $x^p + y^p = z^p$, which has positive integer solutions for x, y, z only if $p = 1$ or $p = 2$. The latter is known as "Fermat's Last Theorem" (see footnote 14).

[14] Pierre de Fermat (17 August 1601–12 January 1665) was a French lawyer and a mathematician who contributed to the development of infinitesimal calculus. He is most famous for his conjecture known as "Fermat's Last Theorem", a long standing open problem until *Andrew Wiles* (see footnote 15) eventually proved it in 1994 after years of hard work and numerous failures.

[15] Sir Andrew John Wiles (born 11 April 1953) is a British mathematician at the University of Oxford. He won numerous awards for proving Fermat's Last Theorem.

Fig. 23.1 Bloch sphere

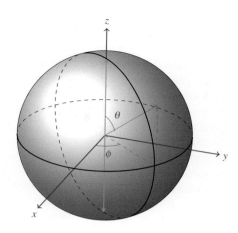

More details, and a discussion of actual linear transformations and how they can represent quantum gates can be found in a standard textbook on this topic like [17]. Other very accessible introductions (including the mathematics) are [1, 7].

23.3 Quantum Computability and Complexity

A few obvious questions arise in the context of this book:

1. Is quantum computing equivalent to all the other notions of computations (once it has been precisely defined[16])? In other words, does the Church-Turing thesis still hold once we have quantum computers?
2. How do we measure time usage for quantum computing? What are quantum complexity classes?
3. What is the relation between quantum complexity classes and the ones we already know (like **P**, **NP**, **BPP**)?
4. In particular, does quantum computing squash the Cook-Karp thesis? In other words, with a quantum computer can we effectively compute problems that are not in **P**?

First of all, a quantum computer can still be simulated on a conventional computer, even if quite some mathematics is needed to describe the superposition. So a quantum computer would not violate the Church-Turing thesis. However, such a simulation can be slow, very slow, exponentially slow. Yet, quantum computer simulators have been built using some clever optimisations reducing the state space (keeping track of entanglements for instance). Most prominently, LIQUi|> (pronounced 'liquid'), short

---

[16]Which we are not able to do here.

for "Language-Integrated Quantum Operations",[17] that *includes state-of-the-art circuit simulation of up to 30 qubits on a single machine with 32 GB RAM, limited only by memory and computing threads. The largest number factored to date on the simulator is a 13-bit number, which required 27 qubits, half a million gates, and 5 days runtime. The circuit was based on Beauregards circuit for Shors algorithm"* [16]. For larger simulations it is necessary to use clusters of machines that run concurrently. Moreover, *certain* quantum computations apparently can be simulated classically in polynomial time.

On the other hand, in general (besides examples like Factorisation) we do not know to what degree quantum computers can speed-up tasks. As for DNA computing, many daily operations would be probably slower on a quantum computer than on a conventional one. Browsing the web is not very likely going to be faster on a quantum computer. But search and optimisation problems might benefit from the (right kind of) superposition. Quantum computers may therefore squash the Cook-Karp thesis for good, but we don't know this yet. Cook's thesis as we phrased it (in Sect. 14.2) is (luckily) not affected by this as its wording cleverly uses the term "sequential" before "notion of computation", and quantum computing is not a sequential computation model.

According to the above, it is actually possible to define a corresponding complexity class and formulate some theorems and prove them. Quantum complexity has first been studied in [3].

Definition 23.1 ([3]) The complexity class **BQP** (bounded error quantum polynomial time) is the class of decision problems solvable by a quantum computer—e.g. a *Quantum Turing machine*—in polynomial time, with an error probability of at most $\frac{1}{3}$ for all instances.[18] It is the quantum analogue of the (classical) probabilistic class **BPP**.

The obvious question is: *what is a Quantum Turing Machine?* It's a Turing machine using quantum states.[19] Any such state will be a superposition of possible configurations which makes the Turing machine "quantum". The (mathematical) details can be found e.g. in [4]. One can alternatively use *quantum circuits* to define quantum computing.

Proposition 23.1 *It holds that* $P \subseteq BPP \subseteq BQP$.

Proof This is because quantum gates can simulate classic gates and quantum computation is probabilistic, so a quantum computer can solve problems that can be solved deterministically or probabilistically on classical hardware.

One also can see that **BQP** $\subseteq$ **EXP** because, as already pointed out, quantum computations can be simulated on a classic computer with exponential slow down. One can

---

[17] Developed by the *Quantum Architectures and Computation Group* (QuArC) at Microsoft Research, Redmond USA.

[18] The concrete value of the probability $\frac{1}{3}$ is actually irrelevant and can be changed.

[19] Which are unit vectors in the Hilbert space generated by the configuration space.

actually do better and show that $\mathbf{BQP} \subseteq \mathbf{PP}$, where $\mathbf{PP}$ (for **p**robabilistic **p**olynomial time) is the following class:

Definition 23.2 ([11]) Let $\mathbf{PP}$ be the class of decision problems accepted by a *probabilistic Turing machine M* (see Definition 21.7) in polynomial time, such that for all inputs x we have

$$\Pr[M(x) = (x \in A)] > \frac{1}{2}$$

Note that $\mathbf{NP} \subseteq \mathbf{PP}$ (Exercise 3), so unlike $\mathbf{BPP}$, $\mathbf{PP}$ *does* contain problems that are not known to be feasible. The difference in definition between $\mathbf{BPP}$ (see Definition 21.9) and $\mathbf{PP}$ is subtle, but crucial, and shall be discussed in Exercise 5.

"*Quantum theory is, in some respects, both superior to and more natural than classical probability theory … since it can describe evolution for finite systems in a continuous way. Since nature is quantum, not classical, it is to be expected that quantum theory is ultimately the more reasonable theory*" [13].

As usual, there is some bad news. It is unknown whether $\mathbf{NP} \subseteq \mathbf{BQP}$, although it is widely believed that this is not true. Neither do we know whether $\mathbf{P} \neq \mathbf{BQP}$.

23.4 Quantum Algorithms

In order to use the power of superposition that quantum computing provides, one must find ways to harness it for one's own purposes. In other words, one must find algorithms that can use superposition to do something useful. It turns out that this is not so easy (and needs proper mathematics). We briefly discuss the most famous quantum algorithms in this section.

23.4.1 Shor's Algorithm

David Deutsch[20] was the first to design a quantum algorithm that is more efficient than a classic counterpart [4]. Then *Peter Shor*[21] published an algorithm in 1994 (see [20, 21]) that uses quantum superposition to factorize primes (and thus solve the Factorisation problem of Definition 20.15) in polynomial time on a (theoretical) quantum computer.[22] This shook the computing world to its core, because it means

---

[20]David Elieser Deutsch, (born 18 May 1953) is a British physicist at the University of Oxford.

[21]Peter Williston Shor (born August 14, 1959) is an American professor of applied mathematics at MIT. For the discovery of the quantum factorisation algorithm, he was awarded the Gödel Prize in 1999.

[22]Shor's algorithm even appeared in the TV sitcom "Big Bang Theory", episode "The Bat Jar Conjecture". It was the answer to a question about factorisation by quantum computers in a "Physics Bowl" quiz.

that if one had a quantum computer, one could solve the Factorisation problem in polynomial time. Recall that the intractability of the Factorisation problem is the basis of public key cryptography. So Shor's result must have also shaken up the security agencies of all major superpowers. However, polynomial time does not necessarily mean fast (as we have discussed earlier), and so we don't really know whether cryptography can actually be broken by a quantum computer unless we have one. Moreover, we do not know whether the Factorisation problem is **NP**-complete. So Shor's result does *not* prove that quantum computers can solve a **NP**-complete problem in polynomial time.

By the way, Shor's algorithm is very clever and uses some non-trivial mathematics.[23]

23.4.2 Grover's Algorithm

There is one other famous quantum algorithm, *Grover's algorithm*, which performs search. Ordinary computers can do search in a (unordered) list of length n no better (worst case) than in $\mathcal{O}(n)$. A quantum computer can perform the same search in $\mathcal{O}(\sqrt{n})$. Even better, Bennett et al. [2] have shown in 1997 that it cannot be done more quickly. So Grover's algorithm is already optimal for a quantum computer. This means that "brute-force search" on a quantum computer won't solve **NP**-complete problems in polynomial time. At best it will provide a quadratic speed-up compared to ordinary computers.

Scott Aaronson formulated the "No Super-Search Postulate"[24] which says that *"there is no physical means to solve NP-complete problems in polynomial time."* First of all, this implies that $\mathbf{P} \neq \mathbf{NP}$, but is *stronger*. Secondly, it connects the $\mathbf{P} \neq \mathbf{NP}$ question to physics. According to Aaronson, $\mathbf{P} \neq \mathbf{NP}$ would "explain" certain physical phenomena, for instance that the Schrödinger equation is linear.

It is worth remembering that:

A quantum computer is NOT like a massively-parallel classical computer.
Scott Aaronson

In particular, we don't know yet whether we can can solve any **NP**-complete problem in polynomial time with a quantum computer. And there is not much evidence right now to believe it.

---

[23] The quantum Fourier transform and algebra [20].

[24] Yes, this is another thesis.

23.5 Building Quantum Computers

In order to build a quantum computer, qubits must be implemented on a very small scale, so that the quantum effects take over. There are many ways to build a qubit, but the main idea is to trap an atom in a magnetic field, either in vacuum or in solids you can send light through. Only very few qubits are actually needed to build powerful quantum computers.

Quantum computers are not expected to be built within the next 10–20 years—if at all—but Canadian firm *D-Wave* claims to have already built a 512 qubit quantum processor (based on "quantum annealing" and not quantum gates). This is, however, highly controversial. It is safe to say though that it costs a large amount of money, allegedly £9 million [18]. The renowned professional organisation IEEE[25] still disputes the fact that this is a quantum computer, but some researchers are convinced it does some sort of quantum computing. A study published in the equally renowned journal *Science* [19] in 2014 could not find evidence of quantum speedup but did not rule out the possibility either. The discussion between various experts will keep going for some time to come. One has to wait and see. Fact is that *D-Wave* has huge financial backing from venture capital firms.

While quantum computers are not available for general households, quantum cryptography is already in use and common. Banks use quantum entanglement to safely communicate keys. That way it can be observed whether anybody has eavesdropped.

23.6 Quantum Computing Challenges

Regarding quantum computing the main challenges appear to be the following:

- in general: building quantum gates, quantum computer hardware;
- in particular: dealing with *decoherence*, the interference, entanglement, of the quantum states with the environment (how to shut out the environment so to speak);
- develop quantum algorithmics, recipes for how to use superposition cleverly to compute whatever one wishes to compute in a "massively parallel way";
- quantum debugging; how do you debug if measurement makes the state space collapse and variable values may be distributed in many entangled qubits?
- design quantum programming languages that then support the elegant coding of the above quantum algorithms. For a survey on the progress up to 2006 see [10]. Latest suggestions include [12, 25].

---

[25]Institute of Electrical and Electronics Engineers.

23.7 To Boldly Go …

Here's another thought: Deutsch uses Shor's algorithm to question where the computation actually took place:

> When Shor's algorithm has factorized a number, using 10^{500} or so times the computational resources that can be seen to be present, where was the number factorized? There are only about 10^{80} atoms in the visible universe, an utterly minuscule number compared with 10^{500}. So if the visible universe were the extent of physical reality, physical reality would not even remotely contain the resources required to factorize such a large number. Who did factorize it, then? How, and where, was the computation performed?
>
> David Deutsch [6, p. 217]

He uses quantum factorisation as an argument for the existence of *parallel universes*. Quantum computation and mechanics, however, can be explained also without the "many world interpretation" (MWI), therefore Shor's algorithm does not appear to be a proper proof of the existence of many universes. If our (single) universe was a *quantum universe* there would not be any need for the extra universes.

To quote Aaronson once more: "*a quantum computer is not a device that could 'try every possible solution in parallel' and then instantly pick the correct one. If we insist on seeing things in terms of parallel universes, then those universes all have to "collaborate"—more than that, have to meld into each other—to create an interference pattern that will lead to the correct answer being observed with high probability*" [1].

Deutsch also argued that the paradox of backwards time travel is only a paradox in *classical physics* and can be avoided in quantum physics such that, at least in theory, time travel could be used to solve hard problems quickly. Of course, we don't know whether time travel is possible but such theoretical discussions can shed light on the *limitations* of computability and feasibility as well as time travel itself. More details can be found in [1, 5]. The time travelling paradoxes and their connection to complexity theory, or the question whether **P** equals **NP**, will be discussed in the exercises.

We have arrived at some mind-boggling physical and meta-physical questions. And fittingly we are approaching the end of this book. Maybe in a *Computability and Complexity* module of the future, the quantum computer will be "the" computer and the computers we use today will be to the students what Turing machines are to us. But this is unlikely, neither molecular nor quantum computers are expected to replace current computing devices any time soon, even if such computers could be built for the mass market. We have observed that they have both very specific application areas. Like with any technology, however, it is impossible to make predictions. As Lord Kelvin, world famous British physicist and engineer[26] and former president of

---

[26]William Thomson, 1st Baron Kelvin (26 June 1824–17 December 1907) is known for his formulation of the laws of thermodynamics and the temperature scale with 'absolute zero' that bears his name.

the Royal Society, "predicted" in 1902: "*Neither the balloon, nor the aeroplane, nor the gliding machine will be a practical success*" [14].

So who knows what the future of computing will bring us.

Exercises

1. Do we know whether one can solve **NP**-complete problems with quantum computers? Discuss.
2. In which complexity classes is the Factorisation problem *known* to be in?
3. Show that **NP** ⊆ **PP**.
4. Show that **BPP** ⊆ **PP**. Note that we do not know whether this inclusion is proper.
5. Explain the difference between **BPP** (Definition 21.9) and **PP** (Definition 23.2). The difference between the classes is subtle, but obviously crucial as **PP** is known to contain infeasible problems of **NP**, and **BPP** is considered "probabilistic **P**" although there is no proof of that.
6. Consider the following polynomial-time "algorithm" to solve **NP**-complete (or even harder) problems: select a brute-force type algorithm and start it on your computer. Then board a space-ship, accelerate to as near to light speed as you can, thus travelling forward in time (w.r.t Earth and your computer) right to the point (or beyond) when your algorithm has produced a solution. Assuming we had the technology to build space-ships that can reach this kind of speed, discuss other issues why this approach probably will not work.
7. Discuss the distinction between the following two well-known paradoxes[27] known to appear in classical physics:

   a. *The Grandfather paradox*: Travel back in time to prevent your grandfather from meeting your grandmother and thus preventing your own birth. Alternatively, if you do not want to interfere with the love life of your ancestors, hop on a time machine and travel back in time to the point before you departed and prevent yourself from departing.

   b. *The Knowledge Generation paradox* (also called *Predestination paradox*): invent something, then send it backwards through time to yourself (or your friends) to use it.

   Hint: Consider logical consistency and computational complexity.

8. Discuss how the *Grandfather paradox* in Exercise 7a involves the concept of "free will".
9. Explain a computational version of the *Grandfather paradox* in Exercise 7a that refers to computations that involve time travel, i.e. computations that can be fed results sent from the future. *Hint: Establish a problematic feedback loop that has no fixpoint.*

---

[27]Both types occur in science fiction movies about time travel, e.g. "Back to the Future", which is, expectably, full of such paradoxes.

10. Discuss how the *Knowledge Generation paradox* from Exercise 7b can be used to solve an **NP**-complete (or even harder) problem in "no time" (assuming time travel is possible and that the *computational Grandfather paradox* is avoidable).

11. The *Grandfather paradox* (see Exercise 7b and 7a) is theoretically avoidable in quantum physics as proved by [5]. We have not introduced enough mathematics (and quantum computation) to even attempt to explain this but according to [1] this can be shown already classically when working probabilisitcally. This means one works with probability distributions of input values as input and produces probability distributions of output values as output. So a Boolean gate with one input and one output (like negation) would be modelled by a function that maps a vector $(p, 1 - p)$, where p is the probability that the input is 0, to another vector $(q, 1 - q)$ where q is the probability that the output is 0.

   a. Whereas classically negation does not have a fixpoint, using probability distributions it does have one. What is this fixpoint?

   b. Show how this helps to avoid the *Grandfather paradox* by interpreting 1 as "prevent own birth" and 0 as "don't prevent own birth". Recall that, classically, preventing your own birth means that you are not born and so you can't prevent your own birth. If you don't prevent it, then you go back in time and prevent it, creating a logical paradox.

References

1. Aaronson, S.: Quantum Computing Since Democritus. Cambridge University Press, New York (2013)
2. Bennett, C.H., Bernstein, E., Brassard, G., Vazirani, U.: Strengths and weaknesses of quantum computation. SIAM J. Comput. **26**, 1510–1523 (1997)
3. Bernstein, E., Vazirani, U.: Quantum complexity theory. SIAM J. Comput. **26**(5), 1411–1473 (1997)
4. Deutsch, D.: Quantum theory, the Church-Turing principle and the universal quantum computer. Proc. R. Soc. A **400**, 97–117 (1985)
5. Deutsch, D.: Quantum mechanics near closed timelike lines. Phys. Rev. D **44**(10), 3197–3217 (1991)
6. Deutsch, D.: The Fabric of Reality. Viking Adult, New York (1997)
7. Edalat, A.: Quantum Computing. CreateSpace Independent Publishing Platform (2015)
8. Feynman, R.P.: Simulating physics with computers. Int. J. Theo. Phys. **21**(6), 467–488 (1982)
9. Feynman, R.P.: QED: The Strange Theory of Light and Matter. Princeton Science Library, Princeton (1988)
10. Gay, S.J.: Quantum programming languages: survey and bibliography. Math. Struct. Comput. Sci. **16**(4), 581–600 (2006)
11. Gill, J.: Computational complexity of probabilistic turing machines. SIAM J. Comput. **6**(4), 675–695 (1977)
12. Green, A., Lumsdaine, P. L., Ross, N., Selinger, P., Valiron, B.: Quipper: A scalable quantum programming language. In Proceedings of PLDI 13, pp. 333–342, ACM (2013)
13. Hardy, L.: Quantum Theory From Five Reasonable Axioms. arXiv:quant-ph/0101012v4 (2001)
14. Kelvin on Science (Interview with Lord Kelvin). Reprinted in The Newark Advocate, April 26, 1902, p. 4 (1902)

15. Li, T., Jevric, M., Hauptmann, J.R., Hviid, R., Wei, Z., Wang, R., Reeler, N.E.A., Thyrhaug, E., Petersen, S., Meyer, J.A.S., Bovet, N., Vosch, T., Nygård, J., Qiu, X., Hu, W., Liu, Y., Solomon, G.C., Kjaergaard, H.G., Bjørnholm, T., Nielsen, M.B., Laursen, B.W., Nørgaard, K.: Ultrathin reduced graphene oxide films as transparent top-contacts for light switchable solid-state molecular junctions. Adv. Mater. **25**(30), 4064–4070 (2013)

16. Microsoft Research: LIQUi|> Project Page. Available via DIALOG. http://research. microsoft.com/en-us/projects/liquid/. Accessed 23 June 2015

17. Nielsen, M.A., Chuang, I.L.: Quantum Computation and Quantum Information. Cambridge University Press, New York (2000)

18. Rincon, P.: D-Wave: Is $15m machine a glimpse of future computing? BBC News Science & Environment Website. 20 May 2014. Available via DIALOG. http://www.bbc.co.uk/news/ science-environment-27264552 Accessed 22 June 2015

19. Rønnow, T.F., Wang, Z., Job, J., Boixo, S., Isakov, S.V., Wecker, D., Martinis, J.M., Lidar, D.A., Troyer, M.: Defining and detecting quantum speedup. Science **345**(6195), 420–424 (2014)

20. Shor, P.W.: Algorithms for Quantum Computation: Discrete Logarithms and Factoring. In Goldwasser, S. (Ed.) Proceedings of the 35th IEEE Symposium on Foundations of Computer Science, pp. 124–134 (1994)

21. Shor, P.W.: Polynomial-time algorithms for prime factorisation and discrete logarithms on a quantum computer. SIAM J. Comput. **26**(5), 1484–1509 (1997)

22. Shrout, R.: Intel Xeon E5–2600 v3 processor overview: Haswell-EP Up to 18 Cores. PC Perspective, 08/09/2014. Available via DIALOG. http://www.pcper.com/reviews/Processors/ Intel-Xeon-E5-2600-v3-Processor-Overview-Haswell-EP-18-Cores (2015). Accessed 23 June 2015

23. Shulaker, M.M., Hills, G., Patil, N., Wei, H., Chen, H.-Y., Philip Wong, H.-S., Mitra, S.: Carbon nanotube computer. Nature **501**, 526–530 (2013)

24. Song, H., Kim, Y., Jang, J.H., Jeong, H., Reed, M.A., Lee, T.: Observation of molecular orbital gating. Nature **462**, 1039–1043 (2009)

25. Wecker, D., Svore, K.M.: LIQUi|>: A Software Design Architecture and Domain-Specific Language for Quantum Computing. Available via DIALOG: arxiv.org/abs/1402.4467. Accessed 23 June 2015 (2014)

Further Reading—Computability and Complexity Textbooks

What books about Computability and Complexity can I read to learn more?

The following is a list of textbooks in Computability and Complexity theory or any combination thereof. The list is ordered alphabetically by author.

According to usage in lectures (where information available online) particularly popular books appear to be [1, 9, 12, 19, 25, 28, 35].

References

1. Arora, S., Barak, B.: Computational Complexity: A Modern Approach. Cambridge University Press, Cambridge (2009)
2. Bovet, D., Crescenzi, P.: Theory of Computational Complexity. Prentice Hall, New York (1993)
3. Brookshear, J.G.: Theory of Computation: Languages, Automata and Complexity, Pearson (1989)
4. Cockshott, P., Mackenzie, L.M., Michaelson, G.: Computation and its Limits. Oxford University Press, Oxford (2012)
5. Cooper, B.: Computability Theory. Chapman Hall/CRC, Boca Raton (2003)
6. Cutland, N.J.: Computability: An Introduction to Recursive Function Theory. Cambridge University Press, Cambridge (1980)
7. Davis, M.: Computability and Unsolvability. McGraw Hill, New York (1958)
8. Davis, M.: The Undecidable. Raven Press, New York (1960)
9. Davis, M.D., Sigal, R., Weyuker, E.J.: Computability, Complexity and Languages. Academic Press (1994)
10. Du, D.-Z., Ko, K.-I.: Theory of Computational Complexity. Wiley, New York (2000)
11. Fernández, M.: Models of Computation—An Introduction to Computability Theory. Springer (2009)

© Springer International Publishing Switzerland 2016
B. Reus, *Limits of Computation*, Undergraduate Topics in Computer Science,
DOI 10.1007/978-3-319-27889-6

12. Garey, M.R., Johnson, D.S.: Computers and Intractability, A Guide to the Theory of NP-Completeness. Freeman (1986)
13. Goldreich, O.: P, NP, and NP-Completeness: The Basics of Computational Complexity. Cambridge University Press, Cambridge (2010)
14. Greenlaw, R., Hoover, H.J.: Fundamentals of the Theory of Computation— Principles and Practice. Morgan Kaufmann, San Francisco (1998)
15. Greenlaw, R., Hoover, H.J., Ruzzo, W.L.: Limits to Parallel Computation: P-Completeness Theory. Oxford University Press, Oxford (1995)
16. Hedman, S.: A First Course in Logic: An Introduction to Model Theory, Proof Theory, Computability, and Complexity. Oxford University Press, Oxford (2004)
17. Hermes, H.: Enumerability, Decidability, Computability. Springer, Berlin (1965)
18. Hopcroft, J.E., Ullman, J.: Introduction to Automata Theory, Languages, and Computation. Addison-Wesley, Reading (1979)
19. Hopcroft, J.E., Motwani, R., Ullman, J.D.: Introduction to Automata Theory, Languages and Computation, 3rd edn. Addison-Wesley (2007)
20. Hromkovi, J.: Theoretical Computer Science: Introduction to Automata, Computability, Complexity, Algorithmics, Randomization. Springer (2003)
21. Jones, N.D.: Computability and Complexity: From a Programming Perspective. MIT Press (1997). (Also available online at http://www.diku.dk/neil/Comp2book.html)
22. Kfoury, A.J., Moll, R.N., Arbib, M.A.: A Programming Approach to Computability. Springer, Berlin (1982)
23. Kozen, D.: Automata and Computability. Springer, Berlin (1997)
24. Lewis, H., Papadimitriou, C.H.: Elements of the Theory of Computation. Prentice Hall (1998)
25. Martin, J.C.: Introduction to Languages and the Theory of Computation. McGraw Hill, New York (1991)
26. Minsky, M.: Computation: Finite and Infinite Machines. Prentice Hall, Upper Saddle River (1967)
27. Moore, C., Mertens, S.: The Nature of Computation. Oxford University Press, Oxford (2011)
28. Papdimitriou, C.H.: Computational Complexity. Addison-Wesley, Reading (1994)
29. Rayward-Smith, V.J.: A first Course in Computability. Blackwell Scientific Publications, Oxford (1986)
30. Reiter, E.E., Johnson, C.M.: Limits of Computation: An Introduction to the Undecidable and the Intractable. Chapman and Hall/CRC, Boca Raton (2012)
31. Rich, E.A.: Automata, Computability and Complexity: Theory and Applications. Pearson (2007)
32. Savage, J.E.: Models of Computation: Exploring the Power of Computing. Addison Wesley, Reading (1998)
33. Selman, H.: Computability and Complexity Theory. Springer, New York (2011)
34. Shen, A., Vereshchagin, N.K.: Computable Functions. American Mathematical Society, Providence (2003)
35. Sipser, M.: Introduction to the Theory of Computation. PWS Publishing, Boston (1997)

36. Soare, R.I.: Recursively Enumerable Sets and Degrees. Springer, Berlin (1987)
37. Stuart, T.: Understanding Computation From Simple Machines to Impossible Programs. O'Reilly (2013)
38. Sudkamp, T.A.: Languages and Machines. Addison Wesley, Reading (1996)
39. Weber, R.: Computability Theory (Student Mathematical Library). American Mathematical Society, Providence (2012)
40. Wegener, I.: Complexity Theory: Exploring the Limits of Efficient Algorithms. Springer, Heidelberg (2005)

Glossary

Logic

$P \Leftrightarrow Q$	statement P is logically equivalent to Q (equivalence)
P iff Q	if and only if, same as $P \Leftrightarrow Q$
$P \Rightarrow Q$	statement P implies Q (implication)
$P \Leftarrow Q$	statement Q implies P (reverse implication)
$\neg P$	not P (negation)
$P \vee Q$	P or Q (disjunction)
$P \wedge Q$	P and Q (conjunction)
$\forall x \in T. \, P(x)$	universal quantification: for all elements x in set T it is the case that P holds, where P depends on x.

Sets and Functions

$\emptyset$	the empty set
Σ^*	set of words over finite alphabet Σ
$A \cap B$	intersection of sets A and B
$A \cup B$	union of sets A and B
$A \subseteq B$	set A is a subset of B
$A \subsetneq B$	set A is a proper subset of B, i.e. $A \subseteq B$ and $A \neq B$
$A \times B$	cartesian product of sets A and B
$A \setminus B$	set difference between sets A and B
$S_\perp$	S lifted, i.e. the set containing all elements of S plus a distinct element $\perp$ representing undefined
$\{x \in S \mid P(x)\}$	the set of all elements in S that satisfy condition P (set comprehension)
$Set(S)$	powerset, i.e. set of all subsets of S
$\overline{S}$	set complement of S
$\mathbb{B}$	the type of Booleans

© Springer International Publishing Switzerland 2016
B. Reus, *Limits of Computation*, Undergraduate Topics in Computer Science,
DOI 10.1007/978-3-319-27889-6

$\mathbb{D}$	the data type of WHILE (i.e. binary trees that carry labels only at leaves in the form of atoms)
$\mathbb{N}$	the type of natural numbers
$\mathbb{R}$	the type of real numbers
$\mathbb{Z}$	the type of integer numbers
L-program	set of L-programs
L-data	set of data values used by L-programs
L-store	set of stores for L-data values needed for the semantics of L-programs
$s \in S$	elementhood: element s occurs in ("is in") set S.
$s \notin S$	element s does not occur in ("is not in") set S
$\bot$	undefined
ε	the empty word
(s, t)	tuple (pair) of elements s and t
$[s_1, s_2, \ldots, s_n]$	list of elements $s_1, s_2, \ldots, s_n$
$[\,]$	empty list
$f : A \to B$	f is a total function from A to B
$f : A \to B_\bot$	f is a partial function from A to B
$f(a)\downarrow$	the application of partial function f to a is defined and thus produces a result value
$f(a)\uparrow$	the application of partial function f to a is undefined and thus does not produce a result value but $\bot$.

Semantics

nil	a special atom, serves a nullary constructor for binary trees in $\mathbb{D}$ to represent the empty tree
$\langle l.r \rangle$	the binary tree in $\mathbb{D}$ that is constructed from a left hand side tree l and a right hand side tree r (binary tree constructor)
$\ulcorner t \urcorner$	the encoding of an expression t (can be an abstract syntax tree, a number, a list etc.) as binary tree in $\mathbb{D}$
$|t|$	the size of a tree $t \in \mathbb{D}$
$\mathcal{E}[\![E]\!]\sigma$	semantics of WHILE-expression E in store σ denoting a value in $\mathbb{D}$
$S \vdash \sigma_1 \to \sigma_2$	semantic judgement stating that executing WHILE-statement list S in store σ_1 terminates and produces result store σ_2
$[\![p]\!]^{\texttt{While}}$	the semantics of WHILE-program p which is a partial function of type $\mathbb{D} \to \mathbb{D}_\bot$
$\sigma[X := v]$	the store σ in which the value for variable X has been updated to v.
$p \vdash (\ell_1, \sigma_1) \to (\ell_2, \sigma_2)$	semantic judgement stating that executing machine language program p in state (ℓ, σ_1) terminates and produces a result state (ℓ_2, σ_2). The ℓ_i are program labels.
$\dot{-}$	subtraction on natural numbers where the result can never be negative
$Acc(p)$	the set of accepted words by nondeterministic program p.

Complexity

$\Theta(f)$	functions growing asymptotically (up to constants) at the same rate as f
$\mathcal{O}(f)$	functions growing asymptotically not faster (up to constants) than f
$o(f)$	functions growing asymptotically much slower than f
$time_p^L(d)$	time it takes L-program p to terminate on input d ($\perp$ denotes non-termination)
$\mathcal{T}E$	time to evaluate WHILE-expression E
$S \vdash^{time} \sigma \Rightarrow t$	judgment stating that the execution of WHILE-statement list S in store σ takes t time units
$time\_nd_p^L$	the runtime of nondeterministic L-program p
$L^{time(f)}$	class of L-programs with runtime bound f
$L^{lintime}$	class of L-programs with linear runtime bounds
L^{ptime}	class of L-programs with polynomial runtime bounds
$L^{exptime}$	class of L-programs with exponential runtime bounds
$L_1 \preceq^{lintime-pg-ind} L_2$	timed programming language L_2 can simulate timed programming language L_1 up to a program independent linear time difference
$L_1 \preceq^{lintime} L_2$	timed programming language L_2 can simulate timed programming language L_1 up to linear difference in time
$L_1 \preceq^{ptime} L_2$	timed programming language L_2 can simulate timed programming language L_1 up to polynomial difference in time
$L_1 \equiv^{lintime-pg-ind} L_2$	timed programming languages L_1 and L_2 are strongly linearly equivalent
$L_1 \equiv^{lintime} L_2$	timed programming languages L_1 and L_2 are linearly equivalent
$L_1 \equiv^{ptime} L_2$	timed programming languages L_1 and L_2 are polynomially equivalent.

Problem Classes

RE	the class of semi-decidable problems
R	the class of decidable problems
TIME$^L(f)$	the class of problems decidable by a L-program with runtime bound f
LIN^L	the class of problems decidable in linear time (by a L-program)
P^L	the class of problems decidable in polynomial time (by a L-program)

EXP^{L}	the class of problems decidable in exponential time (by a L-program)
NTIME$^{\text{L}}(f)$	the class of problems accepted by a nondeterministic L-program with runtime bound f
NLIN^{L}	the class of problems accepted by a nondeterministic L-program in linear time
NP^{L}	the class of problems accepted by a nondeterministic L-program in polynomial time
NEXP^{L}	the class of problems accepted by a nondeterministic L-program in exponential time
RP	(for **r**andomised **p**olynomial) the class of problems accepted by a probabilistic (Monte Carlo) algorithm in polynomial time with an error probability of accepting a wrong answer less than $\frac{1}{3}$.
BPP	(for **b**ounded error **p**robability **p**olynomial time) the class of problems decided by a probabilistic non-biased algorithm in polynomial time with an error probability at least a constant $\varepsilon > 0$ smaller than $\frac{1}{3}$ for all instances
PCP$[r(n), q(n)]$	(for **p**robabilistically **c**heckable **p**roofs) the class of decision problems probabilistically verified in polynomial time using at most $r(n)$ random bits additionally to the certificate, i.e. the "proof", of which at most $q(n)$ bits are read; correct proofs must be accepted, and incorrect proofs must be rejected with probability greater than $\frac{1}{2}$
PP	(for **p**robabilistic **p**olynomial time) the class of problems decided by a probabilistic Turing machine in polynomial time with an error probability of less than $\frac{1}{2}$ for all instances
BQP	(for **b**ounded error **q**uantum **p**olynomial time) the class of problems accepted by a quantum computer in polynomial time with error probability of at most $\frac{1}{3}$
$\mathscr{C}$-hard	the class of problems that all other problems in class $\mathscr{C}$ can be reduced to (using the appropriate form of reduction for $\mathscr{C}$)
$\mathscr{C}$-complete	the class of problems that are $\mathscr{C}$-hard and are also in $\mathscr{C}$ themselves
$\leq_{\text{rec}}$	effective problem reduction
$\leq_{P}$	polynomial time reduction
NPO	(for **n**ondeterministic **p**olynomial **o**ptimisation) the class of optimisation problems for which an optimal solution can be computed by a nondeterministic program in polynomial time
APX	(for **ap**pro**x**imable) the class of optimisation problems in **NPO** for which an approximation to the optimal solution can be computed in polynomial time that has an approximation ratio bounded by a constant

PTAS   (for **p**olynomial **t**ime **a**pproximation **s**cheme) the class of optimisation problems for which an approximation scheme to the optimal solution exists for any approximation factor $1 + \varepsilon$ that can be computed in polynomial time in the size of input

FPTAS   (for **f**ully **p**olynomial **t**ime **a**pproximation **s**cheme) the class of optimisation problems for which an approximation scheme to the optimal solution exists for any approximation factor $1 + \varepsilon$ that can be computed in polynomial time in both the size of input and $\frac{1}{\varepsilon}$.

Index

© Springer International Publishing Switzerland 2016
B. Reus, *Limits of Computation*, Undergraduate Topics in Computer Science,
DOI 10.1007/978-3-319-27889-6